W0193144

Plant Pigments, Flavors and Textures: The Chemistry and Biochemistry of Selected Compounds

Plant Pigments, Flavors and Textures: The Chemistry and Biochemistry of Selected Compounds

Contributors

Lilian Cristina Baldon Aizza and Marcelo Carnier Dornelas et al.

AURIS
Reference

www.aurisreference.com

Plant Pigments, Flavors and Textures: The Chemistry and Biochemistry of Selected Compounds

Contributors: Lilian Cristina Baldon Aizza and Marcelo Carnier Dornelas et al.

Published by Auris Reference Limited

www.aurisreference.com

United Kingdom

Copyright 2016

Printed in 2017 for Sale in the Indian Subcontinent

The information in this book has been obtained from highly regarded resources. The copyrights for individual articles remain with the authors, as indicated. All chapters are distributed under the terms of the Creative Commons Attribution License, which permit unrestricted use, distribution, and reproduction in any medium, provided the original author and source are credited.

Notice

Contributors, whose names have been given on the book cover, are not associated with the Publisher. The editors and the Publisher have attempted to trace the copyright holders of all material reproduced in this publication and apologise to copyright holders if permission has not been obtained. If any copyright holder has not been acknowledged, please write to us so we may rectify.

Reasonable efforts have been made to publish reliable data. The views articulated in the chapters are those of the individual contributors, and not necessarily those of the editors or the Publisher. Editors and/or the Publisher are not responsible for the accuracy of the information in the published chapters or consequences from their use. The Publisher accepts no responsibility for any damage or grievance to individual(s) or property arising out of the use of any material(s), instruction(s), methods or thoughts in the book.

Plant Pigments, Flavors and Textures: The Chemistry and Biochemistry of Selected Compounds

ISBN: 978-1-78154-866-0

British Library Cataloguing in Publication Data
A CIP record for this book is available from the British Library

Printed in the United Kingdom

Exclusively distributed by CBS Publishers & Distributors Pvt. Ltd.

Sales & Distribution Rights only for India, Pakistan, Bangladesh, Sri Lanka, Nepal and Bhutan. This book is not to be sold outside these territories.

Contents

List of Abbreviations

AA	amino acid
ARI	anthocyanin reflectance index
bHLH	Basic helix-loop-helix
CHI	Chalcone isomerase
CHS	Chalcone synthase
CSS	Coastal sage scrub
ESTs	expressed sequence tags
FID	flame ionization detector
HPLC	high performance liquid chromatography
HBH	Hongbaihuatao
IC	Internal control
IEF	Isoelectric focusing
LC	Liquid chromatography
MG	mature green
MEP	methylerythritol phosphate
MPN	Most probable number
NRC	National Research Centre
PCR	Polymerase chain reaction
PC	Principal Components
RWC	relative water content
SSC	soluble solids content
TA	titratable acidity
UV	ultraviolet

List of Contributors

Lilian Cristina Baldon Aizza
Departamento de Biologia Vegetal. Rua Monteiro Lobato 970, Instituto de Biologia, Universidade Estadual de Campinas, Cidade Universitária Zeferino Vaz, 13083-970 Campinas, SP, Brazil

Marcelo Carnier Dornelas
Departamento de Biologia Vegetal. Rua Monteiro Lobato 970, Instituto de Biologia, Universidade Estadual de Campinas, Cidade Universitária Zeferino Vaz, 13083-970 Campinas, SP, Brazil

Hock-Eng Khoo
Department of Nutrition and Dietetics, Faculty of Medicine and Health Sciences, Universiti Putra Malaysia, 43400 UPM Serdang, Selangor, Malaysia

K. Nagendra Prasad
Department of Nutrition and Dietetics, Faculty of Medicine and Health Sciences, Universiti Putra Malaysia, 43400 UPM Serdang, Selangor, Malaysia

Kin-Weng Kong
Department of Nutrition and Dietetics, Faculty of Medicine and Health Sciences, Universiti Putra Malaysia, 43400 UPM Serdang, Selangor, Malaysia

Yueming Jiang
South China Botanical Garden, Chinese Academy of Sciences, Guangzhou 510650, China

Amin Ismail
Department of Nutrition and Dietetics, Faculty of Medicine and Health Sciences, Universiti Putra Malaysia, 43400 UPM Serdang, Selangor, Malaysia
Laboratory of Analysis and Authentication, Halal Products Research Institute, Universiti Putra Malaysia, 43400 UPM Serdang, Selangor, Malaysia

M. M. Hussein
Water Relations and Irrigation Department, National Research Centre, Cairo, Egypt

A. K. Alva
USDA-ARS, Vegetable and Forage Crops Research Unit, Prosser, WA, USA

Michael L. Schwieterman
Plant Molecular and Cellular Biology Program, University of Florida, Gainesville, Florida, United States of America
Plant Innovation Program, University of Florida, Gainesville, Florida, United States of America

Thomas A. Colquhoun
Plant Molecular and Cellular Biology Program, University of Florida, Gainesville, Florida, United States of America
Department of Environmental Horticulture, University of Florida, Gainesville, Florida, United States of America
Plant Innovation Program, University of Florida, Gainesville, Florida, United States of America

Elizabeth A. Jaworski
Department of Environmental Horticulture, University of Florida, Gainesville, Florida, United States of America
Plant Innovation Program, University of Florida, Gainesville, Florida, United States of America

Linda M. Bartoshuk
College of Dentistry, University of Florida, Gainesville, Florida, United States of America
Plant Innovation Program, University of Florida, Gainesville, Florida, United States of America

Jessica L. Gilbert
Horticultural Sciences Department, University of Florida, Gainesville, Florida, United States of America
Plant Innovation Program, University of Florida, Gainesville, Florida, United States of America

Denise M. Tieman
Horticultural Sciences Department, University of Florida, Gainesville, Florida, United States of America
Plant Innovation Program, University of Florida, Gainesville, Florida, United States of America

Asli Z. Odabasi
Food Science and Human Nutrition Department, University of Florida, Gainesville, Florida, United States of America

Plant Innovation Program, University of Florida, Gainesville, Florida, United States of America

Howard R. Moskowitz
Moskowitz Jacobs Inc., White Plains, New York, United States of America

Kevin M. Folta
Plant Molecular and Cellular Biology Program, University of Florida, Gainesville, Florida, United States of America
Horticultural Sciences Department, University of Florida, Gainesville, Florida, United States of America
Plant Innovation Program, University of Florida, Gainesville, Florida, United States of America

Harry J. Klee
Plant Molecular and Cellular Biology Program, University of Florida, Gainesville, Florida, United States of America
Horticultural Sciences Department, University of Florida, Gainesville, Florida, United States of America
Plant Innovation Program, University of Florida, Gainesville, Florida, United States of America

Charles A. Sims
Food Science and Human Nutrition Department, University of Florida, Gainesville, Florida, United States of America
Plant Innovation Program, University of Florida, Gainesville, Florida, United States of America

Vance M. Whitaker
Food Science and Human Nutrition Department, University of Florida, Gainesville, Florida, United States of America
Gulf Coast Research and Education Center, University of Florida, Wimauma, Florida, United States of America
Plant Innovation Program, University of Florida, Gainesville, Florida, United States of America

David G. Clark
Plant Molecular and Cellular Biology Program, University of Florida, Gainesville, Florida, United States of America
Department of Environmental Horticulture, University of Florida, Gainesville, Florida, United States of America

Plant Innovation Program, University of Florida, Gainesville, Florida, United States of America

Paulina Kuczynska
Faculty of Biochemistry, Biophysics and Biotechnology, Department of Plant Physiology and Biochemistry, Jagiellonian University, Gronostajowa 7, Krakow 30-387, Poland

Malgorzata Jemiola-Rzeminska
Faculty of Biochemistry, Biophysics and Biotechnology, Department of Plant Physiology and Biochemistry, Jagiellonian University, Gronostajowa 7, Krakow 30-387, Poland
Małopolska Centre of Biotechnology, Gronostajowa 7A, Krakow 30-387, Poland

Kazimierz Strzalka
Faculty of Biochemistry, Biophysics and Biotechnology, Department of Plant Physiology and Biochemistry, Jagiellonian University, Gronostajowa 7, Krakow 30-387, Poland
Małopolska Centre of Biotechnology, Gronostajowa 7A, Krakow 30-387, Poland

Jingfeng Huang
Institute of Agricultural Remote Sensing & Information Application, Zijingang Campus, Zhejiang University, Hangzhou, China

Chen Wei
Institute of Agricultural Remote Sensing & Information Application, Zijingang Campus, Zhejiang University, Hangzhou, China
Zhejiang Meteorological Service Center, Hangzhou, China

Yao Zhang
Institute of Agricultural Remote Sensing & Information Application, Zijingang Campus, Zhejiang University, Hangzhou, China

George Alan Blackburn
Lancaster Environment Centre, Lancaster University, Lancaster, United Kingdom

Xiuzhen Wang
Institute of Remote Sensing and Earth Sciences, Hangzhou Normal University, Hangzhou, China

Chuanwen Wei
Institute of Agricultural Remote Sensing & Information Application, Zijin-gang Campus, Zhejiang University, Hangzhou, China

Jing Wang
Institute of Agricultural Remote Sensing & Information Application, Zijin-gang Campus, Zhejiang University, Hangzhou, China

Beverley J. Glover
Department of Plant Sciences, University of Cambridge, Downing Street, Cambridge CB2 3EA, UK

Heather M. Whitney
School of Biological Sciences, University of Bristol, Woodland Road, Bristol BS8 1UG, UK

Guillermo Raúl Pratta
National Council for Scientific and Technical Research, Buenos Aires, Argentina
Chair of Genetics, Agronomic Sciences Faculty, National University of Rosario, Zavalla, Argentine

Gustavo Rubén Rodríguez
National Council for Scientific and Technical Research, Buenos Aires, Argentina
Chair of Genetics, Agronomic Sciences Faculty, National University of Rosario, Zavalla, Argentine

Roxana Zorzoli
Chair of Genetics, Agronomic Sciences Faculty, National University of Rosario, Zavalla, Argentine
Council for Research of the National University of Rosario, Zavalla, Argentine

Liliana Amelia Picardi
Chair of Genetics, Agronomic Sciences Faculty, National University of Rosario, Zavalla, Argentine
Council for Research of the National University of Rosario, Zavalla, Argentine

Estela Marta Valle
National Council for Scientific and Technical Research, Buenos Aires, Argentina

Institute of Molecular and Cell Biology of Rosario, CONICET/Biochemical and Pharmaceutical Sciences Faculty, Suipacha, Rosario, Argentine

Congming Lu
Photosynthesis Research Centre, Institute of Botany, Chinese Academy of Sciences, Beijing 100093, PR China
Department of Biology, Hong Kong Baptist University, Kowloon, Hong Kong, PR China

Qingtao Lu
Photosynthesis Research Centre, Institute of Botany, Chinese Academy of Sciences, Beijing 100093, PR China

Jianhua Zhang
Department of Biology, Hong Kong Baptist University, Kowloon, Hong Kong, PR China

Tingyun Kuang
Photosynthesis Research Centre, Institute of Botany, Chinese Academy of Sciences, Beijing 100093, PR China

Jun Cheng
Key Laboratory of Plant Germplasm Enhancement and Specialty Agriculture, Wuhan Botanical Garden of the Chinese Academy of Sciences, Wuhan, 430074, P.R. China
Graduate University of Chinese Academy of Sciences, 19A Yuquanlu, Beijing, 100049, P.R. China

Liao Liao
Key Laboratory of Plant Germplasm Enhancement and Specialty Agriculture, Wuhan Botanical Garden of the Chinese Academy of Sciences, Wuhan, 430074, P.R. China

Hui Zhou
Key Laboratory of Plant Germplasm Enhancement and Specialty Agriculture, Wuhan Botanical Garden of the Chinese Academy of Sciences, Wuhan, 430074, P.R. China
Graduate University of Chinese Academy of Sciences, 19A Yuquanlu, Beijing, 100049, P.R. China

Chao Gu
Key Laboratory of Plant Germplasm Enhancement and Specialty Agricul-

ture, Wuhan Botanical Garden of the Chinese Academy of Sciences, Wuhan, 430074, P.R. China

Lu Wang
Key Laboratory of Plant Germplasm Enhancement and Specialty Agriculture, Wuhan Botanical Garden of the Chinese Academy of Sciences, Wuhan, 430074, P.R. China

Yuepeng Han
Key Laboratory of Plant Germplasm Enhancement and Specialty Agriculture, Wuhan Botanical Garden of the Chinese Academy of Sciences, Wuhan, 430074, P.R. China

Jeff Velten
Plant Stress and Water Conservation Laboratory, United States Department of Agriculture - Agricultural Research Service, Lubbock, Texas, United States of America

Cahid Cakir
Plant Stress and Water Conservation Laboratory, United States Department of Agriculture - Agricultural Research Service, Lubbock, Texas, United States of America

Christopher I. Cazzonelli
Australian Research Council - Centre of Excellence in Plant Energy Biology, Research School of Biology, Australian National University, Canberra, Australian Capital Territory, Australia

Irina C. Irvine
Department of Ecology and Evolutionary Biology, University of California Irvine, Irvine, California, United States of America
Santa Monica Mountains National Recreation Area, United States National Park Service, Thousand Oaks, California, United States of America

Christy A. Brigham
Santa Monica Mountains National Recreation Area, United States National Park Service, Thousand Oaks, California, United States of America

Katharine N. Suding
Department of Environmental Science, Policy and Management, University of California, Berkeley, California, United States of America

Jennifer B. H. Martiny
Department of Ecology and Evolutionary Biology, University of California Irvine, Irvine, California, United States of America

Zhuo Zhang
Department of Occupational and Environmental Health, School of Public Health, China Medical University, No. 92 Bei Er Road, Heping District, Shenyang 110001, China
Department of Nutrition and Food Hygiene, Shenyang Medical College, No146 Huanghe North Street, Shenyang 110034, China

Bo Zhou
Department of Nutrition and Food Hygiene, Shenyang Medical College, No146 Huanghe North Street, Shenyang 110034, China

Hiaohong Wang
Department of Nutrition and Food Hygiene, Shenyang Medical College, No146 Huanghe North Street, Shenyang 110034, China

Fei Wang
Department of Occupational and Environmental Health, School of Public Health, China Medical University, No. 92 Bei Er Road, Heping District, Shenyang 110001, China

Yingli Song
Department of Occupational and Environmental Health, School of Public Health, China Medical University, No. 92 Bei Er Road, Heping District, Shenyang 110001, China

Shengnan Liu
Department of Occupational and Environmental Health, School of Public Health, China Medical University, No. 92 Bei Er Road, Heping District, Shenyang 110001, China

Shuhua Xi
Department of Occupational and Environmental Health, School of Public Health, China Medical University, No. 92 Bei Er Road, Heping District, Shenyang 110001, China

Preface

Plant Pigments, Flavors and Textures: The Chemistry focuses on the chemistry of compounds responsible for the pigments, flavors, and textures of some fruits and vegetables. Since much of the information presented is scattered in the scientific literature, an attempt has been made to integrate the material into a concise yet comprehensive text. First chapter focuses on a genomic approach to study anthocyanin synthesis and flower pigmentation in passionflowers. Second chapter focuses more on several carotenoids and their isomers present in different fruits and vegetables along with their concentrations. Carotenoids and their geometric isomers also play an important role in protecting cells from oxidation and cellular damages. Third chapter demonstrates that exogenous application of ascorbic acid can enhance foliar growth which may contribute to increased plant biomass and yield. In fourth chapter, genetic and environmentally induced variation is exploited to capture biochemically diverse strawberry fruit for metabolite profiling and consumer rating. Photosynthetic pigments are bioactive compounds of great importance for the food, cosmetic, and pharmaceutical industries. Fifth chapter summarizes current knowledge on diatom photosynthetic pigments complemented by some new insights regarding their physico-chemical properties, biological role, and biosynthetic pathways, as well as the regulation of pigment level in the cell, methods of purification, and significance in industries. Sixth chapter determine whether passive optical hyperspectral remote sensing techniques are sufficiently well developed to quantify individual plant pigments, which operational solutions are available for wider plant science and the areas which now require greater focus. Seventh chapter focuses on structural color and iridescence in plants. The general goal of eighth chapter to study free amino acid and pigment composition in selected tomato germplasm, with the aims of contributing to knowledge on variability of ripening metabolism, identifying more consistently the RILs genetic background, and verifying associations between glutamate content and fruit shelf life. The objectives of ninth chapter were to characterize fully photosynthetic pigment composition, to examine if and how down-regulation of PSII happens in senescent leaves when exposed to excess light energy, and to determine if the xanthophyll cycle plays a role in dissipating excess light energy during leaf senescence. Tenth chapter presents on a small indel mutation in an anthocyanin transporter causes variegated colouration of peach flowers. Eleventh chapter focuses on a spontaneous dominant-negative mutation within Arabidopsis 35S. In last chapter, we investigated the effect of PPFMs on several coastal sage scrub (CSS) plant species. CSS is a low shrubland community that once dominated the Mediterranean-type climate regions of coastal California.

Chapter 1

A GENOMIC APPROACH TO STUDY ANTHOCYANIN SYNTHESIS AND FLOWER PIGMENTATION IN PASSIONFLOWERS

Lilian Cristina Baldon Aizza and Marcelo Carnier Dornelas

Departamento de Biologia Vegetal. Rua Monteiro Lobato 970, Instituto de Biologia, Universidade Estadual de Campinas, Cidade Universitária Zeferino Vaz, 13083-970 Campinas, SP, Brazil

ABSTRACT

Most of the plant pigments ranging from red to purple colors belong to the anthocyanin group of flavonoids. The flowers of plants belonging to the genus Passiflora (passionflowers) show a wide range of floral adaptations to diverse pollinating agents, including variation in the pigmentation of floral parts ranging from white to red and purple colors. Exploring a database of expressed sequence tags obtained from flower buds of two divergentPassiflora species, we obtained assembled sequences potentially corresponding to 15 different genes of the anthocyanin biosynthesis pathway in these species. The obtained sequences code for putative enzymes are involved in the production of flavonoid precursors, as well as those involved in the formation of particular ("decorated") anthocyanin molecules. We also obtained sequences encoding regulatory factors that control the expression of structural genes and regulate the spatial and temporal accumulation of pigments. The identification of some of the putative Passiflora anthocyanin biosynthesis pathway genes provides novel resources for research on secondary metabolism in passionflowers, especially on the elucidation of the processes involved in floral pigmentation, which will allow future studies on the role of pigmentation in pollinator preferences in a molecular level.

INTRODUCTION

Anthocyanins belong to a diverse group of secondary metabolites of the phenylpropanoid class, the flavonoids, which are found in different plant species. They represent some of the most important natural pigments, which are responsible for the wide range of red to purple colors present in many flowers, fruits, seeds, leaves, and stems. Besides having great economical relevance, flower and fruit pigments play an important ecological role in the animal attraction for pollination and seed dispersal, wich is a spectacular example of coevolution between plants and animals [1–3].

The biosynthetic pathway of anthocyanins has been well characterized biochemically and genetically in species with different floral morphology, pigmentation pattern, and pollination syndromes such as Petunia hybrida [4,5], Matthiola [6], Dianthus [7], Eustoma [8], Gerbera [9], Zea mays [10, 11], Antirrhinum majus [12], andIpomoea [13, 14]. A representation of a general anthocyanin biosynthetic pathway is shown in Figure 1.

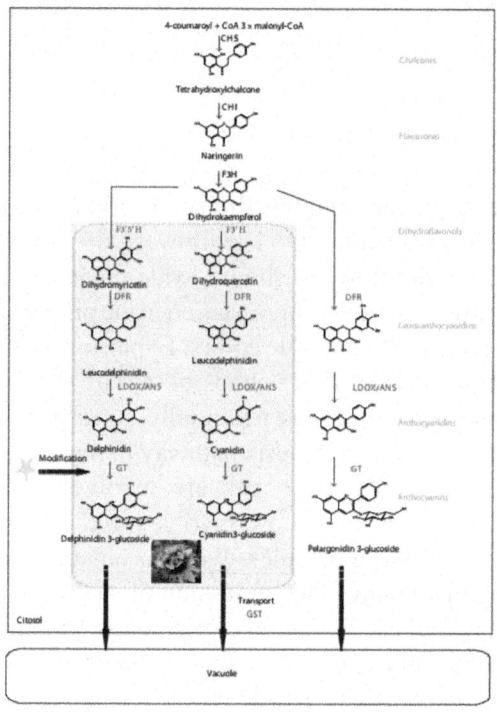

Figure 1: Schematic representation of the anthocyanin biosynthetic pathway (adapted from [16]). Enzymes are indicated in red, and classes of compounds are in green. Anthocyanidin is further modified with glycosyl, acyl, or methyl groups, resulting in the "decorated" anthocyanin. In this case, UF3GT is responsible for the glycosylation of anthocyanidins. The proposed anthocyanin biosynthetic pathway for Passiflora edulis is highlighted by the colored background. CHS: chalcone sintase; CHI: chalcone isomerase; F3H: flavanone 3-hydroxylase; F3'H: flavanone 3'-hydroxylase; F3'5'H:

flavanone 3'5'-hydroxylase; DFR: dihydroflavonol 4-reductase; LDOX/ANS: leuco-anthocyanidin dioxygenase/anthocyanidin synthase; GT: glucosyltransferase; GST: glutathione S-transferase.

Briefly, the pathway is initiated with chalcone synthase (CHS) catalyzing the stepwise condensation of three molecules of acetate residues from malonlyl-CoA with one molecule of 4-coumaroyl-CoA to form the basic structure of flavonoids (tetrahydroxychalcone), which is rapidly isomerized to the colorless naringenin by chalcone isomerase (CHI). Naringenin is then converted to dihydroflavonol by flavanone 3-hydroxylase (F3H). Dihydroflavonol 4-reductase (DFR), which is a specific enzyme for the anthocyanin synthesis, catalyses the production of leucoanthocyanidins from dihydroflavonols, which can be hydroxylated on the 3' or 5' position of the B-ring by flavonoid 3'-hydroxylase (F3'H) to produce dihydroquercetin or by flavonoid 3'5'-hydroxylase (F3'5'H) to form dihydromyricetin. Subsequently, leucoanthocyanidin oxidase/anthocyanidin synthase (LDOX/ANS) is responsible for the formation of the anthocyanidins from the colorless leucoanthocyanidins. GT enzymes (O-glucosyltransferases) represent the final step in anthocyanin biosynthesis: anthocyanidins are converted in differentially "decorated" anthocyanin molecules [15, 16]. Biochemical approaches have demonstrated that all anthocyanin pigments are derived from one of three aglycones: pelargonidin, cyaniding, and delphinidin. The main determinants of the apparent color of these pigments are the hydroxylation and methylation patterns, as well as the number and type of sugars on the beta ring of the flavonoid molecule [1, 3,17–19].

Figure 1 depicts a generalized anthocyanin biosynthesis pathway. At least, two groups of genes are required for anthocyanin biosynthesis: the first group is represented by the structural genes encoding enzymes for the production of the flavonoid precursors, as well as those involved in the formation of particular ("decorated") anthocyanin molecules. The second group includes the genes encoding regulatory factors that control the expression of structural genes which are mainly orenestrated by complexes formed by MYB and basic helix-loop-helix (bHLH) transcription factors that include WDR (WD40 repeats) proteins [2, 4, 15, 16, 20–23].

There are about 600 Passiflora species widely distributed in tropical and subtropical regions. Some Passiflora species have economical importance due to the production of fruits (passionfruit) or use as ornamentals. Nevertheless, a large number of Passiflora species are rare and/or endangered, as the environment of their diversity center has been increasingly degraded by human activities [24]. An enormous floral diversity is observed among Passiflora species, including variation in color, size, morphology, and fusion of floral organs. These and other floral characteristics, including evolutionary innovations such as the presence of coronal filaments and an androgynophore, are indicative of the wide range of pollination syndromes found in the genus [24]. Wide passionflowers may be pollinated by insects (bees and wasps), hummingbirds, and bats [24]. The most striking feature of floral variation among passionflowers is the wide range of pigmentation patterns of the corona filaments. Most of the floral pigments in Passiflora are different types of anthocyanin molecules [25, 26]. Among all Passiflora species, P. edulis Deg and P. suberosa L. are of particular interest, because they are modelPassiflora species for which expressed sequences tags (ESTs) were produced within the frame of the "PASSIOMA" Project [27]. P. edulis Deg flowers are pollinated by large bees of genus Xylocopa. These flowers are about 8–12 cm wide, and their coronas contain multiple series of purplish filaments with white tips. The flowers of P. suberosa L. are small (2-3 cm wide) and show two morphologically distinct series of corona filaments: the outer series is greenish, and the inner series is formed by smaller purple filaments. The flowers of P. suberosa are pollinated by wasps [28].

We are particularly interested in the characterization of genes involved in the anthocyanin biosynthetic pathway of these two Passiflora species. With this aim, we searched for putative Passiflora genes responsible for flower pigmentation, using the key proteins known to be involved in the different enzymatic steps of anthocyanin biosynthesis as baits to search for expressed sequences tags (ESTs) in the PASSIOMA database.

MATERIAL AND METHODS

Searching Passiflora ESTs Homologous to Anthocyanin Biosynthetic Genes

The clustered expressed sequence tags (ESTs) from the PASSIOMA Project database [27] were used as a primary source of data for our analyses. These sequences were assembled from ESTs obtained from the sequencing of several P. edulis or P. suberosa cDNA libraries, made from floral buds at different developmental stages (see [27] for details on library construction, sequencing, and database structure). Nucleotide sequences and their respective deduced amino acid sequences from genes known to be involved in anthocyanin biosynthesis (see Figure 1) were obtained from the National Center for Biotechnology Information (NCBI;http://www.ncbi.nlm.nih.gov/). Searches for putative homolog sequences in the PASSIOMA database were conducted using the tBLASTN module that compares the consensus amino acid sequence with a translated nucleotide sequences database [29]. We generally used Arabidopsis thaliana or Petunia hybrida as query consensus sequences as the anthocyanin biosynthesis pathways in these model species are more thoroughly studied at the molecular level [30–32]. All sequences in the PASSIOMA database that exhibited a significant alignment (-value lower than 10–5) with the query were retrieved from the PASSIOMA database.

The clusterization of all reads identified using a given query sequence was performed using the CAP3 algorithm [33] from the BioEdit software [34]. The novel cluster consensus sequences obtained were reinspected for the occurrence of conserved motives using InterProScan [35] and were compared to NCBI databases using BLAST [29]. Sequences that did not show the main motives present in the query sequence were discarded. Validated sequences were then included in phylogenetic analyses.

Comparison of the Amino Acid Sequences and Phylogenetic Analysis

All amino acid sequences were aligned by CLUSTALX software using default parameters [36]. The obtained alignments were eventually corrected by hand and imported into the molecular evolutionary genetics analysis (MEGA) software [37]. Phylogenetic trees were obtained using parsimony and/or genetic distance calculations (in the later case using pairwise deletion option and with the Poisson correction model). Neighbor-joining [38] and Bootstrap (with 10,000 replicates) trees were also constructed.

RESULTS

The cDNA libraries of the PASSIOMA Project were obtained from mRNA extracted from floral buds at different developmental stages, and it is expected that all EST sequences correspond to genes expressed duringPassiflora flower development [27]. This sequence search detected a total of 75 Passiflora EST sequences, 34 of them corresponding to P. edulis sequences and 41 of them corresponding to sequences derived from P. suberosalibraries. When submitted to the CAP3 algorithm and detailed comparison of their deduced amino acid sequences, the number of valid clusters was reduced to 15, potentially corresponding to 15 different genes. When the validated amino acid sequences obtained from the PASSIOMA database were compared to other plant protein sequences in the public databases, the first BLAST hits generally corresponded to Populus andRicinus sequences. This was expected, as Passiflora and these genera belong to the same order (Malpighiales) and are considered to be closely related [39].

We obtained assembled EST sequences corresponding to genes of the following genes families: CHS, DFR, GT, GST, MYB, and WD40 (see Table 1). Therefore, we used 15 Passiflora assembled sequences from the PASSIOMA database and a selected set of genes from divergent plant species from the public databases to explore their evolutionary relationships. The obtained sequence comparison alignments allowed the construction of phylogenetic trees for each of these families of genes involved in the different enzymatic steps of the anthocyanin pathway.

Table 1: Putative Passiflora homologs of genes encoding elements of the anthocyanin biosynthetic pathway

Enzyme	*Passiflora* AS*	First BLAST hit	e-value	ID/SM
CHS	PACEPE3010G11.g	ABD24222 CHS *Populus alba*	$7e^{-72}$	85/90
	PACEPE3014B06.g	ABC86919 CHS *Populus alba*	$9e^{-67}$	77/84
	PACEPE3007G06.g	XP_002305446 CHS-like *Populus trichocarpa*	$7e^{-121}$	84/92
	PACEPE3023H10.g	XP_002326830 CHS-like *Populus trichocarpa*	$3e^{-110}$	82/91
	PACEPS7017D03.g	AAQ62589 CHS3 *Glycine Max*	$1e^{-98}$	82/88
DFR	PACEPE3003G04.g	XP_002307667 DFR2 *Populus trichocarpa*	$1e^{-93}$	82/94
GT	PACEPE3030G03.g	XP_002532899 UFGT *Ricinus communis*	$6e^{-31}$	53/70
	PACEPS7021H07.g	XP_002518725 UFGT *Ricinus communis*	$8e^{-57}$	56/72
GST	PACEPE3013H01.g	AF048978 GST *Glycine Max*	$7e^{-52}$	79/90
	PACEPE3007A05.g	XM_002519342 GST theta *Ricinus communis*	$4e^{-52}$	77/89
	PACEPE3018F08.g	ADB11335 GSTF7 phi *Populus trichocarpa*	$2e^{-82}$	68/83
	PACEPS4006H06.g	ADB11332 GSTF4 phi *Populus trichocarpa*	$2e^{-61}$	63/78
	PACEPS7023B03.g	AF243378 GST 23 *Glycine max*	$1e^{-51}$	80/88
MYB	PACEPS7022E07.g	XP_002530824 R2R3 MYB *Ricinus communis*	$7e^{-80}$	88/91
WD40	PACEPE3007G07.g	XP_002512788 WD-repeat protein *Ricinus communis*	$3e^{-124}$	92/96

Abbreviations: CHS: chalcone synthase; DFR: dihydroflavonol 4-reductase; GT: glucosyltransferase and GST: glutathione S-transferase.
Using the BLASTp algorithm [29].
*AS: assembled sequence. Codes refer to the longest cDNA clone. PACEPE: *Passiflora edulis*; PACEPS: *Passiflora suberosa*.
ID/SM: identity/similarity (both based on the amino acid sequence) with the first BLAST hit.

The similarities among all genes identified in this study and those reported from other plant species were assembled in Table 1 and ranged from 70% (PACEPE3030G03.g; representing a putative member of the GST, glutathione S-transferase superfamily) to 96% (PACEPE3007G07.g; potentially encoding a WD40 protein).

Some of these gene sequences showed significant similarity to elements required for early or late steps of the pathway; others putatively encode regulatory proteins involved in the control of the spatial and temporal patterns of pigmentation, while others are responsible for intracellular transport of the anthocyanin molecules. The role of each of these genes in the anthocyanin biosynthesis and the probable implications for the understanding of the Passiflora flower pigmentation are presented in the Discussion.

Identification and Phylogenetic Analysis of Passionflower Genes Potentially Involved in Anthocyanin Biosynthesis and Transport

Chalcone Synthases (CHSs)

We have found 5 *Passiflora* assembled sequences (5 putative genes) encoding enzymes of the CHS family: PACEPE3010G11.g, PACEPE3014B06.g, PACEPE3007G06.g, PACEPE3023H10.g and PACEPS7017D03.g. These sequences are expected to encode proteins with 231, 158, 254, 237, and 222 amino acids, respectively. The deduced CHS proteins showed more than 80% similarity to CHSs of other plant species (Table 1). To determine the phylogenetic relationship of different CHSs, we aligned protein sequences from a diverse range of plant species (moss, ferns, gymnosperms and angiosperms), cyanobacterium (*Synechococcus* sp.) and *Passiflora*representatives of the CHS superfamily (Figure 2). The phylogenetic tree was resolved in three clades. These three clades were highly supported with 100% bootstrap values. The *Passiflora* proteins were consistently positioned into different clades. One of these monophyletic clades (highlighted in Figure 2) contains all the anther-specific CHS-like genes (ASCLs; [40, 41]). The remaining sequences, including three *Passiflora*members, were clustered in the other sister clade together with all CHS genes from seed plants.

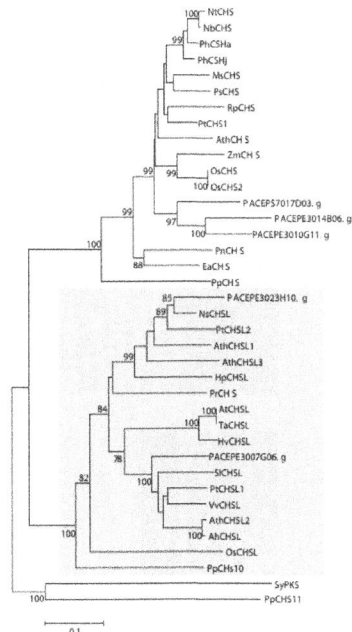

Figure 2: A Neighbor-joining phylogenetic tree of chalcone synthase (CHS) amino acids sequences. The cluster containing all anther-specific CHS-like enzymes is high-

lighted. Bootstrap values from 1,000 replicates were used to assess the robustness of the trees. Only bootstrap values above 75% are indicated at the nodes. Accession numbers for genes from other species are given in Supplementary data.

Dihydroflavonol 4-Reductases (DFR)

A single Passiflora cDNA sequence of 850 bp encoding a predicted protein of 204 amino acids showed significant -value () and 94% similarity to a Populus DFR sequence (Table 1). Figure 3 shows an alignment of the deduced amino acid sequence of the Passiflora DFR with some other plant sequences containing an NADP-binding domain, considered the region of substrate preference of DFR enzymes [42, 43]. Additionally, the Passiflora DFR showed an aspartic acid residue at position 134, as it is observed for thePetunia and Populus proteins, whereas Gerbera and some Lotus DFR show an asparagine residue at the same position (Figure 3). We adopted the terminology suggested by Shimada and coworkers [44] to designate the conserved motifs present in the DFR sequence.

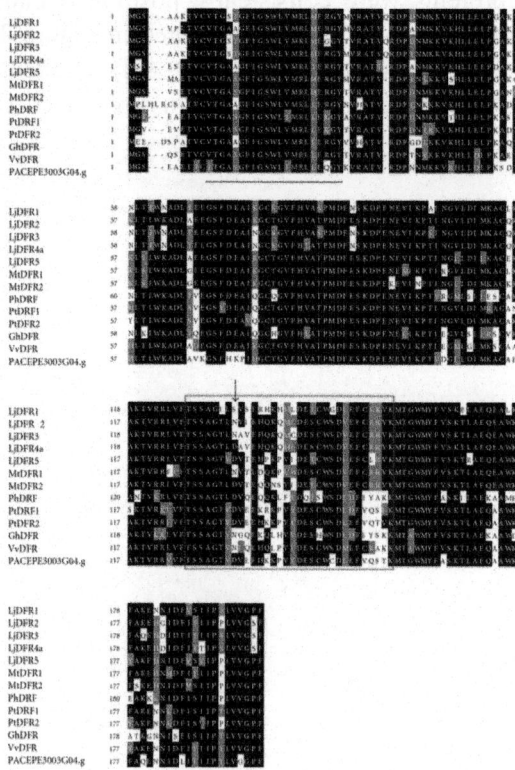

Figure 3: Multiple sequence alignment of the Passiflora sequence with some plant DFR sequences. The identical and similar residues are highlighted on a black and gray

background, respectively. NADP-binding domain is underlined. Boxed amino acids have been considered to control the substrate specificity of DFR enzyme [40], and the amino acid residue (indicated by an arrowhead) is especially important for this specificity [41]. The alignment was performed using CLUSTALX and BOXSHADE program.

A neighbor-joining tree was constructed based on the alignment DFR sequences shown in Figure 3. The monocots and eudicots DFRs were positioned separately. While monocot DFR genes formed one clade, the eudicot DFR sequences diverged into two clades. Clearly, Asn-type DFRs are found in a larger number of species. On the other hand, Asp-type DFRs are restricted to some species, including Passiflora and Populus(Figure 4).

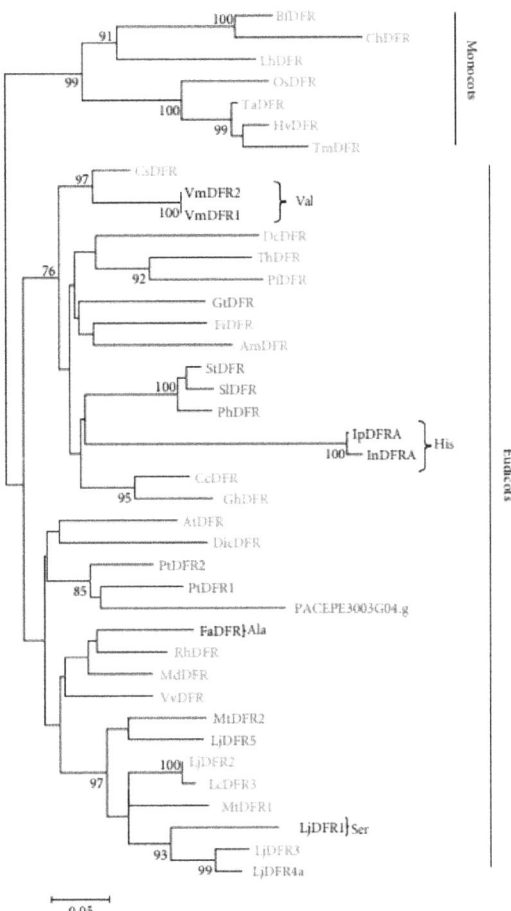

Figure 4: A Neighbor-joining phylogenetic tree of dihydroflavonol 4-reductase (DFR) amino acids sequences. Bootstrap values from 1,000 replicates were used to

assess the robustness of the trees. Only bootstrap values above 75% are indicated at the nodes. Asn-type DFRs, Asp-type DFRs, and DFRs of neither Asn nor Asp-type are indicates in blue, red, and black, respectively [42]. Accession numbers for genes from other species are given in Supplementary data which are available online at doi:10.4061/2011/37157.

Glucosyltransferases (GT)

We identified two Passiflora EST clones, PACEPE3030G03.g and PACEPS7021H02.g, encoding proteins with sequence similarity to Ricinus communis glucosyltransferases (Table 1). The first cDNA sequence contained an ORF specifying a 124 amino acid protein, and the second cDNA encoded a protein of 200 amino acid residues. These putative Passiflora GT proteins were compared with those GT enzymes described by Kovinick and colleagues [45] and retrieved from the NCBI database. The obtained phylogenetic tree resulted in five clades, according to their in vitro substrate specificities [45]. Phylogenetic analysis revealed that the Passiflorasequences were positioned within the Cluster II proteins (Figure 5).

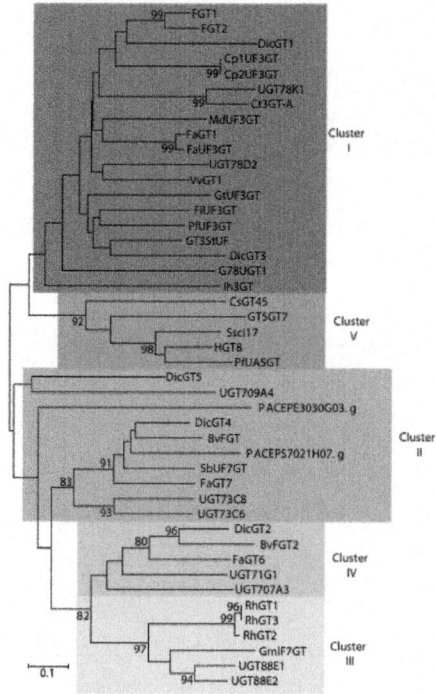

Figure 5: A Neighbor-joining phylogenetic tree of glucosyltransferase (GT) amino acids sequences. Bootstrap values from 1,000 replicates were used to assess the ro-

bustness of the trees. Only bootstrap values above 75% are indicated at the nodes. Accession numbers for genes from other species are given in Supplementary data.

Glutathione S-Transferases (GSTs)

We have identified five Passiflora sequences representing putative members of the GST family. Each member was represented by a single EST sequence. Comparison of these deduced GST protein sequences with those in the GenBank database revealed homology with multifunctional GSTs from Populus, Ricinus, and Glycine spp (see Table 1). Phylogenetic relationships among the putative Passiflora GSTs and family members of other plant species were established (Figure 6). Based on sequence similarity, the five Passiflora putative GSTs were grouped into three clades. PACEPE3018F08.g, PACEPS4006H06.g, and PACEPS7023B03.g are type I GSTs, PACEPE3007A05.g is a type II GST, and PACEPE3013H01.g is a type III GST [46].

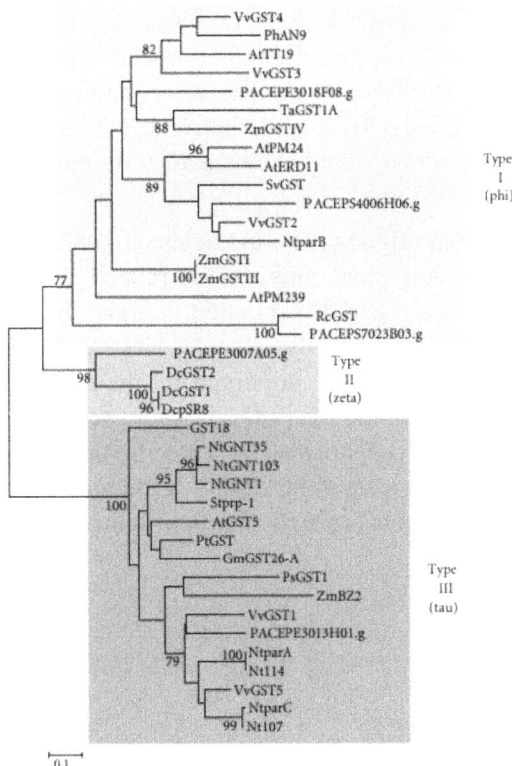

Figure 6: A Neighbor-Joining phylogenetic tree of glutathione S-transferase (GST) amino acids sequences with three types representing phi, tau, and zeta classes. Phi and

tau are plant-specific GSTs. Bootstrap values from 1,000 replicates were used to assess the robustness of the trees. Only bootstrap values above 75% are indicated at the nodes. Accession numbers for genes from other species are given in Supplementary data.

We could not find any putative homologs to chalcone isomerases (CHI), flavanone 3-hydroxylases (F3H), and anthocyanidin synthases (ANS; see Figure 1) in the PASSIOMA database. Three EST sequences were identified corresponding to a putative flavonoid 3-O-hydroxylase (F3'H) gene, and one sequence was found that showed significant homology to genes encoding flavonoid 3-5-O-hydroxylases (F3'5'H; data not show). As these sequences were incomplete at their 5' end, they were not considered in our analyses.

Identification and Phylogenetic Analysis of Passionflower Genes Potentially Involved in Spatially and Temporally Patterning Anthocyanin Deposition

Based on the searches in the PASSIOMA database, we identified one potential homolog for an MYB transcription factor of the R2R3 class. The P. suberosa cDNA clone PACEPS7022E07.g encodes a protein of 132 amino acids showing 91% similarity to the Ricinus communis R2R3 MYB. On the other hand, PACEPE3007G07.g is a putative P. edulis WD40 gene of 886 bp encoding 291 amino acid residues showing 96% similarity to an R. communis, WD40 (Table 1).

Figure 7 shows an alignment of the deduced PACEPS7022E07.g protein sequence with 17 other plant anthocyanin-related R2R3-MYB, indicating the presence of a conserved DNA-binding domain, designated as the R2R3 domain. All sequences analyzed also contained a second conserved amino acid motif in the R3 repeat (red box), important for the interaction between MYB and bHLH proteins in Arabidopsis [48]. The four specific residues required for this interaction in maize [49] are also indicated by the arrows in Figure 7. The third conserved motif appears to be ANDV (blue box) in the R3 repeat of all eudicot R2R3-MYB proteins related to anthocyanin biosynthesis.

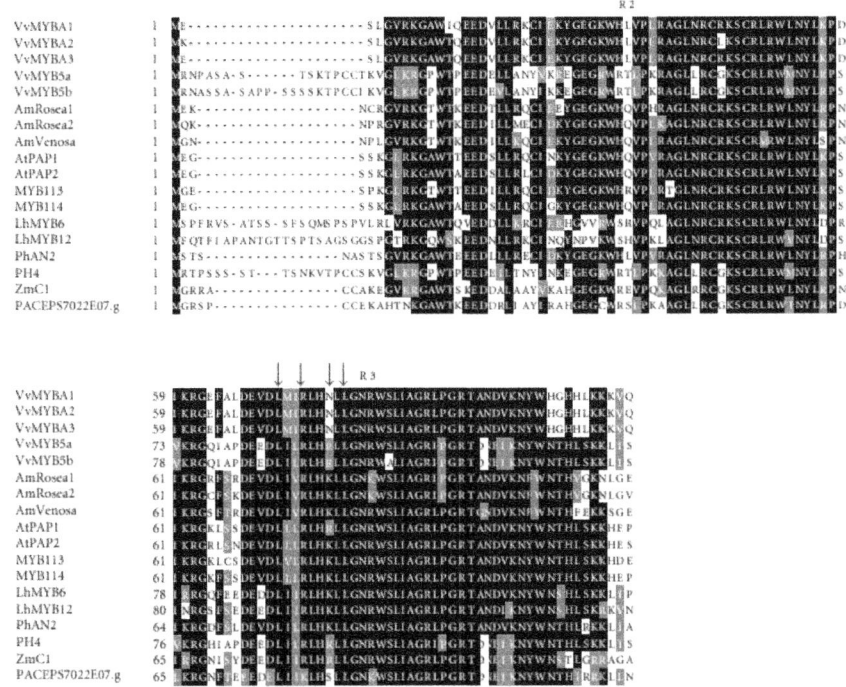

Figure 7: Multiple sequence alignment of the R2R3 MYB domains involved in anthocyanin production including the deduced amino acid sequence of Passiflora suberosa. R2R3 repeats refer to two imperfect repeats of the MYB domain. The identical and similar residues are highlighted on a black and gray background, respectively. Red box shows the R/B like bHLH interacting motif in the R3 repeat [45], and arrows indicate four specific residues of maize C1 required for interaction with a bHLH cofactor R [46]. Blue box shows a conserved motif in the R2R3 repeats for eudicots MYB related to the anthocyanin pigments [47]. The alignment was performed using CLUSTALX and BOXSHADE program.

A phylogenetic tree of selected plant R2R3-MYB transcription factors, including PACEPS7022E07.g, was constructed using the alignment of the conserved R2R3 repeats (Figure 8). The Passiflora sequence was placed within the clade including ZMC1 (Zea mays), PhPH4 (Petunia hybrida), VvMYB5a, and VvMYB5b (Vitis vinifera), which are known to be involved in the regulation of the anthocyanin pathway in these species [49–51].

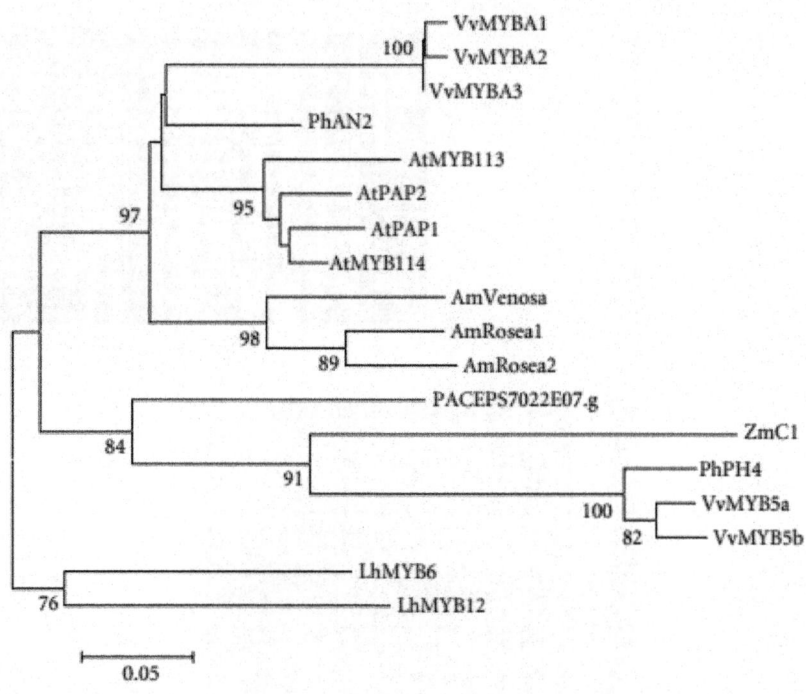

Figure 8: A Neighbor-joining phylogenetic tree of plant R2R3 MYB sequences. Bootstrap values from 1,000 replicates were used to assess the robustness of the trees. Only bootstrap values above 75% are indicated at the nodes. Accession numbers for genes from other species are given in Supplementary data.

Sequence comparison of selected plant WD40 proteins with the sequence obtained from P. edulis indicated that the four WD repeats are highly conserved among all species analyzed (Figure 9). Phylogenetic analysis of these amino acid sequences confirmed that P. edulis WD40 grouped together with Ricinus communis WD40 and found to be more related to other dicot proteins (Figure 10).

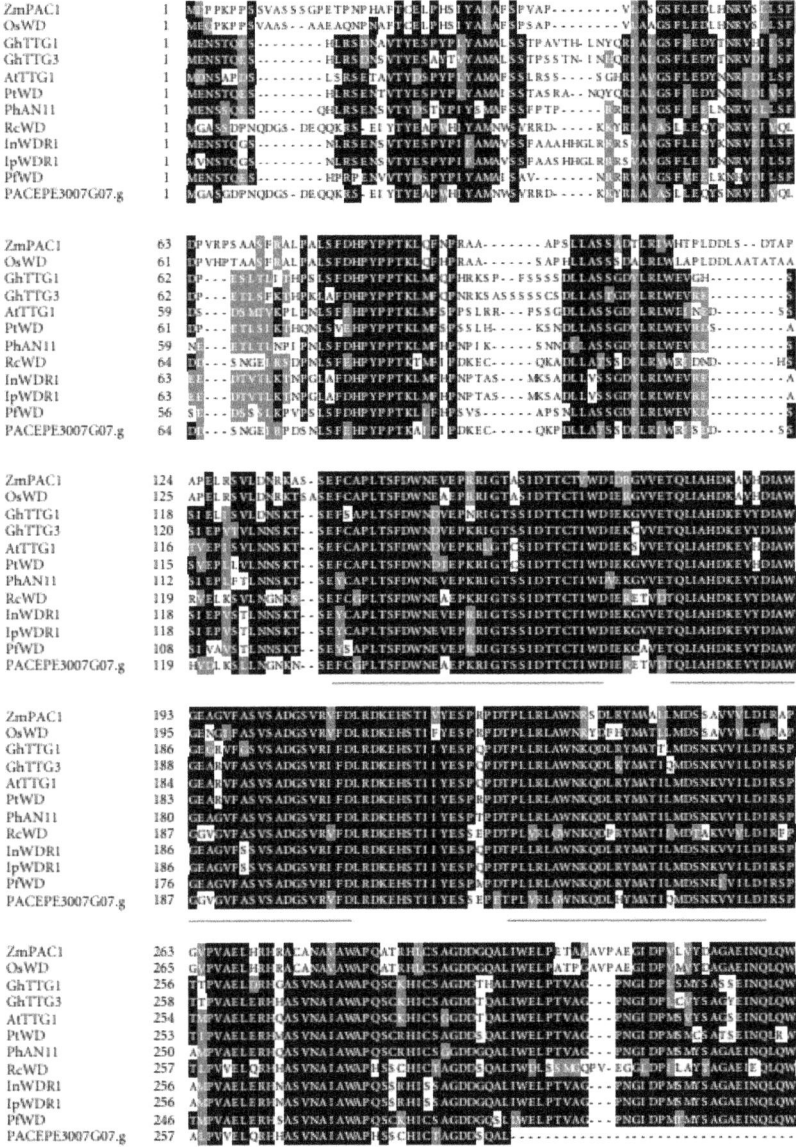

Figure 9: Multiple sequence alignment of the WD40 proteins involved in anthocyanin production, including the deduced amino acid sequence of the Passiflora edulis WD40. The identical and similar residues are highlighted on a black and gray background, respectively. Four conserved WD repeat domain are underlined in red. The alignment was performed using CLUSTALX and BOXSHADE program.

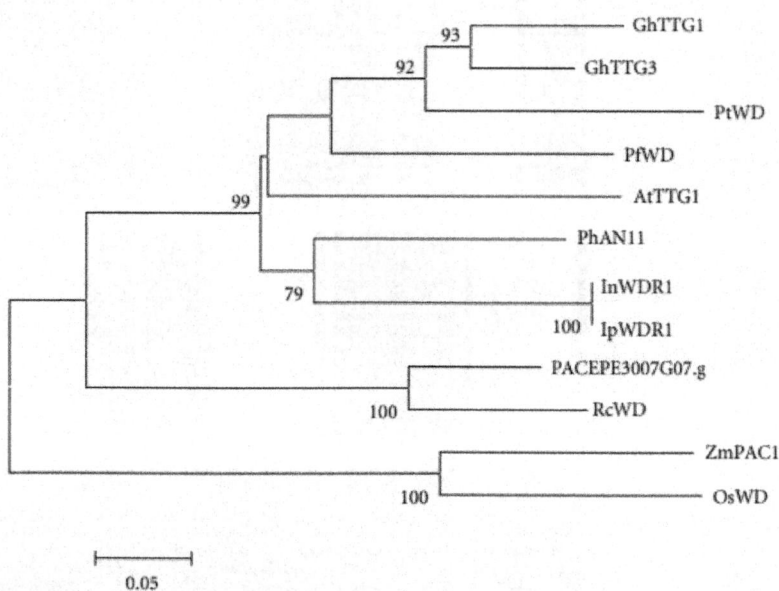

Figure 10: A Neighbor-joining phylogenetic tree of plant WD 40 proteins. Bootstrap values from 1,000 replicates were used to assess the robustness of the trees. Only bootstrap values above 75% are indicated at the nodes. Accession numbers for genes from other species are given in Supplementary data.

No putative homologs to bHLH transcription factors were found in the PASSIOMA database.

DISCUSSION

Flavonoid pathway results in the production of a range of flavonoid compounds, including anthocyanins (Figure 1). CHS is the first enzyme in the phenylpropanoid pathway and is encoded by members of a plant-specific multigene family of polyketide synthases. Nevertheless, genes belonging to the CHS family have been recently described to occur in some microorganisms (Azotobacter vinelandii; [52] and Neurospora crassa; [53]) and, therefore, indicate CHS functions might have evolved previous to the divergence of land plants. Thus, the biological functions of some of the CHS superfamily members are clearly important to plant adaptation. CHS proteins are collectively linked to the biosynthesis of different plant products with diverse functions such as UV protection, defense against pathogens, pigment biosynthesis, and pollen fertility [54, 55].

Sequence analysis indicated that two Passiflora CHS deduced proteins belong to a small distinct group of chalcone synthases that includes angiosperm and gymnosperms homologs to anther-specific chalcone synthase-like genes (ASCLs; highlighted in Figure 2). Furthermore, all ASCLs form a monophyletic clade. Recently, ASCLs transcripts were detected within the tapetum cells during microspore stage in wheat [56]. These genes apparently have important roles in anther development and in pollen fertility [40, 41, 56].

The remaining three Passiflora CHSs were clustered together in a sister clade containing all seed plant CHS genes. Their products are considered key in the biosynthesis of flavonoids. These include CHSA and CHSJgenes, known to be expressed in floral tissues, and involved in floral pigmentation in petunia [30, 31, 57]. Moreover, two nonchalcone genes, divergent from the typical CHSs, formed a separate clade. The SyPKS gene from cyanobacterium encodes an enzyme of the thiolase superfamily [58], whereas the function of thePpCHS11 gene (from Physcomitrella patens) may resemble more the most recent common ancestor of all plant CHSs than do other members of the plant CHS superfamily [55].

We do not have identified putative genes encoding CHI enzymes. Besides the general limitations and drawbacks of the EST-based approach, another possible explanation may be because the rapid isomerization of chalcone to form narigen and the fact that even in the absence of a functional CHI enzyme, chalcone can spontaneously isomerize to form naringenin [15].

DFR is an enzyme catalysing the reduction of three dihydroflavonols: dihydromyricetin (DHM), dihydroquercetin (DHQ), and dihydrokaempferol (DHK) into colorless leucoanthocyanidins. These are further converted to delphinidin, cyaniding, and pelargonidin (Figure 1). The synthesis of three different anthocyanidins is mainly determined by the enzymes activities of two hydroxylases: F3'OH and F3'5'OH. The first converts DHK to DHQ and F3'5'OH converts DHK to DHM [15].

In some plant species, DFR displays distinct substrate specificity in according to the hydroxylation pattern of anthocyanin molecule [30]. A hypothesis to determine substrate specificity was proposed based on the amino acid sequence alignment of Petunia DFR with others plants. The alignment indicated a variable region that controls substrate recognition. Naturally, Petunia hybrida does not produce orange flowers, because the DFR enzyme cannot use dihydrokaempferol as substrate to produce pelargonidin, due to an aspartic acid residue at the 134th position [30, 42], as it was also observed for Passiflora (Figure 3), thus converting dihydroquercetin to leucocyanidin and, more efficiently, the reduction of dihydromyricetin to leucodelphinidin [30, 59]. On the other hand, some Gerbera genotypes have an asparagine residue at this same position and can utilize three dihydroflavonols as substrates of DFR, consequently producing orange to red colored flowers [9, 30]. Thus, the flower color is partly determined by alteration of a single amino acid that changes the substrate specificity of the DFR enzyme.

Almost all anthocyanidins undergo several modifications, which vary across species and involve enzymes of the glucosyltransferase, methyltransferase, and acyltransferase families. The most common is glycosylation of the 3-position of anthocyanidins (represented in Figure 1) to produce stable anthocyanin molecules [15, 30, 31,60]. UDP-glucose:flavonoid 3-O-glucosyltransferase (3GT) belongs to a large multigene glucosyltransferases (GTs) family, representing the final step in anthocyanin biosynthesis.

In this work, we adopted the classification of the GTs into clusters according to Kovinic and colleagues [45]. Cluster I groups includes 3GTs enzymes. Cluster II includes GTs with multiples substrates preferences, generally for chalcones, flavones and flavonols but not anthocyanidins. Enzymes from Cluster III have isoflavone 7-O and anthocyanidin 3,5-O-GT activities. Cluster IV glycosylates flavonol and isoflavonol substrates and Cluster V have anthocyanin 5-O and/or flavonol 7-O-UGT enzymes [45]. Our results indicated that the obtained Passiflora glucosyltransferase gene sequences were grouped in Cluster II, together with other family members that show a high catalytic specificity for more than one class of flavonoid substrates (Figure 5). DicGT5 (from Dianthus caryophyllus) glycosylates a chalcononaringenin 2'-O-glucosyltransferase [61], whereas the Beta vulgaris GT has a favonoid-7, 4'-O-betanidin-5-O-glucosyltransferase activity [62]. Both GTs have non-anthocyanidin substrate specificity. Despite these results, obviously neither GT substrate specificity, nor in vivo function of the Passiflora GTs can be predicted solely based on amino acid sequence similarities and must be experimentally determined.

Anthocyanin biosynthesis has been demonstrated to occur predominantly in the cytosol, but these pigments are exclusively accumulated in the vacuole of epidermal cells [20]. Transport of pigments to the vacuoles requires a glutathione S-transferase and a specific carrier protein localized in the vacuolar membrane. GSTs are multifunctional proteins encoded by a large familiar present in all cellular organisms. Plants GSTs are classified on the basis of sequence identity into four classes: phi, tau, theta, and zeta [46]. The two small zeta and theta classes include GSTs from animals and plants, while the phi and tau classes are plant-specific. Several studies have confirmed the involvement of GSTs in the vacuolar transport of anthocyanins. PhAN2 (from Petunia), ZmBZ2 (from maize), and AtTT19 (from Arabidopsis) are GST proteins involved in anthocyanin transport [30–32, 63–65].

To characterize their phylogenetic relationships, the deduced amino acid sequences from the Passiflora putative GSTs were compared with other plant GST sequences, including the ones mentioned above. Figure 6 shows that the Passiflora GSTs are included into three different clades: three sequences were positioned in the same clade of PhAN9 and AtTT19 (phi class), whereas one sequence was grouped together with ZmBZ2 (tau class; [66]). Although of these known proteins belong to distinct GST clades, they perform similar functions [63–65].

Interestingly, PACEPE3007A05.g was clustered with carnation (Dianthus caryophyllus) GST type II (zeta class) which is associated to petal senescence in response to ethylene [67, 68].

At the moment, we can classify the Passiflora GSTs into type I (phi), type II (zeta), and type III (tau). At least, four of them might be involved in the anthocyanin pathway and PACEPE3007A05.g might be related to other biological processes related to flower development such as those observed for the carnation GST.

In all analyzed species, the spatial and temporal expression of the structural genes of the anthocyanin biosynthetic pathway is controlled by regulatory genes, which interfere with the intensity and pattern of anthocyanin biosynthesis [15]. MYBs, basic helix-loop-helix (bHLH) transcription factors and WD40 proteins form a transcriptional complex for the activation of the structural genes [4, 12, 20, 47, 69, 70]. MYBs and bHLHs proteins are coded by large multigene families, and those associated with anthocyanin biosynthesis are characterized by a conserved DNA-binding domain consisting of two imperfect repeats (named R2R3), and a specific bHLH domain, respectively. These two gene families have been extensively studied in model plants such as Arabidopsis and maize [48, 49, 71].

A multiple sequence alignment of the R2R3 domains of selected MYB proteins known to be involved in anthocyanin biosynthesis regulation, and the deduced amino acid sequence of PACEPS7022E07.g confirmed the presence of the conserved R2R3-MYB domain in this P. suberosa sequence (Figure 7) as well as that of a second conserved domain in the R3 repeat (red box, Figure 7), which is known to be necessary for the interaction between MYB and bHLH transcription factors [48, 49]. Additionally, a third motif in the R3 repeat (ANDV, blue box in Figure 7) represents a conserved motif shared among all eudicot MYBs involved in the anthocyanin biosynthesis [72].

The phylogenetic tree obtained using the alignment shown in Figure 7 is presented in Figure 8 and indicates LhMYB6 and LhMYB12 clustered outside the eudicot clade. These two genes regulate anthocyanin biosynthesis in the flowers of lily (Lilium hybrid), a monocot [73]. One clade is formed exclusively by eudicot anthocyanin regulators (PhAn2, AtPAP1, AtPAP2, AmROSEA1, and AmROSEA2; [12, 71, 74–77]. Curiously, one regulator of the anthocyanin in maize (a monocot), ZmC1 was positioned in the same clade of other dicot members such as PhPH4 (from Petunia), VvMYB5a, and VvMYB5b (from Vitis), as well as the Passiflora R2R3-MYB sequence. PhPH4 is expressed in the petal epidermis and activates vacuolar acidification in petunia [50]. VvMYB5a and VvMYB5b genes are involved in the regulation of anthocyanin biosynthesis during grape berry development [51].

WD40 proteins are highly conserved and can be found in organisms that do not biosynthesize anthocyanins as algae, fungi, and animals [78, 79]. In plants, these proteins are involved in a plethora of developmental and biochemical functions. As an example, the Arabidopsis TRANSPARENT TESTA GLABRA 1 (TTG1), which is a WD40 protein, is involved in regulating trichome formation, anthocyanin biosynthesis, seed coat pigmentation, and seed coat mucilage production. A common feature of WD40 repeat proteins is that they facilitate protein-protein interactions between the MYB and bHLH proteins [22, 79].

The alignment of the Passiflora WD40 protein sequence with other known WD40s from different plant species revealed the presence of conserved WD40 motifs in the C-terminal region (Figure 9). The phylogenetic tree constructed based on this alignment is shown in Figure 10. The results indicated that the monocot sequences ZmPAC1 and OsWD clustered together, whereas the eudicot WD40s known to function as anthocyanin regulators were grouped into a different clade, with Passiflora WD40 being closely related to the Ricinus communis protein (RcWD, Table 1 and Figure 10). Although WD40 proteins are required to regulate anthocyanins and proanthocyanidin together with MYB and bHLH transcription factors, their potential involvement in other biological processes is enormous, therefore, it is premature to say what functions PACEPE3007G07.g might perform in Passiflora.

The fact that no putative homologs to bHLH transcription factors were found in the PASSIOMA database may reflect the high degree of novelty of most of the libraries of the PASSIOMA project indicating that full gene expression spectra was not completely achieved [27]. Perhaps a more deep sequencing effort would reveal that such homologs are indeed expressed in Passiflora flowers, as these elements are generally essential to MYB-WD40 protein complex stability [30–32].

CONCLUSIONS AND PERSPECTIVES

We took the first steps toward the understanding of the molecular processes involved in the biosynthesis of anthocyanins in Passiflora that could account for the differences in pollinator preferences found in the genus. We identified 15 putative coding sequences derived from two distinct Passiflora species (P. edulis and P. suberosa) expressed in developing flower buds and potentially involved in the anthocyanin biosynthetic pathway. Comparisons of deduced amino acid sequences from the 15 Passiflora cDNAs with selected sequences from other plant species revealed strong similarity with genes that encode key elements involved in the biosynthesis (8 sequences), transcriptional regulation (2 sequences), and transport (5 sequences) of anthocyanin molecules.

Needed research concerning the determination of temporal and spatial expression patterns of all thesePassiflora putative anthocyanin-related genes presented here are already ongoing in our group. We expect that future work on the manipulation of their expression patterns, using transgenic approaches, will help us to unravel important aspects relating anthocyanin biosynthesis, flower pigmentation, and flower pollination in rapidly changing tropical environments.

ACKNOWLEDGMENT

The authors acknowledge FAPESP and CNPq (Brazil) for financial support.

REFERENCES

1. J. B. Harborne, The Flavonoids: Advances in Research Since 1986, Chapman & Hall/CRC, New York, NY, USA, 1st edition, 1994.

2. E. Grotewold, "The genetics and biochemistry of floral pigments," Annual Review of Plant Biology, vol. 57, pp. 761–780, 2006.

3. M. D. Rausher, "Evolutionary transitions in floral color," International Journal of Plant Sciences, vol. 169, no. 1, pp. 7–21, 2008.

4. F. Quattrocchio, J. Wing, K. van der Woude et al., "Molecular analysis of the anthocyanin2 gene of Petunia and its role in the evolution of flower color," Plant Cell, vol. 11, no. 8, pp. 1433–1444, 1999.

5. C. Spelt, F. Quattrocchio, J. N. M. Mol, and R. Koes, "Anthocyanin1 of Petunia encodes a basic helix-loop-helix protein that directly activates transcription of structural anthocyanin genes," Plant Cell, vol. 12, no. 9, pp. 1619–1631, 2000.

6. W. Heller, G. Forkmann, L. Britsch, and H. Grisebach, "Enzymatic reduction of (+)-dihydroflavonols to flavan-3,4-cis-diols with

flower extracts from Matthiola incana and its role in anthocyanin biosynthesis,"Planta, vol. 165, no. 2, pp. 284–287, 1985.

7.   K. Stich, T. Eidenberger, F. Wurst, and G. Forkmann, "Enzymatic conversion of dihydroflavonols to flavan-3,4-diols using flower extracts of Dianthus caryophyllus L. (carnation)," Planta, vol. 187, no. 1, pp. 103–108, 1992.

8.   K. M. Davies, J. M. Bradley, K. E. Schwinn, K. R. Markham, and E. Podivinsky, "Flavonoid biosynthesis in flower petals of five lines of lisianthus (Eustoma grandiflorum Grise.)," Plant Science, vol. 95, no. 1, pp. 67–77, 1993.

9.   Y. Helariutta, P. Elomaa, M. Kotilainen, P. Seppänen, and T. H. Teeri, "Cloning of cDNA coding for dihydroflavonol-4-reductase (DFR) and characterization of dfr expression in the corollas of Gerbera hybrida var. Regina (Compositae)," Plant Molecular Biology, vol. 22, no. 2, pp. 183–193, 1993.

10.   E. Grotewold, B. J. Drummond, B. Bowen, and T. Peterson, "The myb-homologous P gene controls phlobaphene pigmentation in maize floral organs by directly activating a flavonoid biosynthetic gene subset," Cell, vol. 76, no. 3, pp. 543–553, 1994.

11.   J. M. Hernandez, G. F. Heine, N. G. Irani et al., "Different mechanisms participate in the R-dependent activity of the R2R3 MYB transcription factor C1," The Journal of Biological Chemistry, vol. 279, no. 46, pp. 48205–48213, 2004.

12.   K. Schwinn, J. Venail, Y. Shang et al., "A small family of MYB-regulatory genes controls floral pigmentation intensity and patterning in the Genus antirrhinum," Plant Cell, vol. 18, no. 4, pp. 831–851, 2006.

13.   Y. Morita, M. Saitoh, A. Hoshino, E. Nitasaka, and S. Iida, "Isolation of cDNAs for R2R3-MYB, bHLH and WDR transcriptional regulators and identification of c and ca mutations conferring white flowers in the Japanese morning glory," Plant and Cell Physiology, vol. 47, no. 4, pp. 457–470, 2006.

14.   K. I. Park, N. Ishikawa, Y. Morita, J. D. Choi, A. Hoshino, and S. Iida, "A bHLH regulatory gene in the common morning glory, Ipomoea purpurea, controls anthocyanin biosynthesis in flowers, proanthocyanidin and phytomelanin pigmentation in seeds, and seed trichome formation," Plant Journal, vol. 49, no. 4, pp. 641–654, 2007.

15.   T. A. Holton and E. C. Cornish, "Genetics and biochemistry of anthocyanin biosynthesis," Plant Cell, vol. 7, no. 7, pp. 1071–1083, 1995.

16. R. Koes, W. Verweij, and F. Quattrocchio, "Flavonoids: a colorful model for the regulation and evolution of biochemical pathways," Trends in Plant Science, vol. 10, no. 5, pp. 236–242, 2005.

17. K. Davies, "Plant pigments and their manipulation," Annual Plant Reviews, vol. 14, p. 368, 2004.

18. H. S. Lee and V. Hong, "Chromatographic analysis of anthocyanins," Journal of Chromatography, vol. 624, no. 1-2, pp. 221–234, 1992.

19. J. B. Harborne, T. J. Mabry, and H. Mabry, The Flavonoids, Academic Press, New York, NY, USA, 1975.

20. J. Mol, E. Grofewold, and R. Koes, "How genes paint flowers and seeds," Trends in Plant Science, vol. 3, no. 6, pp. 212–217, 1998.

21. B. Winkel-Shirley, "Flavonoid biosynthesis. A colorful model for genetics, biochemistry, cell biology, and biotechnology," Plant Physiology, vol. 126, no. 2, pp. 485–493, 2001.

22. N. A. Ramsay and B. J. Glover, "MYB-bHLH-WD40 protein complex and the evolution of cellular diversity," Trends in Plant Science, vol. 10, no. 2, pp. 63–70, 2005.

23. L. Lepiniec, I. Debeaujon, J. M. Routaboul et al., "Genetics and biochemistry of seed flavonoids," Annual Review of Plant Biology, vol. 57, pp. 405–430, 2006.

24. T. Ulmer and J. M. MacDougal, Passiflora, Passion Flowers of the World, Timber Press, Cambridge, UK, 2004.

25. M. M. Halim and R. P. Collins, "Anthocyanins of Passiflora quadrangularis," Bulletin of the Torrey Botanical Club, vol. 97, no. 5, pp. 247–248, 1970.

26. L. KidØy, A. M. Nygård, Ø. M. Andersen, A. T. Pedersen, D. W. Aksnes, and B. T. Kiremire, "Anthocyanins in fruits of Passiflora edulis and P. suberosa," Journal of Food Composition and Analysis, vol. 10, no. 1, pp. 49–54, 1997.

27. M. C. Dornelas, S. M. Tsai, and A. P. M. Rodriguez, "Expressed sequence tags of genes involved in the flowering process of Passiflora spp.," in Floriculture, Ornamental and Plant Biotechnology, J. A. Teixeira da Silva, Ed., pp. 483–488, Global Science Books, London, UK, 2006.

28. G. Varassin, J. R. Trigo, and M. Sazima, "The role of nectar production, flower pigments and odour in the pollination of four species of Passiflora (Passifloraceae) in south-eastern Brazil," Botanical Journal of the Linnean Society, vol. 136, no. 2, pp. 139–152, 2001.

29. S. F. Altschul, W. Gish, W. Miller, E. W. Myers, and D. J. Lipman, "Basic local alignment search tool,"Journal of Molecular Biology, vol. 215, no. 3, pp. 403–410, 1990.

30. T. Nakatsuka, Y. Abe, Y. Kakizaki, S. Yamamura, and M. Nishihara, "Production of red-flowered plants by genetic engineering of multiple flavonoid biosynthetic genes," Plant Cell Reports, vol. 26, no. 11, pp. 1951–1959, 2007.

31. M. Hanumappa, G. Choi, S. Ryu, and G. Choi, "Modulation of flower colour by rationally designed dominant-negative chalcone synthase," Journal of Experimental Botany, vol. 58, no. 10, pp. 2471–2478, 2007.

32. S. Martens, A. Preuß, and U. Matern, "Multifunctional flavonoid dioxygenases: flavonol and anthocyanin biosynthesis in Arabidopsis thaliana L," Phytochemistry, vol. 71, no. 10, pp. 1040–1049, 2010.

33. X. Huang and A. Madan, "CAP3: a DNA sequence assembly program," Genome Research, vol. 9, no. 9, pp. 868–877, 1999.

34. T. A. Hall, "BioEdit: a user-friendly biological sequence alignment editor and analysis program for Windows 95/98/NT," Nucleic Acids Symposium Series, vol. 41, no. 41, pp. 95–98, 1999.

35. S. Hunter, R. Apweiler, T. K. Attwood et al., "InterPro: the integrative protein signature database," Nucleic Acids Research, vol. 37, no. 1, pp. D211–D215, 2009.

36. J. D. Thompson, T. J. Gibson, F. Plewniak, F. Jeanmougin, and D. G. Higgins, "The CLUSTAL X windows interface: flexible strategies for multiple sequence alignment aided by quality analysis tools," Nucleic Acids Research, vol. 25, no. 24, pp. 4876–4882, 1997.

37. S. Kumar, K. Tamura, and M. Nei, "MEGA3: integrated software for molecular evolutionary genetics analysis and sequence alignment," Briefings in Bioinformatics, vol. 5, no. 2, pp. 150–163, 2004.

38. N. Saitou and M. Nei, "The neighbor-joining method: a new method for reconstructing phylogenetic trees," Molecular Biology and Evolution, vol. 4, no. 4, pp. 406–425, 1987.

39. M. J. Moore, P. S. Soltis, C. D. Bell, J. G. Burleigh, and D. E. Soltis, "Phylogenetic analysis of 83 plastid genes further resolves the early diversification of eudicots," Proceedings of the National Academy of Sciences of the United States of America, vol. 107, no. 10, pp. 4623–4628, 2010.

40. Ageez, Y. Kazama, R. Sugiyama, and S. Kawano, "Male-fertility genes expressed in male flower buds ofSilene latifolia include homologs of

anther-specific genes," Genes and Genetic Systems, vol. 80, no. 6, pp. 403–413, 2005.

41. Jiang, C. K. Schommer, S. Y. Kim, and D. Y. Suh, "Cloning and characterization of chalcone synthase from the moss, Physcomitrella patens," Phytochemistry, vol. 67, no. 23, pp. 2531–2540, 2006.

42. M. Beld, C. Martin, H. Huits, A. R. Stuitje, and A. G. M. Gerats, "Flavonoid synthesis in Petunia hybrida: partial characterization of dihydroflavonol-4-reductase genes," Plant Molecular Biology, vol. 13, no. 5, pp. 491–502, 1989.

43. E. T. Johnson, S. Ryu, H. Yi, B. Shin, H. Cheong, and G. Choi, "Alteration of a single amino acid changes the substrate specificity of dihydroflavonol 4-reductase," Plant Journal, vol. 25, no. 3, pp. 325–333, 2001.

44. N. Shimada, R. Sasaki, S. Sato et al., "A comprehensive analysis of six dihydroflavonol 4-reductases encoded by a gene cluster of the Lotus japonicus genome," Journal of Experimental Botany, vol. 56, no. 419, pp. 2573–2585, 2005.

45. N. Kovinich, A. Saleem, J. T. Arnason, and B. Miki, "Functional characterizationofaUDP-glucose:flavonoid3-O-glucosyltransferasefrom the seed coat of black soybean (Glycine max (L.) Merr.),"Phytochemistry, vol. 71, no. 11-12, pp. 1253–1263, 2010.

46. D. P. Dixon, A. Lapthorn, and R. Edwards, "Plant glutathione transferases," Genome Biology, vol. 3, no. 3, pp. 1–10, 2002.

47. Y. Borovsky, M. Oren-Shamir, R. Ovadia, W. De Jong, and I. Paran, "The A locus that controls anthocyanin accumulation in pepper encodes a MYB transcription factor homologous to Anthocyanin2 of Petunia," Theoretical and Applied Genetics, vol. 109, no. 1, pp. 23–29, 2004.

48. M. Zimmermann, M. A. Heim, B. Weisshaar, and J. F. Uhrig, "Comprehensive identification ofArabidopsis thaliana MYB transcription factors interacting with R/B-like BHLH proteins," Plant Journal, vol. 40, no. 1, pp. 22–34, 2004.

49. E. Grotewold, M. B. Sainz, L. Tagliani, J. M. Hernandez, B. Bowen, and V. L. Chandler, "Identification of the residues in the Myb domain of maize C1 that specify the interaction with the bHLH cofactor R,"Proceedings of the National Academy of Sciences of the United States of America, vol. 97, no. 25, pp. 13579–13584, 2000.

50. F. Quattrocchio, W. Verweij, A. Kroon, C. Spelt, J. Mol, and R. Koes, "PH4 of petunia is an R2R3 MYB protein that activates vacuolar acidification through interactions with basic-helix-loop-helix transcription factors of the anthocyanin pathway," Plant Cell, vol. 18, no. 5, pp. 1274–1291,

2006.

51. L. Deluc, J. Bogs, A. R. Walker et al., "The transcription factor VvMYB5b contributes to the regulation of anthocyanin and proanthocyanidin biosynthesis in developing grape berries," Plant Physiology, vol. 147, no. 4, pp. 2041–2053, 2008.

52. N. Funa, H. Ozawa, A. Hirata, and S. Horinouchi, "Phenolic lipid synthesis by type III polyketide synthases is essential for cyst formation in Azotobacter vinelandii," Proceedings of the National Academy of Sciences of the United States of America, vol. 103, no. 16, pp. 6356–6361, 2006.

53. N. Funa, T. Awakawa, and S. Horinouchi, "Pentaketide resorcylic acid synthesis by type III polyketide synthase from Neurospora crassa," The Journal of Biological Chemistry, vol. 282, no. 19, pp. 14476–14481, 2007.

54. M. L. Durbin, B. McCaig, and M. T. Clegg, "Molecular evolution of the chalcone synthase multigene family in the morning glory genome," Plant Molecular Biology, vol. 42, no. 1, pp. 79–92, 2000.

55. P. K. H. Koduri, G. S. Gordon, E. I. Barker, C. C. Colpitts, N. W. Ashton, and D. Y. Suh, "Genome-wide analysis of the chalcone synthase superfamily genes of Physcomitrella patens," Plant Molecular Biology, vol. 72, no. 3, pp. 247–263, 2010.

56. S. Wu, S. J. B. O›Leary, S. Gleddie, F. Eudes, A. Laroche, and L. S. Robert, "A chalcone synthase-like gene is highly expressed in the tapetum of both wheat (Triticum aestivum L.) and triticale (x Triticosecale Wittmack)," Plant Cell Reports, vol. 27, no. 9, pp. 1441–1449, 2008.

57. R. E. Koes, C. E. Spelt, P. J. M. van den Elzen, and J. N. M. Mol, "Cloning and molecular characterization of the chalcone synthase multigene family of Petunia hybrida," Gene, vol. 81, no. 2, pp. 245–257, 1989.

58. C. Jiang, S. Y. Kim, and D. Y. Suh, "Divergent evolution of the thiolase superfamily and chalcone synthase family," Molecular Phylogenetics and Evolution, vol. 49, no. 3, pp. 691–701, 2008.

59. G. Forkmann and B. Ruhnau, "Distinct substrate specificity of dihydroflavonol-4-reductase from flowers of Petunia hybrida," Zeitschrift für Naturforschung—Section C: Biosciences, vol. 42, pp. 1146–1148, 1987.

60. S. Rudd, "Expressed sequence tags: alternative or complement to whole genome sequences?" Trends in Plant Science, vol. 8, no. 7, pp. 321–329, 2003.

61. J. Ogata, Y. Itoh, M. Ishida, H. Yoshida, and Y. Ozeki, "Cloning and heterologous expression of cDNAs encoding flavonoid glucosyltransferases from Dianthus caryophyllus," Plant Biotechnology, vol. 21, no. 5, pp. 367–375, 2004.

62. J. Isayenkova, V. Wray, M. Nimtz, D. Strack, and T. Vogt, "Cloning and functional characterisation of two regioselective flavonoid glucosyltransferases from Beta vulgaris," Phytochemistry, vol. 67, no. 15, pp. 1598–1612, 2006.

63. M. R. Alfenito, E. Souer, C. D. Goodman et al., "Functional complementation of anthocyanin sequestration in the vacuole by widely divergent glutathione S-transferases," Plant Cell, vol. 10, no. 7, pp. 1135–1149, 1998.

64. K. A. Marrs, M. R. Alfenito, A. M. Lloyd, and V. Walbot, "A glutathione S-transferase involved in vacuolar transfer encoded by the maize gene Bronze-2," Nature, vol. 375, no. 6530, pp. 397–400, 1995.

65. S. Kitamura, N. Shikazono, and A. Tanaka, "TRANSPARENT TESTA 19 is involved in the accumulation of both anthocyanins and proanthocyanidins in Arabidopsis," Plant Journal, vol. 37, no. 1, pp. 104–114, 2004.

66. L. Loyall, K. Uchida, S. Braun, M. Furuya, and H. Frohnmeyer, "Glutathione and a UV light-induced glutathione S-transferase are involved in signaling to chalcone synthase in cell cultures," Plant Cell, vol. 12, no. 10, pp. 1939–1950, 2000.

67. R. C. Meyer, P. B. Goldsbrough, and W. R. Woodson, "An ethylene-responsive flower senescence-related gene from carnation encodes a protein homologous to glutathione s-transferases," Plant Molecular Biology, vol. 17, no. 2, pp. 277–281, 1991.

68. H. Itzhaki, J. M. Maxson, and W. R. Woodson, "An ethylene-responsive enhancer element is involved in the senescence- related expression of the carnation glutathione-S-transferase (GST1) gene," Proceedings of the National Academy of Sciences of the United States of America, vol. 91, no. 19, pp. 8925–8929, 1994.

69. P. Elomaa, A. Uimari, M. Mehto, V. A. Albert, R. A. E. Laitinen, and T. H. Teeri, "Activation of anthocyanin biosynthesis in Gerbera hybrida (Asteraceae) suggests conserved protein-protein and protein-promoter interactions between the anciently diverged monocots and eudicots," Plant Physiology, vol. 133, no. 4, pp. 1831–1842, 2003.

70. H. Mathews, S. K. Clendennen, C. G. Caldwell et al., "Activation tagging in tomato identifies a transcriptional regulator of anthocyanin

biosynthesis, modification, and transport," Plant Cell, vol. 15, no. 8, pp. 1689–1703, 2003.

71. R. Stracke, M. Werber, and B. Weisshaar, "The R2R3-MYB gene family in Arabidopsis thaliana," Current Opinion in Plant Biology, vol. 4, no. 5, pp. 447–456, 2001.

72. K. Lin-Wang, K. Bolitho, and K. Grafton, "An R2R3 MYB transcription factor associated with regulation of the anthocyanin biosynthetic pathway in Rosaceae," BMC Plant Biology, vol. 10, no. 1, pp. 50–67, 2010.

73. M. Yamagishi, Y. Shimoyamada, T. Nakatsuka, and K. Masuda, "Two R2R3-MYB genes, homologs of petunia AN2, regulate anthocyanin biosyntheses in flower tepals, tepal spots and leaves of asiatic hybrid Lily," Plant and Cell Physiology, vol. 51, no. 3, pp. 463–474, 2010.

74. C. Jiang, X. Gu, and T. Peterson, "Identification of conserved gene structures and carboxy-terminal motifs in the Myb gene family of Arabidopsis and Oryza sativa L. ssp. indica," Genome Biology, vol. 5, no. 7, p. R46, 2004.

75. S. Kobayashi, M. Ishimaru, K. Hiraoka, and C. Honda, "Myb-related genes of the Kyoho grape (Vitis labruscana) regulate anthocyanin biosynthesis," Planta, vol. 215, no. 6, pp. 924–933, 2002.

76. R. Walker, E. Lee, J. Bogs, D. A. J. McDavid, M. R. Thomas, and S. P. Robinson, "White grapes arose through the mutation of two similar and adjacent regulatory genes," Plant Journal, vol. 49, no. 5, pp. 772–785, 2007.

77. Gonzalez, M. Zhao, J. M. Leavitt, and A. M. Lloyd, "Regulation of the anthocyanin biosynthetic pathway by the TTG1/bHLH/Myb transcriptional complex in Arabidopsis seedlings," Plant Journal, vol. 53, no. 5, pp. 814–827, 2008.

78. N. de Vetten, F. Quattrocchio, J. Mol, and R. Koes, "The an11 locus controlling flower pigmentation in petunia encodes a novel WD-repeat protein conserved in yeast, plants, and animals," Genes and Development, vol. 11, no. 11, pp. 1422–1434, 1997.

79. J. Brueggemann, B. Weisshaar, and M. Sagasser, "A WD40-repeat gene from Malus × domestica is a functional homologue of Arabidopsis thaliana TRANSPARENT TESTA GLABRA1," Plant Cell Reports, vol. 29, no. 3, pp. 285–294, 2010.

Chapter 2

CAROTENOIDS AND THEIR ISOMERS: COLOR PIGMENTS IN FRUITS AND VEGETABLES

Hock-Eng Khoo [1] , K. Nagendra Prasad [1] , Kin-Weng Kong [1] , Yueming Jiang [2] and Amin Ismail [1,3]

[1] Department of Nutrition and Dietetics, Faculty of Medicine and Health Sciences, Universiti Putra Malaysia, 43400 UPM Serdang, Selangor, Malaysia

[2] South China Botanical Garden, Chinese Academy of Sciences, Guangzhou 510650, China

[3] Laboratory of Analysis and Authentication, Halal Products Research Institute, Universiti Putra Malaysia, 43400 UPM Serdang, Selangor, Malaysia

ABSTRACT

Fruits and vegetables are colorful pigment-containing food sources. Owing to their nutritional benefits and phytochemicals, they are considered as 'functional food ingredients'. Carotenoids are some of the most vital colored phytochemicals, occurring as all-trans and cis-isomers, and accounting for the brilliant colors of a variety of fruits and vegetables. Carotenoids extensively studied in this regard include β-carotene, lycopene, lutein and zeaxanthin. Coloration of fruits and vegetables depends on their growth maturity, concentration of carotenoid isomers, and food processing methods. This article focuses more on several carotenoids and their isomers present in different fruits and vegetables along with their concentrations. Carotenoids and their geometric isomers also play an important role in protecting cells from oxidation and cellular damages.

INTRODUCTION

Increasing interest in nutrition, fitness and beauty consciousness has enhanced concerns over a healthy diet. Fruits and vegetables have assumed the status of 'functional' foods, capable of providing additional health benefits, like prevention or delaying onset of chronic diseases, as well as meeting basic nutritional requirements. Appropriate intake of a variety of fruits and vegetables ensures sufficient supply of nutrients and phytochemicals such as carotenoids. Low consumption of fruit and vegetable is among the top ten risk factors resulting in the global mortality. Annually, 2.7 million lives could be saved with sufficient consumption of various kinds of fruits and vegetables [1]. Nowadays, food scientists have collaborated with nutrition researchers to develop plant-based functional foods to promote healthy eating habits. In food research, carotenoids from fruits and vegetables have attracted a great deal of attention, mainly focused on the analysis of geometric carotenoid isomers. Carotenoids found in fruits and vegetables have also attracted great attention for their functional properties, health benefits and prevention of several major chronic diseases [2-4]. Carotenoids are synthesized in plants but not in animals. In nature, more than 600 types of carotenoid have been determined. Carotenoids are localized in subcellular organelles (plastids), i.e. chloroplasts and chromoplasts. In chloroplasts, the carotenoids are chiefly associated with proteins and serve as accessory pigments in photosynthesis, whereas in chromoplasts they are deposited in crystalline form or as oily droplets [5]. Some of the carotenoids such as the xanthophylls are involved in photosynthesis by participating in energy transfer in the presence of chlorophyll in plants [6]. Studies have shown that carotenoids contribute to the yellow color found in many fruits and vegetables [5,7]. The colors of fruits and vegetables depend on conjugated double bonds and the various functional groups contained in the carotenoid molecule [8]. A study also reported that the greater the number of conjugated double bonds, the higher the absorption maxima (λmax) [9]. As a result, the color ranges from yellow, red to orange in many fruits and vegetables [5,10]. Besides, esterification of carotenoids with fatty acids can also occur during fruit ripening, which may affect the color intensity [11]. Naturally, most of the carotenoids occur as trans-isomer in plants. However, cis-isomers may increase due to the isomerization of the trans-isomer of carotenoids during food processing [12]. Many studies have involved in the analysis of dietary carotenoids and their potential isomers [13-15], with much attention given to the geometric isomerization of carotenoids [16-21]. The investigation of carotenoid contents in fresh, frozen and canned foods has been carried out [22]. However, a recent review on contents of carotenoids and their isomers from diverse fruits and vegetables has not been made. The data collected from published literatures will be useful for food researchers, nutritionists and health practitioners in promoting right diets to minimize vitamin A deficiency and maintaining a healthy dietary practice.

CAROTENOIDS AND THEIR ISOMERS

There are many factors influencing the formation and isomerization of carotenoids. Heat, light, and structural differences are the prominent factors that affect the isomerization of carotenoids in foods [23-25]. Various processing methods, such as heating and drying also lead to the isomerization and even degradation of carotenoids [26,27]. De Rigal et al. [24] reported that isomerization of carotenoids in apricot purees was due to enzymatic browning. Oxidative degradation of carotenoids has also led to cis-trans isomerization and formation of carotenoid epoxides [28,29]. Previous studies have shown that cis-isomer of carotenoids can be identified based on the absorption spectrum characteristics, Q ratios, and the relative intensity of the cis peak [8,30]. The UV spectrum of cis carotenoids is characterized with their λ_{max} between 330–350 nm, which has greatest intensity when the double bond is located near or at the center of the chromophore [31]. On the other hand, a hypsochromic shift in the λ_{max} and smaller extinction coefficient is observed. Thus, cis-trans isomerization of carotenoids leads to a decrease of color intensity [12]. Carotenoids that contain more than seven conjugated double bonds were reported to have stronger antioxidant capacity and protection against photo-bleaching of chlorophyll [32]. Di Mascio et al. [33] also reported $^1O_2$ quenching capability of carotenoids is based on the number of conjugated double bonds and not the ionone ring of β-carotene. As the geometric isomers of carotenoids make great contribution to antioxidant activities and health improvement, analyses of carotenoids and their isomers in fruits and vegetables are needed. Liquid chromatography (LC) enables separation and identification of individual carotenoids. Identification of carotenoid isomers can be achieved by high performance liquid chromatography (HPLC). The separation of carotenoid isomers can be done using either polymeric C_{30} or ODS-2 silica columns [34]. However, the identification of carotenoid isomers seemed to be ambiguous. In this review, the analyses of carotenoid geometric isomers and their levels are listed in Table 1, which also should enable researchers to understand the various carotenoid isomers present in different fruits and vegetables.

Table 1: Analyses of carotenoid isomers in fruits and vegetables

Fruit/ vegetable	Analytical method	Carotenoid and its isomer	Ref.
Bambangan (lyophilized pulp) [*Mangifera pajang* Kosterm.]	HPLC: Polymeric C30 column (150 mm × 4.6 mm i.d., 3 μm particle)	cryptoxanthin (mg/100 g): 1.18 α-carotene (mg/100 g) all-*trans* (7.96) β-carotene (mg/100 g) all-*trans* (20.04), 9-*cis* (2.72), *cis*-isomers (3.04–3.07)	[35]
Loquat (fresh) [*Eriobotrya japonica* (Thunb.) Lindl.]	HPLC-PDA-MS/MS: HPLC-MS: YMC C30 column (250 × 4.6 mm i.d., 5 μm particle)	β-cryptoxanthin (μg/100 g): all-*trans* (54.8–715.2); 9- or 90-*cis* (0.8); 13-or 130-*cis* (4.0–20.1); *cis*-5,6:50,60-diepoxy (1.8–3.5); 5,6:50,60-diepoxy (35.0–339.5), 5,8:50,60- or 5.6:50,80-diepoxy (1.8–34.8); *cis*-5,8.50,60- or 5,6:50,80-diepoxy (1.1–10.9), *cis*-5,6:50,60-diepoxy (1.9–213.9); 50,60-epoxy (11.5–109.4); 5,6-epoxy (19.0–213.9); 5,8-Epoxy (1.6–15.3) β-carotene (μg/100 g): all-*trans* (38.1–1441.5), 9-*cis* (1.6–18.0), 13-*cis* (5.0–45.9), 15-*cis* (0.7–4.8)	[36]

Mango (dried pulp) [*Mangifera indica* L.]	HPLC: Polymeric C30 column (250 mm × 4.6 mm i.d., 5 μm particle)	Neoxanthin (μg/g): all-*trans* (0.44–0.71); *cis*-isomers (0.19–0.57) Violaxanthin (μg/g): all-*trans* (0.16–0.32); *cis*-isomers (0.10–4.70) Zeaxanthin (μg/g): all-*trans* (0.89–1.33); *cis*-isomers (0.72–0.96) Lutein (μg/g): 9- or 9'-*cis* (0.53–0.78) β-carotene (μg/g): all-*trans* (9.32–29.34); 13- or 13'-*cis* (0.78–3.79); 15- or 15'-*cis* (0.98–7.20); *cis*-isomers (0.35–0.70)	[37,38]
Peach (fresh) [*Prunus persica* (L.) Batsch]	HPLC: Polymeric C30 column (250 mm × 4.6 mm i.d., 5 μm particle)	β-cryptoxanthin (μg/g): all-*trans* (0.3); 13/13'-*cis* (0.1); 15-*cis* (0.1) β-carotene (μg/g): all-*trans* (2.2); 9-*cis* (0.3); 13-*cis* (0.5); 15-*cis* (trace)	[39]
Tree tomato (yellow) [*Solanum betaceum* Cav.]	HPLC-MS: YMC C30 column (250 × 4.6 mm i.d., 5 μm particle)	β-carotene (% residual carotenoid): all-*trans* (61.1–85.5); 13-*cis* (284.2–518.6) ζ-carotene (% residual carotenoid): *cis*-isomer (46.5–83.9)	[13]
Broccoli (fresh) [*Brassica oleracea* var. Italica]	HPLC: Polymeric C30 column (250 mm × 4.6 mm i.d., 5 μm particle)	β-carotene (μg/g): all-*trans* (29.2); 9-*cis* (5.0); 13-*cis* (3.3), 15-*cis* (1.9); *cis*-isomers (2.0)	[39]
Maize (mutant, fresh) [*Zea mays* L.]	HPLC: Spherisorb ODS-2 silica column (250 × 3.2 mm i.d., 5 μm particle)	ζ-carotene: di-*cis* (55.8) tri-*cis* (17.6–46.3)	[40]
Maize (kernel) - 13 varieties [*Zea mays* L.]	HPLC: Vydac218TP53 column (250 × 3.2 mm i.d.)	β-carotene (μg/100 g): all-*trans* (37–879); *cis*-isomers (<0.1–301)	[41]
Pumpkin (fresh) [*Curcubita moschata* var. Orange]	HPLC: Polymeric C30 column (250 mm × 4.6 mm i.d., 5 μm particle)	β-carotene (μg/g): all-*trans* (61.6); 9-*cis* (2.5); 13-*cis* (2.7)	[15]
Spinach (fresh) [*Spinacia oleracea* L.]	HPLC: Polymeric C30 column (250 mm × 4.6 mm i.d., 5 μm particle)	β-carotene (μg/g): all-*trans* (311.9); 9-*cis* (38.6); 13-*cis* (24.5), 15-*cis* (trace); *cis*-isomers (22.5)	[39]
Tomato (fresh) [*Solanum lycopersicum* L.]	HPLC: Polymeric C30 column (250 mm × 4.6 mm i.d., 5 μm particle)	β-carotene (μg/g): all-*trans* (71.0); 9-*cis* (4.8); 13-*cis* (5.8)	[39]

Carotenes

Carotenes include several related compounds having the general formula $C_{40}H_{56}$. They are a simple type of carotenoid and occur in several isomeric forms, such as alpha (α), beta (β), gamma (γ), delta (δ), epsilon (ε), and zeta (ζ) [42]. Among the various carotenoids, α- and β-carotene are the two primary forms of carotenes. In human body, β-carotene is broken down by β-carotene dioxygenase in the mucosa of small intestine into two retinyl molecules, which is later reduced to vitamin A (retinol) [43]. Carotenes can be found in many dark green and yellow leafy vegetables and appear as fat soluble pigments, while β-carotene can be found in yellow, orange and red colored fruits and vegetables [44]. Naturally, β-carotene is mostly found as all-trans isomers and lesser as cis-isomers (Figure 1), with the relative abundances in the following order: all-trans > 9-cis > 13-cis > 15-cis [45].

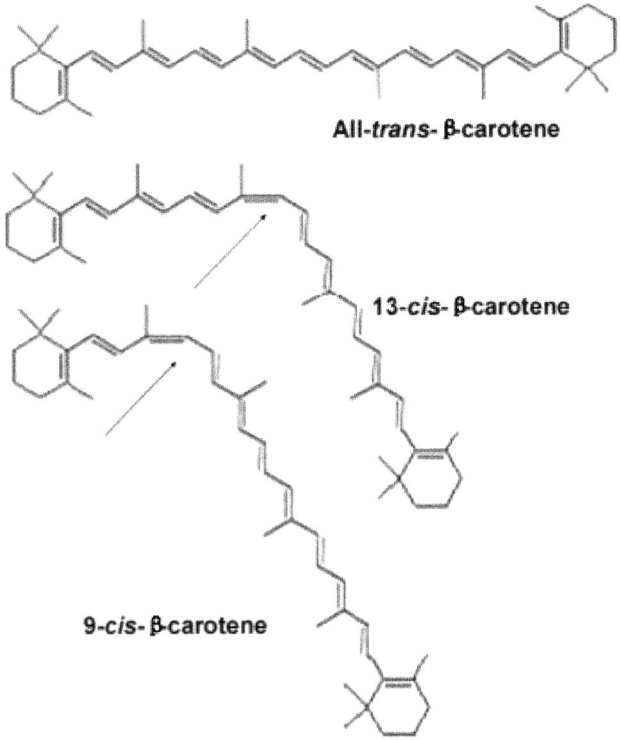

Figure 1: The structure of all-trans-β-carotene and its two geometric isomers [35,46].

All-trans-β-carotene is very unstable and can be easily isomerized into cis-isomers, when exposed to heat and light. Isomerization energy is involved in relocation of the single or double bond of one form of carotenoid into another [46,47]. A study has been carried out to determine the isomerization energy of carotenoids, especially neurosporene, spheroidene and spirilloxanthin [16], but the excited energy stages are not well understood. Besides, processing of fruit could result in significant cis-trans isomerization of β-carotene which was shown by the formation of 13-cis-β-carotene [48]. In regard to the effect of processing and isomerization of carotenoids in fruits and vegetables, 13-cis-β-carotene is the main product of geometric isomerization [49], 9-cis-β-carotene is formed when exposure to light [12,49], while 13-cis-α- and β-carotene isomers are formed during storage [50]. A study on the effect of β-carotene isomerization due to reflux heating has exhibited that degradation occurs to all-trans-β-carotene, with a significant increase in 13-cis-β-carotene [51]. Based on the structures of all-trans-β-carotenes, the double bonds can be relocated during heating and form several isomers (Figure 1) [46]. Marx et al. [52] have revealed that in pasteurized and sterilized samples, 13- cis-β-carotene was the only isomer formed during pasteurization and sterilization of carrot juice, while 9-cis-β-carotene was probably formed during blanching of sterilized carrot juice. Moreover, 9-cis- and 13-cis-β-carotenes were thought to originate independently from cis precursors by non-enzymatic isomerization of all-trans forms [53]. On the other hand, cis-β-carotene has been shown to isomerize into all-trans-isomer when heated and exposed to air [17]. It shows that isomerization of β-carotene occurs instead of degradation. The isomerization process was also known to occur when a crystaline β-carotene is heated at 90 °C and 140 °C in a nitrogen environment, which might be due to the partially melted β-carotene that has increased the probability of cis- to all-trans-β-carotene isomerization [17]. Carotene in all-trans form has higher bioavailability than its cis counterpart, while β-carotene and β-apo-12'-carotenal have the highest bioconversion rate at 100% and 120% (on a weight basis), respectively [54].

Lycopene

Lycopene is an unsaturated acyclic carotenoid with open straight chain hydrocarbon consisting of 11 conjugated and two unconjugated double bonds. Lycopene has no provitamin A activity due to the lack of terminal β-ionic ring as the basic structure for vitamin A [55]. Most of the lycopene occurs naturally in all-trans form [56]. The red color of lycopene is mainly due to many conjugated carbon double bonds, as it absorbs more visible spectrum compared to other carotenes [57]. Lycopene contains seven double bonds which can be isomerized to mono-cis- or poly-cis-isomers [58]. Based on the isomeric conformation of lycopene, 5-cis-lycopene was the most stable isomer, followed by alltrans- and 9-cis-lycopene [59]. Besides, 5-cis-lycopene has the lowest isomerization energy among other lycopene cis-isomers, and its very large rotational barrier restricts it to form all-trans structure [45]. More studies on isomerization energy are needed to explain the rationale on the conversion of alltrans-carotene to its cis-isomers by thermal processing, under low pH condition and exposure to light. Lycopene cis-isomers are more soluble in oil or organic solvents than all-trans-lycopene [60]. There is dissimilarity between the isomerization of β-carotene and lycopene [61]. Lycopene isomerization occurs under the simulated gastric digestion, thermal processing and low pH [62], but the effect of these conditions on lycopene isomerization is unclear. Boileau et al. [56] reviewed that isomerization of lycopene was found to occur in human body due to the effect of gastric juice in the stomach. However, Blanquet-Diot et al. [59] reported that no cis-trans isomerization of lycopene has occurred using gastrointestinal tract model. Heating at 60 °C and 80 °C favored the isomerization of lycopene [63]. The formation of 9-cis-lycopene is more favorable at low pH condition while 13-cis-lycopene is the major degradation product formed from thermal processing [62]. The uptake of cis-lycopene by intestinal cells is known to exceed those of all-trans-lycopene, which was in agreement with the study by Tyssandier et al. [64] that cis-lycopene had greater bioaccessibility compared to its all-trans form. Lycopene cis-isomers also found to have greater bioactivity and bioavailability than their all-trans counterpart [65]. Besides, lycopene is less bioavailable than β-carotene and lutein [66]. Processing method could help to release the lycopene from the matrix in fruits and vegetables, and thus increases bioavailability [12].

Xanthophylls

Xanthophylls are the oxidized derivatives of carotenes. Xanthophylls, with a general chemical formula $C_{40}H_{56}O_2$, contain hydroxyl groups and are more polar than carotenes [67]. In Nature, xanthophylls are found in the leaf of most plants and are synthesized within the plastids [68], which occur as yellow to red colored pigments. They are also considered accessory pigments, along with anthocyanins, carotenes, and sometimes phycobiliproteins [69]. Commonly found xanthophylls include lutein, zeaxanthin, and cryptoxanthin. In plant, violaxanthin, antheraxanthin and zeaxanthin participate in xanthophyll cycle, which involves the conversion of pigments from a non-energy-quenching form to energy-quenching forms [6]. Lutein is one kind of xanthophyll found abundantly in fruits and vegetables [44,65]. It is a fat soluble compound and very stable in emulsion [66]. Although, lutein and zeaxanthin are isomers but they are not stereoisomers. In addition, lutein is one of the xanthophyll discovered in egg yolk [67]. As animals cannot produce xanthophylls, xanthophylls found in animals are known to be ingested from food [68]. The isomers of xanthophyll are not well studied. Since the development of the C30 HPLC analytical column, the determination of xanthophyll isomers is becoming a hot issue. Identification of xanthophyll isomers has been carried out using different polymeric columns [69]. Study also reported that cis-isomers of xanthophyll determined using a C30 stationary phase were relatively higher than accessed using C18 column [70]. Tóth and Szabolcs [71] had identified 9-cis- and 9'-cis-isomers of antheraxanthin, capsanthin, lutein and lutein epoxide in several higher plants. They found that 9-cisisomers of antheraxanthin and lutein epoxide occurred without their 9'-cis counterparts in nonphotosynthetic tissues. This could be explained by the non-stereoselective biosynthesis or stereomutation, while the 9-cis form is protected stereoselectively against photoisomerization. Isomers of violaxanthin namely, 9-cis-, 13-cis- and di-cis-violaxanthin have been identified in orange juice [72,73]. Besides, lutein epoxide has been identified in dandelion petal, with high amounts of the 9-cis- and 9'-cis-isomers, with the all-trans form as the major carotenoid [74]. Moreover, 13-ciszeaxanthin was found as the major isomerization products of all-trans form, which was induced by light and temperature (35–39 °C) [75]. In one study by Kishimoto et al. [76], sixteen xanthophylls were isolated from the petals of chrysanthemum. These xanthophylls were mainly the isomers of violaxanthin, luteoxanthin, lutein, and also lutein epoxides. They also concluded that chrysanthemum petals have a unique carotenoid characteristic compared to the flowers of other species. Furthermore, Yahia et al. [77] reported that the saponified crude extract of mango fruit has all-trans-violaxanthin and 9-cis-violaxanthin present in the esterified form. In ripening fruit, esterification of xanthophylls occurs [10], but the mechanism and biosynthetic pathways of esterification are still to be explored.

CAROTENOID PIGMENTS IN FRUITS AND VEGETABLES

Carotenoids are widely distributed in the cellular tissues of plants [78]. The distribution of carotenoids in human tissues is originated from plant sources. Therefore, fruits and vegetables constitute the major source of carotenoids in human diet [79,80]. In plant, carotenoids are found as fat soluble and colored-pigments [81,82]. Carotenoids can be isolated from the grana of chloroplasts in the form of carotenoprotein complexes, which give various colors to the outer surfaces of the plants [83]. The visible colors of the plant are due to the conjugated double bonds of carotenoids that absorb light. The more number of double bonds results in the more absorbance of red color wavelength. The occurrences of carotenoids in plants are not as a single compound. Most of the carotenoids are bound with chlorophyll, and a combination of carotene-chlorophyll and xanthophyll-chlorophyll occurs often.

The binding of carotenoids to chlorophylls can give rise to a variety of colors in plants, fruits and vegetables. However, as fruit matures, the chlorophyll content decreases, and results in coloredcarotenoid pigments [84]. Besides, study had carried out to improve carotenoids color retention during ripening [85]. In nature, fruits have lesser xanthophyll contents compared to vegetables. Some fruits such as papaya (Carica papaya L.) and persimmon (Diospyros sp.) have high amount of xanthophylls (lutein and zeaxanthin), like that found in vegetables [44]. In fruits and vegetables, β-carotene is found to be bound to either chlorophylls or xanthophylls, forming chlorophyll-carotenoid complexes, which absorb light in the orange or red light spectrum and give rise to green, purple or blue coloration [86], These complexes could decrease the bioavailability of β-carotene and further weaken its bioefficacy for the conversion to vitamin A. However, this setback can be resolved by saponifying the plant extract to yield all-trans-β-carotene in a free-state form [77]. In vegetables, provitamin A carotenoids have lower bioavailability as compared to fruits [87], which may be due to their protein-complex structures in chloroplasts [54]. In this review, a comprehensive data for the typical carotenoids content in fruits and vegetables are given in Tables 2 and 3, where the carotenoids contents in fruits and vegetables are summarized. The data from this compilation are useful for comparison of the ongoing study with other previous reports.

Orange and Yellow Pigment Carotenoids

Naturally occurring β-carotene, with 11 double bonds, is orange in color [55]. Takyi [83] reported β-carotene occurs as an orange pigment, while α-carotene is a yellow pigment, which can be found in fruits and vegetables. Yellow colored fruits that contain low or trace amounts of β-carotene are mainly from the genera Ananas, Averrhoa, Citrus, Durio, Malus, Musa, Nephelium, Pyrus, Rubus and Vitis or vegetables from the genera Apium, Cucumis, Manihot, Vigna and Maranta (Tables 2 and 3). Besides, yellow maize (Zea mays L.) is a good source of β-carotene [88]. Several vegetables are known to contain β-carotene. For example, β-carotene is present in carrot, sweet potato and tomato which are from the genera of Daucus, Ipomea and Solanum, respectively. Carrot is the major contributor of β-carotene in the diet, along with green leafy vegetables. Rajyalakshmi et al. [89] reported that the β-carotene contents in 70 edible wild green leafy vegetables ranged from 0.4–4.05 mg per 100 gram edible portion. A few underutilized green leafy vegetables from India were also found to have 0.68–12.6 mg/100 g β-carotene [90]. Therefore, other than carotene-rich yellow-orange colored vegetables (e.g., carrot, pumpkin and sweet potato), green leafy vegetables are good sources of β-carotene. The β-carotene contents of some green leafy vegetable grown in the wild such as black nightshade (Solanum nigrum) and Mulla thotakura (Amaranthus spinosus) are comparable to carrot or sweet potato (Table 3).

Table 2: Carotenoid contents (mg/100 g fresh weight) of some common fruits

Family	Genus	Species	Common name	α-Carotene	β-Carotene	Lycopene	References
Anacardiaceae	Mangifera	indica L.	Mango	–	0.553	0.353	[91]
				–	1.71(0.95)	–	[90]
				0.017[0.001]	0.445[0.016]	–	[44]
		var. Black-gold		ND	0.615	ND	[65]
		var. Gedong		0.061(0.086)	3.267(2.075)	–	[92]
		var. Manalagi		ND	0.19(0.123)	–	[92]
		var. Indramayn		0.067(0.005)	1.606(0.166)	–	[92]
		var. Harum manis		0.055(0.001)	1.08(0.264)	–	[92]
		var. Golek		0.055(0.003)	1.237(0.626)	–	[92]
	Spondias	dulcis L.	Hog plum	–	0.201	0.364	[91]
Actinidiaceae	Actinidia	deliciosa C.F.Liang.& A.R.Ferguson.	Kiwifruit				
		var. Hayward		ND	0.074[0.021]	ND	[93]
		var. Zespri gold		ND	0.092[0.008]	ND	[93]
Bromeliaceae	Ananas	comosus (L.) Merr.	Pineapple	ND	0.056[0.005]	ND	[93]
				ND	0.17	ND	[94]
Caricaceae	Carica	papaya L.	Papaya	ND	0.23-1.981	1.477-5.75	[65,91,95]
				–	1.05(0.44)	–	[90]
				ND	0.276[0.245]	–	[44]
		var. Fruit tower		ND	0.409[0.027]	2.481[0.692]	[93]
		var. Sun rise		ND	1.981[0.059]	1.477[0.302]	[93]
		var. Yellow sweet		ND	1.048[0.026]	1.987[0.851]	[93]
		var. Hawaiian		ND	0.5	1.7	[94]
Cucurbitaceae	Citrullus	lanatus (Thunb.) Matsum. & Nakai	Watermelon	0-0.76	0.14-6.806	0.071-11.389	[65,91,94,95]
				ND	0.59[0.033]	6.184[0.152]	[93]
Ebenaceae	Diospyros	sp.	Persimmon	–	0.253	–	[44]
				ND	0.129[0.003]	0.415[0.013]	[93]
Ericaceae	Vaccinium	spp.	Blueberries	ND	0.035	ND	[44]
				ND	0.027[0.005]	ND	[93]
Malvaceae	Durio	zibethinus L.	Durian	0.006	0.023	–	[44]
Moraceae	Artocarpus	heterophyllus Lam.	Jackfruit	ND	0.026-0.36	0.037	[65,91,94,96]
				–	0.16(0.06)	–	[90]

Family	Genus	Species	Common name	α-Carotene	β-Carotene	Lycopene	References
Musaceae	Musa	spp.	Banana	0.005[0.005]	0.021[0.014]	–	[44]
				0.058[0.007]	0.058[0.006]	ND	[93]
		paradisiaca L. var. Ambon	Banana	–	0.097	0.114	[91]
		sapientum Linn.					
		var. Emas		–	0.04	–	[65]
		var. Tanduk		–	0.092	–	[65]
Myrtaceae	Psidium	guajava L.	Guava	–	0.001	0.114	[91]
				–	0.001 (0.0001)	–	[90]
		var. Pink	Pink guava	–	–	5.4	[95]
				ND	0.359[0.015]	2.307[0.058]	[93]
					5.027[0.08]	4.383[0.371]	
Oxalidaceae	Averrhoa	carambola L.	Starfruit	ND	0.028-0.042	0-0.042	[65,91,96]
				ND	ND	ND	[93]
Passifloraceae	Passiflora	edulis Sims	Passion fruit	0.035	0.53	–	[44]
				ND	0.156[0.02]	0.057[0.003]	[93]
Rosaceae	Eriobotrya	japonica (Thunb.) Lindl.	Loquat	–	0.207	–	[97]
	Fragaria	ananassa Duchesne	Strawberry	0.005	–	–	[44]
	Malus	domestica Borkh.	Apple	0.001-0.03	0.031-0.072	0.209	[96,98,99]
		var. Fuji		ND	0.036[0.003]	ND	[93]
Rosaceae	Prunus	armeniaca L.	Apricot	ND	2.554	0.005	[44]
		salicina Lindl.	Nectarine				
		var. Red Jim		–	0.073(0.016)	–	[100]
		var. August red		–	0.128(0.005)	–	[100]
		var. Spring bright		–	0.085(0.006)	–	[100]
		var. May glo		–	0.058(0.005)	–	[100]
		var. September red		–	0.131(0.023)	–	[100]
		persica (L.) Batsch	Peach	0.001[0.001]	0.097[0.013]	–	[44]
		var. Summer sweet		–	0.04(0.01)	–	[100]
		var. Snow king		–	0.008(0.002)	–	[100]

Family	Genus	Species	Common name	β-cryptoxanthin	Lutein	Zeaxanthin	References
		var. Snow giant		–	0.006(0.001)	–	[100]
		var. Champagne		–	0.007(0.001)	–	[100]
		var. September snow		–	0.004(0.001)	–	[100]
		var. Hakuto		ND	0.048[0.032]	ND	[93]
		var. Kanto 5 go		ND	0.036[0.006]	ND	[93]
		var. Mochizuki		ND	ND	ND	[93]
		var. Nishiki		ND	0.16[0.005]	ND	[93]
		var. Ogonto		ND	0.121[0.008]	ND	[93]
		domestica L.	Plum	–	0.098	–	[44]
		var. Red		ND	0.127	ND	[65]
		var. Wickson		–	0.04(0.004)	–	[100]
		var. Black Beaut		–	0.188(0.017)	–	[100]
		var. Red Beaut		–	0.064(0.012)	–	[100]
		var. Santa Rosa		–	0.049(0.012)	–	[100]
		var. Angeleno		–	0.057(0.009)	–	[100]
		var. Ponteroza		ND	0.218[0.019]	ND	[93]
		var. Soldam		ND	0.439[0.029]	ND	[93]
Rosaceae	*Prunus*	*spp.*	Cherry	ND	ND	ND	[94]
				–	0.14(0.06)	–	[90]
				–	0.028	–	[44]
		var. Domestic		0.018(0.004)	0.071(0.004)	ND	[93]
		var. USA		ND	0.037(0.004)	ND	[93]
	Pyrus	*sp.*	Pear	0.006	0.027	ND	[44]
	Rubus	*sp.*	Raspberry	0.012	0.008	–	[44]
Rutaceae	*Citrus*	*aurantium* L.	Orange	–	0.17(0.08)	–	[90]
		maxima Merr.	Pummelo	0.014	0.32	–	[44]
		microcarpa Bunge	Musk lime	–	0.012	–	[65]
		nobilis L.	Orange	–	0.025	–	[65]
		paradisiaca Macfad.	Grapefruit				
		var. Star ruby		ND	0.452[0.019]	1.869[0.654]	[93]
		var. Pink		0.005[0.005]	0.603[0.152]	–	[44]
				–	–	3.36	[95]
		var. White		0.008	0.014	–	[44]
		sinensis (L.) Osbeck	Orange	0.016	0.051	–	[44]
		var. Navel		0.019[0.002]	0.139[0.014]	ND	[93]
		var. Valencia		0.015[0.001]	0.051[0.004]	ND	[93]
		reticulata Blanco	Mandarin	–	0.081	–	[65]
			orange	ND	0.03	ND	[94]
Sapindaceae	*Nephelium*	*lappaceum* L.	Rambutan	–	ND	0.148	[91]
Vitaceae	*Vitis*	*vinifera* Linnaeus	Grape	–	0.039	–	[44]
		var. Deraware		ND	0.058[0.004]	ND	[93]

Taxonomy			Common name	β-cryptoxanthin	Lutein	Zeaxanthin	References
Family	**Genus**	**Species**					
Anacardiaceae	*Mangifera*	*indica* L.	Mango	0.137	–	–	[91]
				0.011[0.009]	–	–	[44]
		var. Black-gold		ND	ND	–	[65]
	Spondias	*dulcis* L.	Hog plum	0.309	–	–	[91]
Actinidiaceae	*Actinidia*	*deliciosa* L.	Kiwifruit				
		var. Hayward		ND	0.153(0.005)	ND	[93]
		var. Zespri gold		ND	0.156(0.005)	0.113(0.006)	[93]
Bromeliaceae	*Ananas*	*comosus* (L.) Merr.	Pineapple	0.089	ND	ND	[93]
Caricaceae	*Carica*	*papaya* L.	Papaya	0.18-3.182	0.016-0.063	0.165-0.564	[65,91,95]
				0.076[0.225]	0.075^c	–	[44]
		var. Fruit tower		0.725[0.012]	0.016[0.001]	0.165[0.001]	[93]
		var. Sun rise		3.182[0.117]	0.063[0.001]	0.564[0.01]	[93]
		var. Yellow sweet		1.629[0.064]	0.029[0.001]	0.303[0.007]	[93]
		var. Hawaiian		–	–	–	[94]
Cucurbitaceae	*Citrullus*	*lanatus* (Thunb.) Matsum. & Nakai	Watermelon	0.09-0.48	0, 0.017^c	ND	[65,91,95]
				ND	ND	ND	[93]
Ebenaceae	*Diospyros*	*sp.*	Persimmon	1.45	0.834^c	0.49	[44]
				0.52[0.02]	ND	0.238[0.01]	[93]
Ericaceae	*Vaccinium*	*spp.*	Blueberries	–	–	–	[44]
				0.011[0.006]	0.042[0.011]	ND	[93]
Malvaceae	*Durio*	*zibethinus* L.	Durian	ND	–	–	[44]

Moraceae	Artocarpus	heterophyllus Lam.	Jackfruit	0.017-0.036	0.095	–	[65,91,96]
Musaceae	Musa	spp.	Bananas	ND	ND^c	–	[44]
				ND	0.113(0.008)	ND	[93]
		paradisiaca L. var. Ambon		0.003	–	–	[91]
Myrtaceae	Psidium	guajava L.	Guava	0.012, 0.464	0.044	ND	[91,95]
		var. Pink	Pink guava	0.012[0.003], 0.464[0.015]	0.044[0.002]	ND	[93]
Oxalidaceae	Averrhoa	carambola L.	Starfruit	0.036-1.066	0.066	ND	[65,91,96]
				ND	ND	ND	[93]
Passifloraceae	Passiflora	edulis Sims.	Passion fruit	0.046	–	–	[44]
				0.027[0.001]	–	0.042[0.002]	[93]
Rosaceae	Eriobotrya	japonica (Thunb.) Lindl.	Loquat	0.518	–	–	[97]
	Fragaria	ananassa Duchesne	Strawberry	–	–	–	[44]
	Malus	domestica Borkh.	Apple	0.001-0.106	0.017	0.0019	[91,96,99]
		var. Fuji		ND	ND	ND	[93]
	Prunus	armeniaca L.	Apricot	ND	–	–	[44]
		salicina Lindl.	Nectarine				
		var. Red Jim		0.014(0.005)	–	–	[100]
		var. August red		0.014(0.003)	–	–	[100]
		var. Spring bright		0.021(0.002)	–	–	[100]
		var. May glo		0.008(0)	–	–	[100]
		var. September red		0.015(0.006)	–	–	[100]
		persica (L.) Batsch	Peach	0.024	0.057^c	–	[44]
		var. Summer sweet		0.012(0)	–	–	[100]
		var. Snow king		ND	–	–	[100]
		var. Snow gaint		ND	–	–	[100]
		var. Champagne		ND	–	–	[100]
		var. September snow		ND	–	–	[100]

Rosaceae	Prunus	persica (L.) Batsch	Peach				
		var. Hakuto		ND	ND	ND	[93]
		var. Kanto 5 go		0.283[0.003]	ND	0.51[0.015]	[93]
		var. Mochizuki		0.081[0.011]	ND	0.028[0.002]	[93]
		var. Nishiki		0.074[0.003]	0.051[0.005]	0.116[0.005]	[93]
		var. Ogonto		0.025[0.008]	0.029[0.002]	0.104[0.002]	[93]
		domestica L.	Plum	0.016	–	–	[44]
		var. Red		0.04	0.149	–	[65]
		var. Wickson		0.05(0.01)	–	–	[100]
		var. Black Beaut		0.13(0.01)	–	–	[100]
		var. Red Beaut		0.03(0.01)	–	–	[100]
		var. Santa Rosa		0.07(0.03)	–	–	[100]
		var. Angeleno		0.03(0)	–	–	[100]
		var. Ponteroza		0.05[0.008]	0.133[0.024]	0.049[0.006]	[93]
		var. Soldam		0.077[0.009]	0.207[0.011]	0.026[0.002]	[93]
		spp.	Cherry	–	–	–	[44]
		var. Domestic		0.021[0.001]	0.112[0.008]	0.042[0.005]	[93]
		var. USA		0.014[0.002]	0.091[0.004]	0.027[0.001]	[93]
Rutaceae	Citrus	maxima Merr.	Pummelo	0.103	–	–	[44]
		paradise Macfad.	Grapefruit				
		var. Star ruby		ND	ND	ND	[93]
		var. Pink		0.012[0.009]	–	–	[44]
		var. White		–	–	–	[44]
		nobilis L.	Orange	–	0.275	ND	[91]
		sinensis (L.) Osbeck	Orange				
		var. Navel		0.462[0.031]	0.059[0.006]	0.164[0.013]	[93]
		var. Valencia		0.278[0.001]	0.071[0.002]	0.019[0.001]	[93]
				0.122	0.187^c	–	[44]
Sapindaceae	Nephelium	lappaceum L.	Rambutan	ND	–	–	[91]
		var. Deraware		ND	0.103[0.014]	0.028[0.004]	[93]

^a ND, Not detected; – data not available; var., variety; ^b mean(standard deviation), mean[standard error]; ^c content of lutein + zeaxanthin

Table 3: Carotenoid contents (mg/100 g fresh weight) of common leafy and non-leafy vegetables

Family	Taxonomy Genus	Species	Common name	α-Carotene	β-Carotene	Lycopene	References
		Leafy Vegetables					
Alliaceae	*Allium*	*fistulosum* L.	Spring onion leaves	–	1.28	–	[65]
		sativum L.	Garlic leaves	–	5.0	–	[42]
		cepa L.	Onion leaves	–	4.9(0.15)	–	[90]
Apiaceae	*Apium*	*graveolens* L.	Celery	ND	0.77	ND	[100]
				ND	0.15	–	[43]
	Coriandrum	*sativum* L.	Coriander leaves	ND	3.17	ND	[65]
			Coriander	–	4.8(0.16)	–	[90]
	Foeniculum	*vulgare* Mill.	Fennel common	–	4.4	–	[101]
Amaranthaceae	*Amaranthus*	*spp.*	Amaranth	–	1.96-8.6	–	[42,101,102]
		spinosus L.	Mulla thotakura	–	10.9(1.25)	–	[90]
		sp.	Yerramolakakaura	–	11.9(1.48)	–	[90]
	Spinacia	*oleracea* L.	Spinach	ND	3.177, 36.53(6.4)	ND	[65,103]
		var. Red		ND	5.088	ND	[65]
				–	1.1(0.36)	–	[90]
				ND	5.597[0.561]	–	[44]
Asteraceae	*Lactuca*	*sativa* L.	Lettuce	ND	0.097	ND	[65]
				–	1.4(0.28)	–	[90]
		var. Cos or Romaine		ND	1.272	–	[44]
		var. Iceberg		0.002	0.192[0.069]	–	[44]
Brassicaceae	*Brassica*	*juncea* (L.) Czern.	Chinese mustard leaves	ND	2.93	ND	[65]
		oleracea L.					
		var. Acephala	Kale	ND	9.23	ND	[44]
		var. Alboglabra	Chinese kale	ND	4.09	ND	[65]
		var. Capitata	Cabbage	ND	0.01-3.02	ND	[44,101]
		var. Chinensis		–	2.703	–	[65]
		var. Pekinensis		–	0.01(0.01)	–	[104]
		papaya L.	Papaya leaves	0.424(0.355)	5.229(2.195)	–	[92]
		aquatica Forssk.	Swamp cabbage	ND	1.895	ND	[61]
			Water spinach	0.014(0.026)	2.73 (1.013)	–	[92]
Cucurbitaceae	*Momordica*	*Charantia* Descourt.	Bitter melon leaves	–	3.4	–	[101]
Euphorbiaceae	*Manihot*	*esculenta* Crantz	Cassava leaves	0.038(0.054)	9.912(2.503)	–	[92]
Fabaceae	*Sesbania*	*grandiflora* (L.) Poiret	Sesbania	ND	13.61, 13.28(3.2)	ND	[65,103]
	Trigonella	*foenum-graecum* L.	Fenugreek	–	9.2(1.48), 12.13(4.1)	–	[91,103]
Lamiaceae	*Mentha*	*arvensis* L.	Pudina	–	4.3(2.0)	–	[90]
Meliaceae	*Azadirachta*	*indica* L.	Neem tree leaves	–	0.92	–	[101]
Moringaceae	*Moringa*	*oleifera* Lam.	Drumstick leaves	ND	5.2, 7.54	ND	[65,102]
				–	19.7(5.55), 22.89(6.8)	–	[91,103]
Phyllanthaceae	*Sauropus*	*androgynus* L.	Sweet shoot leaves	ND	13.35	ND	[65]
				1.335(0.878)	10.01(2.189)	–	[92]
Solanaceae	*Solanum*	*nigrum* L.	Black nightshade	ND	7.05	ND	[65]
Rutaceae	*Murraya*	*koenigii* (L.) Sprengel	Curry leaves	–	7.1(2.36)	–	[90]
		Non-leafy Vegetables					
Alliaceae	*Allium*	*schoenoprasum* L.	Chive	ND	0.83, 3.51	ND	[42,65]
Apiaceae	*Daucus*	*carota* L.	Carrot	3.41-6.2	6.5-21	ND	[44,65,93]
Araceae	*Colocasia*	*esculenta* (L.) Schott	Taro	–	–	–	–
Asparagaceae	*Asparagus*	*officinalis* L.	Asparagus	0.012	0.493	–	[44]
Brassicaceae	*Brassica*	*oleracea* L.					
		var. Calabrese	Broccoli	–	0.898	–	[97]
				0.001[0.001]	0.779[0.19]	–	[44]
Brassicaceae	*Brassica*	var. Italica Plenck.		–	0.81(0.2)	–	[103]
		var. Gemmiferae	Brussels sprout	ND	0.14	ND	[42]
				0.006	0.45[0.057]	–	[44]

Family	Genus	Species	Common name	β-Cryptoxanthin	Lutein	Zeaxanthin	References
				-	0.14(0.02)	-	[104]
		var. Botrytis	Cauliflower	-	0.08	-	[42]
				-	0.08(0.03)	-	[104]
				-	6.5(1.46)	-	[90]
Convolvulaceae	Ipomea	batatas (L.) Lam	Sweet potato	0.002	0.058, 9.18	ND	[44,92]
				-	1.87(0.14)	-	[90]
				ND	9.18[1.272]	-	[44]
Cucurbitaceae	Coccinea	grandis (L.) J. Voigt	Ivy gourd	-	3.2-4.1	-	[42]
	Cucumis	sativus L.	Cucumber	0.008	0.031-0.14	ND	[44]
				ND	ND	ND	[94]
	Cucurbita	maxima Duch.	Pumpkins	0.03-7.5	0.06-14.85	ND	[65,94]
		(12 varieties)		0-7.5	1.4-7.4	-	[105]
		minima L.		-	1.16(0.057)	-	[90]
		moschata Duch.		-	9.29(7.5)	-	[106]
		(4 varieties)		0.98-5.9	3.1-7.0	-	[105]
		pepo L. (5 varieties)		0.03-0.17	0.06-2.3	-	[105]
	Momordica	charantia Descourt.	Bitter gourd	ND	ND	ND	[94]
Euphorbiaceae	Manihot	esculenta Crantz	Cassava	ND	0.008	-	[44]
		var. Monroe		ND	0.52	ND	[94]
		var. Beqa		ND	0.43	ND	[94]
		var. Common		ND	<0.02	ND	[94]
		utilissima Pohl.	Tapioca shoot	ND	5.72	ND	[65]
	Phaseolus	vulgaris L.	French bean	ND	0.24	ND	[65]
		var. Red	Common Bean	0.28	0.8	ND	[94]
		var. Yellow		ND	ND	-	[94]
		var. French		0.72	0.78	ND	[94]
	Vigna	unguiculata (L.) Walp.					
		subsp. unguiculata	Cow pea	-	-	-	[107]
		subsp. sesquipedalis	Long bean	ND	0.41-0.57	ND	[65]
Malvaceae	Abelmoschus	esculentus (L.) Moench	Okra	0.028	0.43	-	[44]
Marantaceae	Maranta	arundinacea L.	Arrowroot	ND	0.01	-	[44]

Family	Genus	Species	Common name	β-Cryptoxanthin	Lutein	Zeaxanthin	References
Poaceae	Zea	mays L.	Maize	-	0.014	-	[44]
		(13 varieties)		0.003-0.086	0.037-0.879	-	[92]
				(0-0.009)	(0-0.028)		
Solanaceae	Capsicum	annuum L.					
		var. Cayenne	Chilies	3.41	0.47-6.77	ND	[65]
		var. Grossa	Capsicum	-	1.13(0.8)	-	[90]
				ND	0.27	ND	[65]
		sp.	Pepper	0.022-0.059	0.2-2.38	ND	[44]
				-	0.11 (0.04)	-	[91]
	Solanum	betaceum Cav.	Tree tomato	ND	0.6	ND	[65]
		lycopersicum L.	Tomatoes	2.5	0.365-1.3	0.009-2.0	[65,94,95]
				-	0.62(0.19)	-	[90]
		melongena L.	Red eggplant	ND	ND	ND	[94]
		tuberosum L.	Potato	-	0.006	-	[44]

Taxonomy			Common name	β-Cryptoxanthin	Lutein	Zeaxanthin	References
Family	Genus	Species					
		Leafy Vegetables					
Alliaceae	Allium	fistulosum Linnaeus	Spring onion leaves	ND	0.323	-	[65]
Amaranthaceae	Spinacia	oleracea L.	Spinach	-	77.58(6.6)	1.51(0.4)	[103]
Apiaceae	Apium	graveolens L.	Celery	ND	0.23^e	0.003	[44]
	Spinacia	oleracea L.	Spinach	ND	4.175	-	[65]
		var. Red		ND	2.047	-	[65]
				ND	11.938^e	-	[44]
Asteraceae	Lactuca	sativa L.	Lettuce	ND	0.073	-	[65]
		var. Cos or Romaine		ND	2.635^e	-	[44]
		var. Iceberg		ND	0.352^e	-	[44]
Brassicaceae	Brassica	juncea (L.) Czern.	Chinese mustard	ND	1.02	-	[65]
		oleracea L.					
		var. Acephala	Kale	ND	39.55^e	-	[44]
		var. Alboglabra	Chinese kale	ND	1.54	-	[65]

Family	Genus	Species/variety	Common name				References
		var. Capitata	Cabbage	ND	0.02, 0.31[c]	–	[44,101]
		rapa L.					
		var. Chinensis	Chinese cabbage	–	2.703	–	[65]
		var. Pekinensis		–	0.02(0.01)	–	[104]
Convolvulaceae	Ipomoea	aquatica Forssk	Swamp cabbage	ND	0.335	–	[65]
Fabaceae	Sesbania	grandiflora (L.) Poiret	Sesbania	ND	20.21, 16.9(3.7)	0.57(0.7)	[65,103]
Moringaceae	Moringa	oleifera Lam.	Drumstick leaves	ND	7.13, 50.4(0.8)	4.13(0.7)	[102,103]
Phyllanthaceae	Sauropus	androgynus L.	Sweet shoot leaves	ND	29.91	–	[65]
Rutaceae	Murraya	koenigii (L.) Sprengel	Curry leaves	ND	5.25	–	[65]
Solanaceae	Solanum	nigrum L.	Black nightshade	ND	2.89	–	[65]
		Non-leafy Vegetables					
Alliaceae	Allium	schoenoprasum L.	Chive	ND	1.08	–	[42,65]
Apiaceae	Daucus	carota L.	Carrot	ND	ND	–	[44]
Araceae	Colocasia	esculenta (L.) Schott	Taro	–	0.16	0.006	[107]
Brassicaceae	Brassica	oleracea L.					
		var. Calabrese	Broccoli	ND	1.28, 2.45[c]	–	[44,97]
		var. Italica Plenck		–	0.68(0.22)	–	[104]
		var. Gemmiferae	Brussels sprout	ND	0.43	–	[42]
				ND	1.59[c]	–	[44]
				–	0.43(0.06)	–	[104]
		var. Botrytis	Cauliflower	–	0.13	–	[42]
				–	0.05(0.02)	–	[104]
Convolvulaceae	Ipomea	batatas (L.) Lam	Sweet potato	ND	ND	–	[44]
Cucurbitaceae	Coccinea	grandis (L.) J. Voigt	Ivy gourd	–	0.99	ND	[107]
	Cucumis	sativus L.	Cucumber	–	0.544	0.009	[107]
	Cucurbita	maxima Duch.	Pumpkins	ND	0.94-17.0	0.278	[65,107]
		(12 varieties)		–	0.8[c]-17.0[c]	–	[105]
		minima L.		–	1.16(0.057)	–	[90]
		moschata Duch.		–	9.29(7.5)	–	[106]
		(4 varieties)		–	0.08[c]-1.1[c]	–	[105]
		pepo L. (5 varieties)		–	0[c]-1.8[c]	–	[105]
Euphorbiaceae	Manihot	esculenta Crantz	Cassava	ND	–	–	[44]
		var. Monroe		–	–	–	[94]
		var. Beqa		–	–	–	[94]
		var. Common		–	–	–	[94]
		utilissima Pohl	Tapioca shoot	ND	1.68[c]	–	[65]
Fabaceae	Phaseolus	vulgaris L.	French bean	ND	0.171-0.46	0.02	[65,107]
	Vigna	unguiculata (L.) Walp.					
		subsp. unguiculata	Cow pea	–	0.24	0.009	[107]
		subsp. sesquipedalis	Long bean	ND	0.3-0.42	–	[65]
Malvaceae	Abelmoschus	esculentus (L.) Moench	Okra	–	0.347	0.008	[44]
Marantaceae	Maranta	arundinacea L.	Arrowroot	ND	–	–	[44]
Poaceae	Zea	mays L. (13 varieties)	Maize	0.037-0.988 (0.001-0.015)	0-2.047 (0-0.075)	0.173-2.07 (0.004-0.073)	[92]
Solanaceae	Capsicum	annum L.					
		var. Cayenne	Chilies	1.75	0.39-1.902	0.063	[65,107]
		var. Grossa	Capsicum	–	0.425	0.005	[107]
		spp.	Pepper	2.21	0.22	–	[44,65]
	Solanum	betaceum Cav.	Tree tomato	1.24	ND	–	[65]
		lycopersicum L.	Tomatoes	ND	0.13-0.289	0.014	[65,107]
		melongena L.	Red eggplant	–	0.065-1.8	0.005-0.016	[107]
		tuberosum L.	Potato	–	–	–	–
	Phaseolus	vulgari L.	French bean	ND	0.171-0.46	0.02	[65,107]

[a] ND, Not detected; – data not available; var., variety; [b] mean(standard deviation), mean[standard error]; [c] content of lutein + zeaxanthin

Orange colored fruits such as apricot (Prunus armeniaca L.), grapefruit (Citrus paradise Macfad.), mango (Mangifera indica L.), papaya (Carica papaya L.), persimmon (Diospyros sp.), pink guava (Psidium guajava L. var. Pink) and watermelon [Citrullus lanatus (Thunb.) Matsum. & Nakai] are rich in β-carotene. Khoo et al. [108] reported that orange colored underutilized fruits contained high amount of β-carotene. Although papaya is orange in color, certain cultivars have shown to contain low β-carotene [93,94]. Furthermore,

Levy et al. [98] reported some of the orange colored fruits had low amount of β-carotene. Naturally, most of xanthophylls are yellow-orange colored pigments, especially lutein and zeaxanthin which can be found in most of the fruits and vegetables [82]. As lutein can absorb blue light, it appears as yellow color; while zeaxanthin appears yellow-orange color. Cryptoxanthins are other types of yellow-orange colored carotenoids. Takyi [82] reported α-cryptoxanthin appears as yellow colored pigment, while β-cryptoxanthin is orange in color. As shown in Table 3, lutein is found to be in higher amounts in green leafy and yellow colored non-leafy vegetables as compared to fruits. Green leafy vegetables that contain high amount of xanthophylls are mainly from the genera of Brassica, Coriandrum, Lactuca, Moringa, Murraya, Sauropus, Sesbania, Solanum and Spinacia, while the non-leafy vegetables are from the genera of Allium, Brassica, Capsicum, Cucurbita and Zea (Table 3). These green vegetables contain mainly lutein and zeaxanthin [44,65]. Kale (Brassica oleracea L. var. Acephala), lettuce (Lactuca sativa L. var. Cos or Romaine), Sesbania (Sesbania grandiflora L. Poiret), spinach (Spinicia oleracea L. var. Red) and sweet shoot leaves (Sauropus androgynus L.) are the example of the leafy vegetables that have high lutein content (Table 3); while other lutein-rich non-leafy vegetables are red eggplant (Solanum melongena L.), chili (Capsicum annum L. var. Cayenne), ivy gourd [Coccinea grandis (L.) J. Voigt], and pumpkin (Cucurbita maxima Duch.) [107]. Muzhingi et al. [88] reported that 36 genotypes of yellow maizes (Zea mays L.) contained lutein and zeaxanthin, whereas saponification significantly decreased the xanthophyll contents. In some cases, zeaxanthin-rich fruits [e.g., papaya (Carica papaya L.) and persimmon (Diospyros sp.)] and zeaxanthin-rich non-leafy vegetables [e.g. pumpkin (Cucurbita maxima Duch.) and maize (Zea mays L.)] were found to have high amount of β-carotene (Tables 2 and 3). Cryptoxanthin is another yellow colored carotenoid, which is closely related to carotene [9,82]. Cryptoxanthin has approximately half of provitamin A activity as compared to β-carotene [109]. Cryptoxanthins have been identified in various types of fruits and vegetables [9,82,110]. Besides, the level of β-cryptoxanthin is high in fruits such as papaya (Carica papaya L.), persimmon (Diospyros sp.) and starfruit (Averrhoa carambola L.) (Table 2) and non-leafy vegetables such as chili (Capsicum annum L. var. Cayenne), maize (Zea mays L.), pepper (Capsicum sp.) and tree tomato (Solanum betaceum Cav.) (Table 3).

Red Pigment Carotenoids

Lycopene is one of the naturally occurring red colored carotenoids [58]. The all-trans-isomer of lycopene is the most predominant geometrical isomer in fruits and vegetables [111]. Lycopene has two more double bonds than β-carotene, hence it appears red. Beside lycopene, δ-carotene pigment is redorange in color, while astaxanthin is a red colored pigment [82].

This review shows that red lycopene pigment is abundant in fruits such as papaya (Carica papaya L.), pink grapefruit (Citrus paradise Macfad. var. Pink), pink guava (Psidium guajava L. var. Pink) and watermelon [Citrullus lanatus (Thunb.) Matsum. & Nakai] [93]. For non-leafy vegetables, USDA database [112] showed that raw red cabbage (Brassica oleracea var. Capitata) and boiled asparagus (Asparagus officinalis L.) contained 20 and 30 µg lycopene per 100 g edible portions. Red colored pigments in fruits and vegetables are believed to be originated from lycopene, which might also account for xanthophylls [113-115] and anthocyanins [116,117]. In fruits, dry persimmon (Diospyros sp.) contains the highest amount of lycopene (53.21 mg/100 g dry weight), which is two times higher than dry tomato [118]. A review by Bramley [95] has shown that pink guava and watermelon had comparable amounts of lycopene, which are even higher than fresh tomato (Tables 2 and 3). Lycopene content in tomato products such as tomato ketchup is 5.5-time higher than in fresh ripe tomato [44]. Lycopene content in fresh tomato is influenced by the cultivars, agricultural practices, maturity and environmental factors [61]. Besides, mutant tomatoes have an almost two-fold increase in lycopene content [119].

CONCLUSIONS

Carotenoids are colorful pigments found in fruits and vegetables. The geometric isomers of carotenoids are present in all-trans and cis forms, together with carotenoid epoxides. Although alltrans-isomer is the major form of carotenoid, the cis isomers are available in small quantities. Heating and thermal processing could increase the amount of carotenoid cis-isomers. The degradation of carotenoids in fruits and vegetables is a major issue due to carotenoid loss. More attention should be given to the control of carotenoid geometry isomer degradation, and to improve the quality of dietary carotenoids. The color changes during geometric isomerization of carotenoids, thermal processing and even fruit ripening have been widely studied. However, there is still lack of information on the chemical and kinetic pathways of color changes during carotenoid degradation and isomerization. In the future, more studies are needed to focus on the isomerization of carotenoids in relation to the colorful pigments in the biodiversity of fruits and vegetables.

REFERENCES

1. WHO. Fruit, Vegetables and NCD Disease Prevention; World Health Organization: Geneva, Switzerland, 2003; http://www.who.int/hpr/NPH/ fruit_and_vegetables/fruit_vegetables_fs.pdf, (accessed on 3 May 2010).

2. Cooper, D.A. Carotenoids in health and disease: Recent scientific evaluations, research recommendations and the consumer. J. Nutr. 2004, 134, 221S–224S.

3. Young, C.Y.F.; Yuan, H.Q.; He, M.L.; Zhang, J.Y. Carotenoids and prostate cancer risk. MiniRev. Med. Chem. 2008, 8, 529–537.

4. Shahidi, F. Nutraceuticals and functional foods: Whole versus processed foods. Trends Food Sci. Technol. 2009, 20, 376–387.

5. Bartley, G.E.; Scolnik, P.A. Plant carotenoids: Pigments for photoprotection, visual attraction, and human health. Plant Cell 1995, 7, 1027–1038.

6. Janik, E.; Grudziński, W.; Gruszecki, W.I.; Krupa, Z. The xanthophyll cycle pigments in Secale cereale leaves under combined Cd and high light stress conditions. J. Photochem. Photobiol. B 2008, 90, 47–52.

7. Lancaster, J.E.; Lister, C.E.; Reay, P.F.; Triggs, C.M. Influence of pigment composition on skin color in a wide range of fruit and vegetables. J. Am. Soc. Hortic. Sci. 1997, 122, 594–598.

8. Rodriguez-Amaya, D.B.; Kimura, M. HarvestPlus Handbook for Carotenoid Analysis. In HarvestPlus Technical Monograph, Series 2; International Food Policy Research Institute and International Center for Tropical Agriculture: Washington, DC, USA, 2004.

9. Rodriguez-Amaya, D.B. A Guide to Carotenoid Analysis in Foods; International Life Sciences Institute, ILSI Press: Washington, DC, USA, 2001.

10. Hornero-Méndez, D.; Mínguez-Mosquera, M.I. Xanthophyll esterification accompanying carotenoid overaccumulation in chromoplast of Capsicum annuum ripening fruits is a constitutive process and useful for ripeness index. J. Agric. Food Chem. 2000, 48, 1617–1622.

11. Minguez-Mosquera, M.I.; Hornero-Mendez, D. Changes in carotenoid esterification during the fruit ripening of Capsicum annuum Cv. Bola. J. Agric. Food Chem. 1994, 42, 640–644.

12. Schieber, A.; Carle, R. Occurrence of carotenoid cis-isomers in food: Technological, analytical, and nutritional implications. Trends Food Sci. Technol. 2005, 16, 416–422.

13. Mertz, C.; Brat, P.; Caris-Veyrat, C.; Gunata, Z. Characterization and thermal lability of carotenoids and vitamin C of tamarillo fruit (Solanum betaceum Cav.). Food Chem. 2010, 119, 653–659.

14. Zepka, L.Q.; Mercadante, A.Z. Degradation compounds of carotenoids formed during heating of a simulated cashew apple juice. Food Chem. 2009, 117, 28–34.

15. Shi, J.; Yi, C.; Ye, X.; Xue, S.; Jiang, Y.; Ma, Y.; Liu, D. Effects of supercritical CO2 fluid parameters on chemical composition and yield of carotenoids extracted from pumpkin. LWT – Food Sci. Technol. 2010, 43, 39–44.

16. Niedzwiedzki, D.M.; Sandberg, D.J.; Cong, H.; Sandberg, M.N.; Gibson, G.N.; Birge, R.R.; Frank, H.A. Ultrafast time resolved absorption spectroscopy of geometric isomers of carotenoids. Chem. Phys. 2009, 357, 4–16.

17. Qiu, D.; Chen, Z.-R.; Li, H.-R. Effect of heating on solid β-carotene. Food Chem. 2009, 112, 344–349.

18. Liu, R.S.H.; Asato, A.E. The primary process of vision and the structure of bathorhodopsin: a mechanism for photoisomerization of polyenes. Proc. Natl. Acad. Sci. USA 1985, 82, 259–263.

19. Britton, G. Overview of carotenoid biosynthesis. In Carotenoids: Biosynthesis and Metabolism; Britton, G., Liaaen-Jensen, S., Pfander, H., Eds.; Birkhäuser Verlag: Basel, Switzerland, 1998; pp. 13–147.

20. Vásquez-Caicedo, A.L.; Sruamsiri, P.; Carle, R.; Neidhart, S. Accumulation of all-trans-β- carotene and its 9-cis and 13-cis stereoisomers during postharvest ripening of nine Thai mango cultivars. J. Agric. Food Chem. 2005, 53, 4827−4835.

21. Aman, R.; Schieber, A.; Carle, R. Effects of heating and illumination on trans-cis isomerization and degradation of β-carotene and lutein in isolated spinach chloroplasts. J. Agric. Food Chem. 2005, 53, 9512–9518.

22. Rickman, J.C.; Bruhn, C.M.; Barrett, D.M. Nutritional comparison of fresh, frozen, and canned fruits and vegetables II. Vitamin A and carotenoids, vitamin E, minerals and fiber. J. Sci. Food Agric. 2007, 87, 1185–1196.

23. Parker, R. Absorption, metabolism, and transport of carotenoids. FASEB J. 1996, 10, 542–551.

24. De Rigal, D.; Gauillard, F.; Richard-Forget, F. Changes in the carotenoid content of apricot (Prunus armeniaca, var Bergeron) during enzymatic browning: β-carotene inhibition of chlorogenic acid degradation. J. Sci. Food Agric. 2000, 80, 763–768.

25. Bohm, V.; Puspitasari-Nienaber, N.L.; Ferruzzi, M.G.; Schwartz, S.J. Trolox equivalent antioxidant capacity of different geometrical isomers of α-carotene, β-carotene, lycopene, and zeaxanthin. J. Agric. Food Chem. 2002, 50, 221–226.

26. Chen, B.H.; Tang, Y.C. Processing and stability of carotenoid powder from carrot pulp waste. J. Agric. Food Chem. 1998, 46, 2312–2318.

27. Goula, A.M.; Adamopoulos, K.G.; Chatzitakis, P.C.; Nikas, V.A. Prediction of lycopene degradation during a drying process of tomato pulp. J. Food Eng. 2006, 74, 37–46.

28. Mordi, R.C.; Walton, J.C.; Burton, G.W.; Hughes, L.; Keith, I.U.; David, L.A.; Douglas, M.J. Oxidative degradation of β-carotene and β-apo-8'-carotenal. Tetrahedron 1993, 49, 911–928.

29. Wacheä, Y.; Bosser-Deratuld, A.L.; Lhuguenot, J.-C.; Belin, J.-M. Effect of cis/trans isomerism of β-carotene on the ratios of volatile compounds produced during oxidative degradation. J. Agric. Food Chem. 2003, 51, 1984–1987.

30. Lin, C.H.; Chen, B.H. Determination of carotenoids in tomato juice by liquid Chromatography. J. Chromatogr. A 2003, 1012, 103–109.

31. Britton, G. UV/Visible Spectroscopy. In Carotenoids Spectroscopy; Brotton, G., Liaaen-Jensen, S., Pfander, H., Eds.; Birkhäuser Verlag: Basel, Switzerland, 1995; Volume 1B, pp. 57–61.

32. Krinsky, N.I. Carotenoid protection against oxidation. Pure Appl. Chem. 1979, 51, 649–660.

33. Di Mascio, P.; Kaiser, S.; Sies, H. Lycopene as the most efficient biological carotenoid singlet oxygen quencher. Arch. Biochem. Biophys. 1989, 274, 532–538.

34. Marx, M.; Schieber, A.; Carle, R. Quantitative determination of carotene stereoisomers in carrot juices and vitamin supplemented (ATBC) drinks. Food Chem. 2000, 70, 403–408.

35. Khoo, H.-E.; Prasad, K.N.; Ismail, A.; Mohd-Esa, N. Carotenoids from Mangifera pajang and their antioxidant capacity. Molecules 2010, 15, 6699–6712.

36. De Faria, A.F.; Hasegawa, P.N.; Chagas, E.A.; Pio, R.; Purgatto, E.; Mercadante, A.Z. Cultivar influence on carotenoid composition of loquats from Brazil. J. Food Compos. Anal. 2009, 22, 196–203.

37. Chen, J.P.; Tai, C.Y.; Chen, B.H. Improved liquid chromatographic method for determination of carotenoids in Taiwanese mango (Mangifera indica L.). J. Chromatogr. A 2004, 1054, 261–268.

38. Chen, J.P.; Tai, C.Y.; Chen, B.H. Effects of different drying treatments on the stability of carotenoids in Taiwanese mango (Mangifera indica L.). Food Chem. 2007, 100, 1005–1010.

39. Lessin, W.J.; Schwartz, S.J. Quantification of cis-trans isomers of provitamin A carotenoids in fresh and processed fruits and vegetables. J. Agric. Food Chem. 1997, 45, 3728–3732.

40. Li, F.; Murillo, C.; Wurtzel, E.T. Maize Y9 encodes a product essential for 15-cis-ζ-carotene isomerization. Plant Physiol. 2007, 144, 1181–1189.

41. Hulshof, P.J.M.; Kosmeijer-Schuil, T.; West, C.E.; Hollman, P.C.H. Quick screening of maize kernels for provitamin A content. J. Food Compos. Anal. 2007, 20, 655–661.

42. Rodriguez-Amaya, D.B. Carotenoids and Food Preparation. In The Retention of Provitamin A Carotenoids in Prepared, Processed, and Stored Foods; John Snow Inc: Rio de Janeiro, Brazil, 1997.

43. During, A.; Smith, M.K.; Piper, J.B.; Smith, J.C. β-Carotene 15,15'-dioxygenase activity in human tissues and cells: Evidence of an iron dependency. J. Nutr. Biochem. 2001, 12, 640–647.

44. Holden, J.M.; Eldridge, A.L.; Beecher, G.R.; Buzzard, I.M.; Bhagwat, A.S.; Davis, C.S.; Douglass, L.W.; Gebhardt, E.S.; Haytowitz, D.; Schakel, S. Carotenoid content of U.S. foods: An update of the database. J. Food Compos. Anal. 1999, 12, 169–196.

45. Guo, W.-H.; Tu, C.-Y.; Hu, C.-H. Cis-trans isomerizations of β-carotene and lycopene: A theoretical study. J. Phys. Chem. 2008, 112, 12158–12167.

46. ESA. Carotenoid Isomers. ESA Application Note, 5600A; ESA Inc: Chelmsford, MA, USA, 2009; http://www.esainc.com (accessed on 7 October 2009).

47. Kuki, M.; Koyama, Y.; Nagae, H. Triplet-sensitized and thermal isomerization of all-trans, 7-cis, 9-cis, 13-cis and 15-cis isomers of β-carotene: Configurational dependence of the quantum yield of isomerization via the T1 state. J. Phys. Chem. 1991, 95, 7171–7180.

48. Vásquez-Caicedo, A.L.; Schilling, S.; Carle, R.; Neidhart, S. Effects of thermal processing and fruit matrix on beta-carotene stability and enzyme inactivation during transformation of mangoes into purée and nectar. Food Chem. 2007, 102, 1172–1186.

49. Lozano-Alejo, N.; Carrillo, G.V.; Pixley, K.; Palacios-Rojas, N. Physical properties and carotenoid content of maize kernels and its nixtamalized snacks. Innov. Food Sci. Emerg. Technol. 2007, 8, 385–389.

50. Tang, Y.C; Chen, B.H. Pigment change of freeze-dried carotenoid powder during storage. Food Chem. 2000, 69, 11–17.

51. Chen, B.H.; Huang, J.H. Degradation and isomerization of chlorophyll a and β-carotene as affected by various heating and illumination treatments. Food Chem. 1998, 62, 299–307.

52. Marx, M.; Stuparic, M.; Schieber, A.; Carle, R. Effects of thermal processing on trans–cisisomerization of β-carotene in carrot juices and carotene-containing preparations. Food Chem. 2003, 83, 609–617.

53. Breitenbach, J.; Sandmann, G. ζ-Carotene cis isomers as products and substrates in the plant polycis carotenoid biosynthetic pathway to lycopene. Planta 2005, 220, 785–793.

54. Castenmiller, J.J.M.; West, C.E. Bioavailability and bioconversion of carotenoids. Annu. Rev. Nutr. 1998, 18, 19–38.

55. Rao, A.V.; Rao, L.G. Carotenoids and human health. Pharmacol. Res. 2007, 55, 207–216.

56. Boileau, T.W.M.; Boileau, A.C.; Erdman, J.W., Jr. Bioavailability of all-trans and cis-isomers of lycopene. Exp. Biol. Med. 2002, 227, 914–919.

57. Schulz, H.; Baranska, M.; Baranski, R. Potential of NIR-FT-Raman spectroscopy in natural carotenoid analysis. Biopolymers 2005, 77, 212–221.

58. Kong, K.-W.; Khoo, H.-E; Prasad, K.N.; Ismail, A.; Tan, C.-P.; Rajab, N.F. Revealing the power of the natural red pigment lycopene. Molecules 2010, 15, 959–987.

59. Blanquet-Diot, S.; Soufi, M.; Rambeau, M.; Rock, E.; Alric, M. Digestive stability of xanthophylls exceeds that of carotenes as studied in a dynamic in vitro gastrointestinal system. J. Nutr. 2009, 139, 876–883.

60. Failla, M.L.; Chitchumroonchokchai, C.; Ishida, B.K. In vitro micellarization and intestinal cell uptake of cis isomers of lycopene exceed those of all-trans lycopene. J. Nutr. 2008, 138, 482–486.

61. Shi, J.; Maguer, M.L.; Bryan, M. Lycopene from Tomatoes. In Functional Food: Biochemical & Processing Aspects; Shi, J.; Mazza, G., Maguer, M.L., Eds.; CRC Press LCC: Danvers, MA, USA, 2002; Volume 2, pp. 135–167.

62. Moraru, C.; Lee, T.-C. Lycopene isomerization at gastric pH. In Nutraceuticals & Functional Foods Session of IFT Annual Meeting, Las Vegas, NV, USA, 12–16 July 2004.

63. Lee, M.T.; Chen, B.H. Stability of lycopene during heating and illumination in a model system. Food Chem. 2002, 78, 425–432.

64. Tyssandier, V.; Reboul, E.; Dumas, J.-F.; Bouteloup-Demange, C.; Armand, M.; Marcand, J.; Sallas, M.; Borel, P. Processing of vegetable-borne carotenoids in the human stomach and duodenum. Am. J. Physiol-Gastr. L. 2003, 284, G913–G923.

65. Tee, E.-S.; Lim, C.-L. Carotenoid composition and content of Malaysian vegetables and fruits by the AOAC and HPLC methods. Food Chem. 1991, 41, 309–339.

66. Losso, J.N.; Khachatryan, A.; Ogawa, M.; Godber, J.S.; Shih, F. Random centroid optimization of phosphatidylglycerol stabilized lutein-enriched oil-in-water emulsions at acidic pH. Food Chem. 2005, 92, 737–744.

67. Matsuno, T.; Hirono, T.; Ikuno, Y.; Maoka, T.; Shimizu, M.; Komori, T. Isolation of three new carotenoids and proposed metabolic pathways of carotenoids in hen's egg yolk. Comp. Biochem. Physiol. B 1986, 84, 477–481.

68. Handelman, G.J. The evolving role of carotenoids in human biochemistry. Nutrition 2001, 17, 818–822.

69. Emenhiser, C.; Sande, L.C.; Schwartza, S.J. Capability of a polymeric C30 stationary phase to resolve cis-trans carotenoid isomers in reversed-phase liquid chromatography. J. Chromatogr. A 1995, 707, 205–216.

70. Sander, L.C.; Sharpless, K.E.; Pursch, M. C30 stationary phases for the analysis of food by liquid chromatography. J. Chromatogr. A 2000, 880, 189–202.

71. Tóth, G.; Szabolcs, J. Occurrence of some mono-cis-isomers of asymmetric C40-carotenoids. Phytochem. 1981, 20, 2411–2415.

72. Meléndez-Martínez, A.J.; Vicario, I.M.; Heredia, F.J. Geometrical isomers of violaxanthin in orange juice. Food Chem. 2007, 104, 169–175.

73. Gross, J. Pigments in Fruits: Food Science and Technology; Academic Press: Orlando, FL, USA, 1987.

74. Meléndez-Martínez, A.J.; Britton, G.; Vicario, I.M.; Heredi, F.J. HPLC analysis of geometrical isomers of lutein epoxide isolated from dandelion (Taraxacum officinale F. Weber ex Wiggers). Phytochemistry 2006, 67, 771–777.

75. Milanowska, J.; Gruszecki, W.I. Heat-induced and light-induced isomerization of the xanthophyll pigment zeaxanthin. J. Photochem. Photobiol. B. 2005, 80, 178–186.

76. Kishimoto, S.; Maoka, T.; Nakayama, M.; Ohmiya, A. Carotenoid composition in petals of chrysanthemum (Dendranthema grandiflorum (Ramat.) Kitamura). Phytochemistry 2004, 65, 2781–2787.

77. Yahia, E.M.; Ornelas-Paz, J.J.; Gardea, A. Extraction, separation and partial identification of 'Ataulfo' mango fruit carotenoids. Acta Hortic. 2006, 712, 333–338.

78. Furr, H.C.; Clark, R.M. Intestinal absorption and tissue distribution of carotenoids. J. Nutr. Biochem. 1997, 8, 364–377.

79. Scott, K.J.; Thurnham, D.I.; Hart, D.J.; Bingham, S.A.; Day, K. The correlation between the intake of lutein, lycopene and β-carotene from vegetables and fruits, and blood plasma concentrations in a group of women aged 50-65 years in the UK. Brit. J. Nutr. 1996, 75, 409–418.

80. Krinsky, N.I.; Johnson, E.J. Carotenoid actions and their relation to health and disease. Mol. Aspects Med. 2005, 26, 459–516.

81. Simpson, K.L. Chemical Changes in Natural Food Pigments. In Chemical Changes in Food during Processing; Richardson, T.; Finley, J.W., Eds.; Van Nostrand Reinhold Company, Inc.: New York, NY, USA, 1985; pp. 409–437.

82. Minguez-Mosquera, M.I.; Gandul-Rojas, B.; Garrido-Fernandez, J.; Gallardo-Guerrero, L. Pigments present in virgin olive oil. J. Am. Oil Chem. Soc. 1990, 67, 192–196.

83. Takyi, E.E.K. Bioavailability of Carotenoids from Vegetables versus Supplements. In Vegetables, Fruits, and Herbs in Health Promotion; Watson, R.R., Ed.; CRC Press LCC: Danvers, MA, USA, 2001; pp. 19–31.

84. Marín, A.; Ferreres, F.; Tomas-Barberan, F.A.; Gil, M.I. Characterization and quantitation of antioxidant constituents of sweet pepper (Capsicum annuum L.). J. Agric. Food Chem. 2004, 52, 3861–3869.

85. Markus, F.; Daood, H.G.; Kapitany, J.; Biacs, P.A. Change in the carotenoid and antioxidant content of spice red pepper (paprika) as a function of ripening and some technological factors. J. Agric. Food Chem. 1999, 47, 100–107.

86. Wieruszewski, J.B. Astaxanthin bioavailabity, retention efficiency and kinetics in Atlantic salmon (Salmo salar) as influenced by pigment concentration and method of administration (kinetics only). Master Thesis, Simon Fraser University, Ottawa, Canada, 2002.

87. De Pee, S.; West, C.E.; Permaesih, D.; Martuti, S.; Muhilal; Hautvast, J.G.A.J. Increasing intake of orange fruits is more effective than increasing intake of dark-green leafy vegetables in increasing serum concentrations of retinol and β-carotene in schoolchildren in Indonesia. Am. J. Clin. Nutr. 1998, 68, 1058–1067

88. Muzhingi, T.; Yeum, K.J.; Russell, R.M.; Johnson, E.J.; Qin, J.; Tang, G. Determination of carotenoids in yellow maize, the effects of saponification and food preparations. Intl. J. Vit. Nutr. Res. 2008, 78, 112–120.

89. Rajyalakshmi, P.; Venkatalaxmi, K.; Venkatalakshmamma, K.; Jyothsna, Y.; Devi, K.B.; Suneetha, V. Total carotenoid and beta-carotene contents of forest green leafy vegetables consumed by tribals of south India. Plant Food Hum. Nutr. 2001, 56, 225–238.

90. Bhaskarachary, K.; Rao, D.S.S.; Deosthale, Y.G.; Reddy, V. Carotene content of some common and less familiar foods of plant origin. Food Chem. 1995, 54, 189–193.

91. Setiawan, B.; Sulaeman, A.; Giraud, D.W.; Driskell, J.A. Carotenoid content of selected Indonesian fruits. J. Food Compos. Anal. 2001, 14, 169–176.

92. Hulshof, P.J.M.; Chao, X.; Van De Bovenkamp, P.; Muhilal; West, C.E. Application of a validated method for the determination of provitamin A carotenoids in Indonesia foods of different maturity and origin. J. Agric. Food Chem. 1997, 45, 1174–1179.

93. Yano, M.; Kato, M.; Ikoma, Y.; Kawasaki, A.; Fukazawa, Y.; Sugiura, M.; Matsumoto, H.; Oohara, Y.; Nagao, A.; Ogawa, K. Quantitation of carotenoids in raw and processed fruits in Japan. Food Sci. Technol. Res. 2005, 11, 13–18.

94. Lako, J.; Trenerry, V.C.; Wahlqvist, M.; Wattanapenpaiboon, N.; Sotheeswaran, S.; Premier, R. Phytochemical flavonols, carotenoids and the antioxidant properties of a wide selection of Fijian fruit, vegetables and other readily available foods. Food Chem. 2007, 101, 1727–1741.

95. Bramley, P.M. Is lycopene beneficial to human health? Phytochem. 2000, 54, 233–236.

96. Charoensiri, R.; Kongkachuichai, R.; Suknicom, S.; Sungpuag, P. Beta-carotene, lycopene, and alpha-tocopherol contents of selected Thai fruits. Food Chem. 2009, 113, 202–207.

97. Granado-Lorencio, F.; Olmedilla-Alonso, B.; Herrero-Barbudo, C.; Blanco-Navarro, I.; PérezSacristán, B.; Blázquez-García, S. In vitro bioaccessibility of carotenoids and tocopherols from fruits and vegetables. Food Chem. 2007, 102, 641–648.

98. Levy, A.; Harcl, S.; Palevitch, D.; Akiri, B.; Menagem, E.; Kanner, J. Carotenoid pigments and β- carotene in paprika fruits (Capsicum spp.) with different genotypes. J. Agric. Food Chem. 1995, 43, 362–366.

99. Dias, M.G.; Filomena, M.; Camões, G.F.C.; Oliveira, L. Carotenoids in traditional Portuguese fruits and vegetables. Food Chem. 2009, 113, 808–815.

100. Gil, M.I.; Tomás-Barberán, F.A.; Hess-Pierce, B.; Kader, A.A. Antioxidant capacities, phenolic compounds, carotenoids, and vitamin C contents of nectarine, peach, and plum cultivars from California. J. Agric. Food Chem. 2002, 50, 4976–4982.

101. Speek, A.J.; Speek-Saichua, S.; Schreurs, W.H.P. Total carotenoids and β-carotene contents of Thai vegetables and effects of processing. Food Chem. 1988, 27, 245–251.

102. Begum, A.; Pereira, S.M. The β-carotene content of Indian edible green leaves. Trop. Geogr. Med. 1977, 29, 47–50.

103. Lakshminarayana, R.; Raju, M.; Krishnakantha, T.P.; Baskaran, V. Determination of major carotenoids in a few Indian leafy vegetables by high-performance liquid chromatography. J. Agric. Food Chem. 2005, 53, 2838–2842.

104. Singh, J.; Upadhyay, A.K.; Prasad, K.; Bahadur, A.; Rai, M. Variability of carotenes, vitamin C, E and phenolics in Brassica vegetables. J. Food Compos. Anal. 2007, 20, 106–112.

105. Murkovic, M.U.; Mülleder, U.; Neunteufl, H. Carotenoid content in different varieties of pumpkins. J. Food Comp. Anal. 2002, 15, 633–638.

106. Pandey, S.; Singh, J.; Upadhyay, A.K.; Ram, D.; Rai, M. Ascorbate and carotenoid content in an Indian collection of pumpkin (Cucurbita moschata Duch. Ex Poir.). Cucurbit Gen. Coop. Rep. 2003, 26, 51–53.

107. Aruna, G.; Mamatha, B.S.; Baskaran, V. Lutein content of selected Indian vegetables and vegetable oils determined by HPLC. J. Food Compos. Anal. 2009, 22, 632–636.

108. Khoo, H.E.; Ismail, A.; Mohd-Esa, N.; Idris, S. Carotenoid content of underutilized fruits. Plant Food Hum. Nutr. 2008, 63, 170–175.

109. Yuan, J.-M.; Gao, Y.-T.; Ong, C.-N.; Ross, R.K.; Yu, M.C. Prediagnostic level of serum retinol in relation to reduced risk of hepatocellular carcinoma. J. Natl. Cancer I. 2006, 98, 482–490.

110. Homnava, A.; Payne, J.; Koehler, P.; Eitenmiller, R. Provitamin A (alpha-carotene, beta-carotene and beta-cryptoxanthin) and ascorbic acid content of Japanese and American persimmons. J. Food Quality 1990, 13, 85–95.

111. Xianquan, S.; Shi, J.; Kakuda, Y.; Yueming, J. Stability of lycopene during food processing and storage. J. Med. Food 2005, 8, 413–422.

112. USDA database. USDA National Nutrient Database for Standard Reference, Release 23; Agricultural Research Service, United States Department of Agriculture: Washington, DC, USA, 2011; http://www. ars.usda.gov (accessed on 20 January 2011).

113. Camara, B.; Hugueney, P.; Bouvier, F.; Kuntz, M.; Monéger, R.; Kwang, W.J.; Jonathan, J. Biochemistry and molecular biology of chromoplast development. Int. Rev. Cytol. 1995, 163, 175–247.

114. Bouvier, F.; d'Harlingue, A.; Hugueney, P.; Marin, E.; Marion-Poll, A.; Camara, B. Xanthophyll biosynthesis. J. Biol. Chem. 1996, 271, 28861–28867.

115. Prasad, K. N.; Lyee, C.; Khoo, H.E.; Bao. Y.; Azlan, A.; Amin. I. Carotenoids and antioxidant capacities from Canarium odontophyllum Miq. fruits. Food Chem. 2011, 124, 1549–1555.

116. Tanaka, Y. Flower colour and cytochromes P450. Phytochem. Rev. 2006, 5, 283–291.

117. Bakker, J.; Timberlake, C.F. Isolation, identification, and characterization of new color-stable anthocyanins occurring in some red wines. J. Agric. Food Chem. 1997, 45, 35–43.

118. Ben-Amotz, A.; Fishler, R. Analysis of carotenoids with emphasis on 9-cis β-carotene in vegetables and fruits commonly consumed in Israel. Food Chem. 1998, 62, 515–520.

119. Levin, I.; De Vos, C.H.R.; Tadmor, Y.; Bovy, A.; Lieberman, M.; Oren-Shamir, M.; Segev, O.; Kolotilin, I.; Keller, M.; Ovadia, R. High pigment tomato mutants—more than just lycopene (a review). Isr. J. Plant Sci. 2006, 54, 179–190.

Chapter 3

EFFECTS OF ZINC AND ASCORBIC ACID APPLICATION ON THE GROWTH AND PHOTOSYNTHETIC PIGMENTS OF MILLET PLANTS GROWN UNDER DIFFERENT SALINITY

M. M. Hussein[1], A. K. Alva[2]

[1]Water Relations and Irrigation Department, National Research Centre, Cairo, Egypt

[2]USDA-ARS, Vegetable and Forage Crops Research Unit, Prosser, WA, USA

ABSTRACT

Salinity stress impacts crop growth as well as production. The need for increased food production to feed the increasing population and the limited resources, i.e. optimal quality land and water, require developing strategies to mitigate marginal stresses, including salinity stress, for reasonable expectation of crop production. A pot experiment was conducted in a greenhouse at the National Research Centre, Dokki, Cairo, Egypt in the summer season of 2005 to evaluate the effects of foliar application of ascorbic acid alone or in combination with zinc sulfate on the growth and photosynthetic pigments of millet plants irrigated by tap water (250 ppm, 0.39 dS $\cdot$ m^{-1}) or moderate to high salinity irrigation water [2500 ppm (3.9 dS $\cdot$ m^{-1}) and 5000 ppm (7.8 dS $\cdot$ m^{-1})]. Increased salinity in the irrigation water decreased the plant growth, biomass, and carotenoid content. Foliar application of ascorbic acid alone increased number of leaves and leaf area, while in combination with zinc sulfate increased the plant height and total plant biomass. However, these treatments had no significant effects on the photosynthetic pigments. This study demonstrates that exogenous application of ascorbic acid can enhance foliar growth which may contribute to increased plant biomass and yield.

INTRODUCTION

Soil salinity is a major abiotic stress for crop production in many parts of the world. Approximately one third of the irrigated area in the world (227 million hectares) is already affected by varying degree of excess salinity/sodicity [1] , primarily caused by inadequate drainage. About 23% and 37% of the world's cultivated lands (1.5×10^9 ha) are characterized as saline and sodic, respectively.

The negative effects of salinity on plant growth and metabolism were reported on millet [2] , wheat [3] -[5] , barley [6] [7] , and rice [1] .

Zinc (Zn) is required for plant growth as an activator of several enzymes and is directly involved in the biosynthesis of growth regulators such as auxin, which promotes production of more plant cells and biomass that will be stored in the plant organs especially in seeds [8] .

Ascorbic acid is an organic acid with antioxidant properties. Many oxidants, typically reactive oxygen species such as the hydroxyl radical (formed from hydrogen peroxide), contain an unpaired electron and thus are highly reactive and damage plant cells at molecular level. This is due to their interaction with nucleic acid, proteins, and lipids [9] .

The problems of salinity can be mitigated by developing crop cultivars with improved tolerance to salts and/ or by altering the growth and physiology of the crop by ways of amendments such as fertilizers and/or phyto- chemicals [10] -[12] . The objective of this study was to evaluate the potential for mitigation of salt stress in millet plants by foliar application of Zn sulfate with or without combination of ascorbic acid.

MATERIALS AND METHODS

A pot experiment was conducted in a greenhouse at the National Research Centre (NRC), Dokki, Cairo, Egypt, during 2005 summer season. A clay soil was sampled (0 - 15 cm depth) from Kirdasa village, Giza governorate. The bulk soil was air-dried, sieved to pass 2 mm sieve and five replicate samples were taken for analyses of some basic physiochemical properties (Table 1). Metallic pots (35 cm diameter, and 50 cm depth) were used with 30 kg air-dried soil. The inner surface of the pots was coated with three layers of bitumen to prevent direct contact between the soil and the metal. The base of the pot was filled with 2 kg gravel (about 2 - 3 cm diameter) prior to filling the pots with soil.

Each pot received 3 g calcium super phosphate (6.8% P) and 1.5 g potassium sulfate (40.3% K) and mixed with top 10 cm depth soil. These rates were equivalent to 106 and 212 kg/ha P and K, respectively. The soil was moistened to field capacity (22% water by weight) prior to planting. Millet (Pennisetum glaucum (L.) R. Br.) seeds were sown (10 seeds/plot) on July, 5, 2005. The plants were thinned 15 and 25 days after seedling emergence to leave three uniform plants/pot. Nitrogen was broadcasted using ammonium sulfate (20.5% N), 6.86 g per pot [equivalent to 488 kg $\cdot$ N $\cdot$ ha^{-1}] in two equal doses (244 kg $\cdot$ ha^{-1} each) i.e. before planting and 2 weeks after the seedling emergence. The treatments included:

1)   Main Treatments: 3 levels of irrigation water quality, i.e. tap water and 2 dilute sea water with salinity levels of 250, 2500, and 5000 ppm, respectively (i.e. 0.39, 3.9, and 7.8 dS $\cdot$ m^{-1}).

2)   Sub treatments: Two foliar sprays (21 and 36 days after sowing), using: i) tap water as Control; ii) Ascorbic acid (150 ppm); iii) as in treatment ii) plus 200 ppm zinc sulphate.

Table 1: Physical and chemical analysis of the soil used in this experiment

	Sand (%)		Silt (%) 20 - 2 μm	Clay (%) <2 μm	Soil Texture
Course > 200 μm		Fine 200 - 20 μm			
7.20		14.25	30.22	48.33	clay

Soil chemical analysis											
pH (1:2.5)	EC (dS·m^{-1} 1:5)	CaCO$_3$ (%)	CEC (C mol·Kg^{-1})	OM (%)	Soluble cations and anions (meq/100 g soil)						
					Na$^+$	K$^+$	Ca^{2+}	Mg^{2+}	HCO$_3^-$	Cl$^-$	SO$_4^{2-}$
7.15	1.3	2.53	33.5	1.3	1.82	0.23	2.38	1.27	0.91	1.9	1.89

Available macro-nutrients (%)			Available micro-nutrients (ppm)			
N	P		K	Zn	Fe	Mn	Cu
0.47	0.25		0.95	3.1	4.8	7.3	5.2

Soil Chemical Analysis

High salinity water irrigation began 30 days after sowing. Each high salinity water irrigation was alternated by tap water irrigation through the entire duration of the experiment.

Each of the above 9 treatments were replicated 6 times. Thus the total number of pots was: 3 × 3 × 6 = 54. The mean minimum and maximum temperatures during the course of the experiment were 22°C and 37°C, respectively. Relative humidity range was 51% - 64%. The range of day length was 11 to 14 h.

On 45 days after the seedling emergence, five leaves were sampled per plant from two plants per pot for analyses of photosynthetic pigments i.e. chlorophyll_a (Chl_a), chlorophyll_b (Chl_b) and total carotenoids concentrations using the procedures described by von Wetestien et al. [13] . Plants from all replicate pots of all treatments were cut 2 cm above the soil surface. The plant height was measured by a ruler. Number of leaves was counted. The stem diameter was measured using a caliper. Leaves were separated from the stem, and leaf area was measured using a Li-Cor portable leaf area meter. The fresh biomass weights of leaves and stem were measured. The biomass of leaves and stem were dried in an oven at 72˚C for 48 h and the dry weights were recorded.

The significance of the treatments effects on the response parameters was evaluated by analysis of variance (ANOVA) and mean separation tests [14] .

RESULTS AND DISCUSSION

Increased salinity in irrigation water decreased the growth of millet plants (Table 2). Significant decrease in growth parameters occurred at the high salinity level as compared to the growth of the plants irrigated by tap water. The stem or total plant dry weights decreased by 52% in plants irrigated with 7.8 dS $\cdot$ m^{-1} salinity water as compared to those of the plants irrigated by tap water. These results are in agreement with those of other researchers [12] [15] -[17] .

Olmos and Hellín [18] reported that adaptation of cell line of Pisumsativum germplasm to NaCl depends on modification of the osmotic adjustment together with physiological and biochemical modifications. When cultivar calli, with tolerance to high salinity, was grown in high NaCl medium (85.5 mM), intracellular levels of Na, Cl, reducing sugars were increased. Total free amino acids and ascorbic acid contents also increased. Veeranagamallaiah et al. [19] reported reduction in millet (Setariaitalica L. cv Prasad) seedling growth and biomass when subjected to 100 - 200 mM NaCl. Furthermore, Beltagi [11] reported no significant negative effects of salinity (20 and 40 mM) on stem and root length of pure strain of Chickpea. However, stem and root fresh weights and the root dry weight were significantly decreased at 40 mM NaCl. The number of leaves per plant was decreased at 20 and 40 mM NaCl.

The negative effects of salt stress were reported on water and mineral absorption [20] -[22] , water adjustment [23] [24] , protein synthesis [19] [25] , photosynthesis and carbohydrate accumulation [23] [26] , enzymes activities [27] [28] , growth regulators [7] [29] and antioxidant defense mechanism [30] [31] .

Salinity stress had no significant effects on the Chl_a and Chl_b content (Table 3). Carotenoid content, however, decreased significantly in the plants subjected to irrigation with 3.9 dS $\cdot$ m^{-1} salinity water as compared to that of the plants received tap water irrigation. Further increase in salinity to 7.8 dS $\cdot$ m^{-1} had no significant effects on the carotenoid content. The ratio of (Chl_a + Chl_b): carotenoid increased with an increase in salinity levels.

Beltaji [11] reported no negative effects of salinity on Chlorophyll content in chick pea plants. Pinheiro et al. [23] reported that Chl_a and carotenoids contents in castor bean increased with increased salinity in the range of 0 to 30 mM NaCl salinity on 38 days after germination, but decreased on 59 days after germination. Chl_b content,

Table 2: Effects of salinity on growth and biomass of millet plants

Salinity (dS·m⁻³)	Plant height (cm)	No. of leaves	Stem diameter (cm)	Leaf area (cm²)	Fresh weight (g/plant)			Dry weight (g/plant):		
					Stem	Leaves	Total	Stem	Leaves	Total
0.39[1]	129	8.0	1.89	1299	54.6	18.3	72.9	23.0	11.5	34.5
3.9	108	7.2	1.70	992	43.5	19.5	65.0	19.5	10.4	29.9
7.8	68	7.0	1.54	880	33.1	17.6	50.7	10.9	10.0	20.9
LSD (P ≤ 0.05)[2]	42.3	1.01	0.29	291	NS[3]	NS	NS	10.7	N.S	18.3

[1]Tap water; [2]LSD: Least significant difference; [3]NS: Non-significant.

[1]Tap water; [2]LSD: Least significant difference; [3]NS: Non-significant.

however, decreased at both stages and the reverse was true for Chl_a: Chl_b ratio. On the other hand, Sairam and Srivstava [32] reported a decrease in chlorophyll content in wheat genotypes subjected to 6.85 dS $\cdot$ m^{-1} salinity using NaCl.

RESPONSE TO FOLIAR APPLICATION OF ASCORBIC ACID AND ZINC SULFATE

Foliar application of only ascorbic acid significantly increased number of leaves and leaf area as compared to those of the plants which were sprayed with tap water (Table 4). All other response parameters were non-sig- nificant. Plant height, number of leaves, and total plant dry biomass were significantly increased with foliar application of ascorbic acid plus zinc sulfate. Foliar application of Zinc has contributed to increased plant growth and yield of peanuts [33] [34] and sunflower [35] .

Beltagi [11] indicated that the addition of ascorbic acid (4 mM) significantly increased the stem dry weight of chickpea plants. Abd El-Moniem et al. [36] reported foliar application of zinc on orange trees' improved leaf N, K, and Zn concentrations. Abd El-Aziz et al. [37] also reported increased growth and nutrient uptake by Kaya sengalensis with foliar Zinc application.

Zinc is an important activator of several enzymes in plants and is directly involved in the biosynthesis of growth substances, such as auxin which produces more plant cells which result in increased dry matter. Darwish et al. [33] reported the highest seed and oil yields, and protein percentage of peanuts grown with foliar application of 96 kg · m⁻¹ K. Gobarah et al. [34] also reported an increase in peanut yield and quality with foliar application of 2% zinc solution. Similar responses were also reported for sunflower [35] . No significant response was evident on photosynthetic pigments by foliar application of ascorbic acid alone or in combination with zinc sulfate (Table 5).

The interactions between the salinity and foliar application of ascorbic acid and zinc sulfate were mostly insignificant on all growth parameters, except leaf area and stem dry weight (Table 6), and photosynthetic pigments, except Chl_a + Chl_b (Table 7). A significant increase in stem dry weight of chickpea by application of ascorbic acid (4 mM) has been reported only at the low salinity level (20 mM NaCl) [11] . Beneficial effects of the exogenous application of ascorbic acid in partially mitigating the adverse effects of salt stress on growth of Chickpea plants (Cicer arietinum L.), cell division and cell enlargement have been reported [11] [38] . Shalata and Neumann [39] reported that salt-stress increased the accumulation of lipid peroxidation products produced

Table 3: Effects of salinity on the concentrations of photosynthetic pigments of millet plants

Salinity (dS·m⁻¹)	Chl_a (ppm)	Chl_b (ppm)	Carotenoid (ppm)	Chl_a + Chl_b (ppm)	Chl_a: Chl_b	(Chl_a + Chl_b) Carotenoids
0.39 (Tap water)	5.96	2.10	2.45	8.06	2.84	3.29
3.9	4.80	2.27	1.96	7.07	2.12	3.61
7.8	5.33	2.06	1.93	7.39	2.59	3.82
LSD (P ≤ 0.05)	NS	NS	0.77	0.354	-	-

LSD: Least significant difference; NS: Non-significant; Chl_a: Cholorphyll_a; Chl_b: Cholorphyll_b.

LSD: Least significant difference; NS: Non-significant; Chl_a: Cholorphyll_a; Chl_b: Cholorphyll_b.

Table 4; Effects of foliar application of ascorbic acid without or with zinc sulfate (ZnSO₄) on growth and biomass production of millet plants

Foliar spray	Plant height (cm)	No. of leaves	Stem diameter (cm)	Leaf area (cm²)	Fresh weight (g/plant)			Dry weight (g/plant)		
					Stem	Leaves	Total	Stem	Leaves	Total
1. Tap water	88	6.4	1.36	891	37.4	15.2	52.6	14.5	9.0	23.5
2. Ascorbic acid (150 ppm)	102	7.8	1.71	1162	41.2	19.7	60.9	16.9	11.4	28.3
3. Trt. 2 + ZnSO₄ (200 ppm)	116	8.0	1.80	1089	52.5	20.3	72.8	22.0	11.5	33.5
LSD (P ≤ 0.05)	19	1.1	N.S	250	N.S	N.S	N.S	N.S	N.S	9.8

LSD: Least significant difference; NS: Non-significant.

LSD: Least significant difference; NS: Non-significant.

Table 5: Effects of foliar application of ascorbic acid with or without zinc sulfate on the concentrations of photosynthetic pigments in millet plants

Foliar spray	Chl_a (ppm)	Chl_b (ppm)	Carotenoid (ppm)	Chl_a + Chl_b (ppm)	Chl_a: Chl_b	(Chl_a + Chl_b) Carotenoids
1.Tap water	5.61	2.29	2.36	7.90	2.45	3.31
2. Ascorbic acid (150 ppm)	5.03	2.02	1.89	7.05	2.49	3.73
3. Trt. 2 + ZnSO₄ (200 ppm)	5.46	2.13	2.09	7.59	2.56	3.63
LSD (P ≤ 0.05)	N.S	N.S	N.S	N.S	-	-

LSD: Least significant difference; NS: Non-significant; Chl_a: Cholorphyll_a; Chl_b: Cholorphyll_b.

LSD: Least significant difference; NS: Non-significant; Chl_a: Cholorphyll_a; Chl_b: Cholorphyll_b.

Table 6: Effects of foliar application of ascorbic acid and zinc sulfate on growth of millet plants under different salinity levels

Salinity (dS·m⁻¹)	Foliar spray	Plant height (cm)	No. of leaves	Stem diameter (cm)	Leaf area (cm²)	Fresh weight (g/plant) Stem	Leaves	Total	Dry weight (g/plant) Stem	Leaves	Total
0.39 (Tap water)	1. Tap water	107	7	1.7	1109	57.4	13.5	70.9	23.2	8.8	32.0
	2. Ascorbic acid (150 ppm)	140	6	1.7	1505	53.2	20.3	73.5	21.6	14.2	35.8
	3. Trt. 2 + ZnSO₄ (200 ppm)	140	9	2.0	1283	53.3	20.4	73.7	24.2	11.5	35.7
3.9	1. Tap water	94	6	1.2	833	27.3	14.7	42.0	11.7	9.3	21.0
	2. Ascorbic acid (150 ppm)	63	8	1.5	1065	38.8	21.9	60.7	16.3	11.2	27.5
	3. Trt. 2 + ZnSO₄ (200 ppm)	127	8	1.7	992	65.0	21.8	86.8	30.2	10.8	40.9
7.8	1. Tap water	62	5	1.2	731	39.7	17.5	57.2	8.5	8.9	17.4
	2. Ascorbic acid (150 ppm)	63	7	1.7	917	31.7	16.4	48.1	8.9	8.9	17.8
	3. Trt. 2 + ZnSO₄ (200 ppm)	80	8	1.7	991	40.0	18.8	58.8	11.8	12.2	23.9
	LSD at P ≤ 0.05	N.S	N.S	N.S	481	N.S	N.S	N.S	17.4	N.S	N.S

LSD: Least significant difference; NS: Non-significant.

LSD: Least significant difference; NS: Non-significant.

Table 7: Effects of foliar application of ascorbic acid and zinc sulfate on concentrations of photosynthetic pigments in millet plants grown under different salinity levels

Salinity (dS·m⁻¹)	Foliar spray	Chl_a (ppm)	Chl_b (ppm)	Carotenoid (ppm)	Chl_a + Chl_b (ppm)	Chl_a: Chl_b	(Chl_a + Chl_b) Carotenoids
0.39 (Tap water)	1. Tap water	6.2	2.0	2.5	8.2	3.1	3.3
	2. Ascorbic acid (150 ppm)	5.9	2.2	2.4	8.1	2.7	3.3
	3. Trt. 2 + ZnSO₄ (200ppm)	5.8	2.2	2.5	8.0	2.6	3.2
3.9	1. Tap water	4.5	2.5	2.3	7.0	1.8	3.0
	2. Ascorbic acid (150 ppm)	4.4	2.0	1.7	6.5	2.2	3.8
	3. Trt. 2 + ZnSO₄ (200ppm)	5.5	2.3	1.9	7.8	2.4	4.1
7.8	1. Tap water	6.1	2.4	2.3	8.5	2.5	3.7
	2. Ascorbic acid (150 ppmt does)	4.8	1.9	1.6	6.7	2.5	4.2
	3. Trt. 2 + ZnSO₄ (200ppm)	5.1	1.9	1.9	7.0	2.7	3.7
	LSD at P ≤ 0.05	N.S	N.S	N.S	0.742	-	-

LSD: Least significant difference; NS: Non-significant

LSD: Least Significant Difference; NS: Non-Significant.

by interactions with damaging active oxygen species in roots, stems and leaves. Exogenous application of ascorbic acid partially mitigated the above response, but did not significantly reduce sodium uptake or plasma membrane leakiness. Verma and Mishra [40] reported that salinity caused reduction in seedling growth and biomass accumulation, which was parallel to that caused by increased superoxide (O_2^-), hydrogen peroxide (H_2O_2) levels, lipid peroxidation and electrolyte leakage in leaf tissues.

Beltagi [41] reported that application of ascorbic acid (4 mM) increased Chl_a content in cowpea plants under high salinity conditions. However, Chl_b and Chl_a + Chl_b contents were not influenced by exogenous ascorbic acid. Salt stress can lead to oxidative stress through an increase in reactive oxygen species which are highly reactive and may cause cellular damage. Ascorbic acid acts as an antioxidant for scavenging hydrogen peroxide [41]. Sairam and Srivastava [32] revealed that NaCl salinity caused decrease in relative water content (RWC), and chlorophyll content. Results of this study demonstrate that salinity-induced growth suppression of millet plants can be mitigated by foliar application of ascorbic acid in combination with zinc sulfate.

REFERENCES

1.   Khan A.A. and Abdullah, Z. (2003) Salinity-Sodicity Induced Changes in Reproductive Physiology of Rice (Oryza sativa L.) under Dense Soil. Environment and Experimental Botany, 47, 145-157. http://dx.doi.org/10.1016/S0098-8472(02)00066-7

2.   Heidari, M. and Jamshidi, P. (2011) Effects of Salinity and Potassium Application on Antioxidant Enzyme Activities and Physiological Parameters in Pearl Millet. Agricultural Sciences in China, 10, 228-237. http://dx.doi.org/10.1016/S1671-2927(09)60309-6

3.   Sastry, E.V.D., Sharma, H. and Sharma, H. (2000) Effect of Temperature and Salinity on the Germination and Seedling Growth in Wheat. Indian Journal of Agricultural Science, 70, 117-118.

4.   Hu, Y., Schmidhalter, U. and Hu, Y.C. (2001) Reduced Cellular Cross-Section Area in the Leaf Elongation Zone of Wheat Causes a Decrease in Dry Weight Deposition under Saline Conditions. Australian Journal of Plant Physiology, 28, 165-170.

5.   Akram, N., Hussein, M., Akhtar, S. and Rasul, E. (2002) Impact of NaCl Salinity on Yield Components of Some Wheat Accessions/Varieties. International Journal of Agriculture Biology, 4, 156-158.

6.   Hussein, M.M., El-Geratly, N.H. and Abo El-Khier, M.S. (2002) Endogenous Hormones, Growth and Yield of Barley Plants as Affected by

Benzyl Adenine under Different Salinity Levels. Journal of Agricultural Science, Moshtohor, University, 27, 5283-5292.

7. Hussein, M.M., Balbaa, L.H. and El-Liethy, S. (2007) The Effect of Saline Irrigation, Adenine Spraying and Their Interaction on the Growth and Photosynthetic Pigments in Barley. Egyptian Journal of Applied Science, 22, 173-186.

8. Marschner, H. (1995) Mineral Nutrition of Higher Plants. 2nd Edition, Academic Press, London, 645.

9. Younis, M.E., Hasaneen, M.N. and Kazamel, A.M. (2009) Plant Growth, Metabolism and Adaptation in Relation to Stress Conditions. XXVII. Can Ascorbic Acid Modify the Adverse Effects of NaCl and Mannitol on Amino Acids, Nucleic Acids and Protein Patterns in Viciafaba Seedlings? Protoplasma, 235, 37-47. http://dx.doi.org/10.1007/s00709-008-0025-4

10. Hussein, M.M. and El-Masry, M.F. (2007) Irrigation by Mixed Drainage Water and Micronutrients Spray and Its Effects on Micronutrients in Straw and Grains of Wheat. Proceedings of ICID 22nd European Regional Conference, Pavia, 2-7 September 2007, 89.

11. Beltagi, M.S. (2008) Exogenous Ascorbic Acid (Vitamin C) Induced Anabolic Changes for Salt Tolerance in Chick Pea (Cicer arietinum L.) Plants. African Journal of Plant Science, 2, 118-123.

12. Hussein , M.M. , Shaaban, M.M. and El-Saady, A.M. (2008) Response of PK-Foliar Fertilizations Cowpea Plants Grown under Salinity Stress. American Journal of Plant Physiology, 3, 81-88. http://dx.doi.org/10.3923/ajpp.2008.81.88

13. Von Wetstein, D. (1957) Chlorophyll Letale and Der Sub-Mikroskopishe Formweschselder Plastiden. Experimental Cell Research, 12, 427-506. http://dx.doi.org/10.1016/0014-4827(57)90165-9

14. Snedecor, G.W. and Cochran , W.G. (1990) Statistical Methods. 8th Edition, Iowa State University Press, Ames.

15. Manikandan, K. and Desingh, R. (2009) Effect of Salt Stress on Growth, Carbohydrate and Proline Content of Two Finger Millet Varieties. Recent Research in Science and Technology, 1, 48-51.

16. Hussein , M.M. , Tawfik, M.M. , Ahmed , M.K. and El-Karamany, F. (2013) Effect of Water Stress on Growth and Some Physiological Aspects of Jojoba [Simmondisia chinieses (Link) Schenider] in New Reclaimed Sandy Soil. Elixer Pollution, 55, 12903-12909.

17. Ashraf, M.Y., Akhtar, K., Sarwar, G. and Ashraf, M. (2003) Evaluation of Arid and Semi-Arid Ecotypes of Guar (Cyamopsis tetragonoloba L.)

for Salinity (NaCl) Tolerance. Journal of Arid Environments, 52, 473-482. http://dx.doi.org/10.1006/jare.2002.1017

18. Olmos, E. and Hellín, E. (1996) Mechanisms of Salt Tolerance in a Cell Line of Pisum sativum: Biochemical and Physiological Aspects. Plant Science, 120, 37-45.http://dx.doi.org/10.1016/S0168-9452(96)04483-4

19. Veeranagamallaiah, G., Jyothsnakumari, G., Thippeswamy, M., Reddy, P.C. , Surabhi,G.K. , Srirangayakulu, G., Mahesh, Y., Rajasekhar, B., Madhurarekha, Ch. and Sudhakar, C. (2008) Proteomic Analysis of Salt Stress Responses in Foxtail Millet (Setaria italica L. cv Prasad) Seedlings. Plant Science, 175, 631-641.http://dx.doi.org/10.1016/j.plantsci.2008.06.017

20. Sheldon, A., Menzies , N.W., So, H.B. and Dalal, R. (2004) The Effect of Salinity on Plant Available Water. Proceedings of Australian New Zealand Soil Conference, University of Sydney, Sydney, 5-9 December 2004.

21. Ragab, A.A. , Hellal, F.A. and Abd El-Hady , M. (2008) Water Salinity Impacts on Some Soil Properties and Nutrients Uptake by Wheat Plants in Sandy and Calcareous Soils. Australian Journal of Basic and Applied Science, 2, 225-233.

22. Kaya, C., Ashraf, M., Dikilitas, M. and Tuna, A.L. (2013) Alleviation of Salt Stress-Induced Adverse Effects of Maize Plants by Exogenous Application of Indoleacetic Acid (IAA) and Inorganic Nutrients. A Field Trial. Australian Journal of Crop Science, 7, 249-254.

23. Pinheiro, H.A., Silva, J., Endres, L., Ferreira, V.M., Camara, C.A., Cabral, F.F., Oliveira, V.M., Carvalha, L.W., Santos, J.M. and Filho, B.G. (2008) Leaf Gas Exchange, Chloroplastic Pigments and Dry Matter Accumulation in Castor Bean (Ricinus communis L.) Seedlings Subjected to Salt Stress Conditions. Industrial Crops and Products, 27, 385- 392. http://dx.doi.org/10.1016/j.indcrop.2007.10.003

24. Tuna, A.L. , Kaya, C., Dikilitas, M. and Higgs, D. (2008) The Combined Effects of Gibberelic Acid and Salinity on Some Antioxidant Enzymes Activity, Plant Growth Parameters and Nutritional Status in Maize Plants. Environmental and Experimental Botany, 62, 1-9. http://dx.doi.org/10.1016/j.envexpbot.2007.06.007

25. Maslenkova, L.T . , Miteva, T.S. and Popova, L.P. (1992) Changes in Polypeptide Patterns of Barley Seedlings Exposed to Jasmonic Acid and Salinity. Plant Physiology, 98, 700-707.http://dx.doi.org/10.1104/pp.98.2.700

26. El-Tayeb, M.A. (2005) Response of Barley Grains to the Interactive Effect of Salinity and Salicylic Acid. Plant Growth Regulators, 45, 215-224. http://dx.doi.org/10.1007/s10725-005-4928-1

27. Ahmadi, A. and Baker, D.A. (2001) The Effect of Water Stress on the Activities of Key Regulatory Enzymes of the Sucrose of Starch Pathway in Wheat. Plant Growth Regulation, 35, 81-91. http://dx.doi.org/10.1023/A:1013827600528

28. Raza, S., Athar, H., Ashraf, M. and Hameed, A. (2007) Glycinebetaine-Induced Modulation of Antioxidant Enzymes Activities and Ion Accumulation in Two Wheat Cultivars Differing in Salt Tolerance. Environmental and Experimental Botany, 69, 368-376.http://dx.doi.org/10.1016/j.envexpbot.2006.12.009

29. Ghanem, M.E., Albacete, A., Andújar, C.M. , Acosta, M., Aranda, R.R. , Dodd, I., Lutts, C. and Alfocea, F.P. (2008) Hormonal Changes during Salinity-Induced Leaf Senescence in Tomato (Solanum lycopersicum L.). Journal of Experimental Botany, 59, 3039-3050.http://dx.doi.org/10.1093/jxb/ern153

30. Athar, H., Khan, A. and Ashraf, M. (2008) Exogenously Applied Ascorbic Acid Alleviates Salt-Induced Oxidative Stress in Wheat. Environmental and Experimental Botany, 63, 224-231. http://dx.doi.org/10.1016/j.envexpbot.2007.10.018

31. Abd El-Baky , H.H. , Hussein , M.M. and Baroty, G.S. (2008) Algal Extraction Improve Antioxidants Defense Abilities and Salt Tolerance of Wheat Plant Irrigated with Sea Water. Electronic Journal of Environmental Agriculture and Food Chemistry, 7, 2812-832.

32. Sairam, R.K. and Srivastava, G.C. (2002) Changes in Antioxidant Activity in Sub-Cellular Fractions of Tolerant and Susceptible Wheat Genotypes in Response to Long Term Salt Stress. Plant Science, 162, 897-904. http://dx.doi.org/10.1016/S0168-9452(02)00037-7

33. Darwish, D.S., El-Gharreib, E.G., El-Hawary, M.A. and Rafft, O.A. (2002) Effect of Some Macro and Micronutrients Application on Peanut Production in a Saline Soil in El-Faiyum Governorate. Egyptian Journal of Applied Science, 17, 17-32.

34. Gobarah, M.E., Mohamed, M.H. and Tawfik, M.M. (2006) Effect of Phosphorus Fertilizer and Foliar Spraying with Zinc on Growth, Yield and Quality of Groundnut under ReclaimedSandy Soils. Journal of Applied Science Research, 2, 491-496.

35. Thalooth, A.T. , Badr, N.M. and Mohamed, M.H. (2005) Effect of Foliar Spraying with Zn and Different Levels of Phosphatic Fertilizer on Growth

and Yield of Sunflower Plants Grown under Saline Condition. Egyptian Journal of Agronomy, 27, 11-22.

36.   Abd El-Moniem , A., Emman, A., Abd El-Mageed , M.M. and Omayma, M.M. (2007) GA3 and Zinc Sprays for Improving Yield and Fruit Quality of Washington Novel Orange Trees Grown under Sandy Conditions. Research Journal of Agriculture and Biological Science, 3, 498-395.

37.   Abd El-Aziz , N. Mazhar, A.M. and Habaa, E. (2006) Effect of Foliar Spray with Ascorbic Acid on Growth and Chemical Constituents of Kaya sengalensis Grown under Salt Condition. American-Eurasian Journal of Agriculture & Environmental Science, 1, 207-214.

38.   Ahmed-Hamed, A.M. and Monsaly, H.M. (1998) Seed Soak in Pre-Sowing in Vitamins versus the Adverse Effects of NaCl Salinity on Photosynthesis and Some Related Activities of Maize and Sunflower Plants. Proceedings of the XIth International Photosynthesis Conference, Budapest, August 1998, 17-22.

39.   Shalata, A. and Neumann , P.M. (2007) Exogenous Ascorbic Acid (Vitamin C) Increases Resistance to Salt Stress and Reduces Lipid Peroxidation. Journal of Experimental Botany, 52, 2207-2211.

40.   Verma, S. and Misra, N. (2005) Putrescine Alleviation of Growth in Salt Stressed Brassica juncia by Inducing Antioxidative Defense System. Journal of Plant Physiology, 162, 669-677. http://dx.doi.org/10.1016/j.jplph.2004.08.008

41.   Khan, T., Mazid, M. and Mohammad , F. (2011) A Review of Ascorbic Acid Potentialities against Oxidative Stress in Plants. Journal of Agrobiology, 28, 97-111.http://dx.doi.org/10.2478/v10146-011-0011-x

Chapter 4

STRAWBERRY FLAVOR: DIVERSE CHEMICAL COMPOSITIONS, A SEASONAL INFLUENCE, AND EFFECTS ON SENSORY PERCEPTION

Michael L. Schwieterman,[1,8] Thomas A. Colquhoun,[1,2,8] Elizabeth A. Jaworski,[2,8] Linda M. Bartoshuk,[3,8] Jessica L. Gilbert,[4,8] Denise M. Tieman,[4,8] Asli Z. Odabasi,[5,8] Howard R. Moskowitz,[6] Kevin M. Folta,[1,4,8] Harry J. Klee,[1,4,8] Charles A. Sims,[5,8] Vance M. Whitaker,[5,7,8] and David G. Clark[1,2,8]

[1] Plant Molecular and Cellular Biology Program, University of Florida, Gainesville, Florida, United States of America

[2] Department of Environmental Horticulture, University of Florida, Gainesville, Florida, United States of America

[3] College of Dentistry, University of Florida, Gainesville, Florida, United States of America

[4] Horticultural Sciences Department, University of Florida, Gainesville, Florida, United States of America

[5] Food Science and Human Nutrition Department, University of Florida, Gainesville, Florida, United States of America

[6] Moskowitz Jacobs Inc., White Plains, New York, United States of America

[7] Gulf Coast Research and Education Center, University of Florida, Wimauma, Florida, United States of America

[8] Plant Innovation Program, University of Florida, Gainesville, Florida, United States of America

ABSTRACT

Fresh strawberries (Fragaria x ananassa) are valued for their characteristic red color, juicy texture, distinct aroma, and sweet fruity flavor. In this study, genetic and environmentally induced variation is exploited to capture biochemically diverse strawberry fruit for metabolite profiling and consumer rating. Analyses identify fruit attributes influencing hedonics and sensory perception of strawberry fruit using a psychophysics approach. Sweetness intensity, flavor intensity, and texture liking are dependent on sugar concentrations, specific volatile compounds, and fruit firmness, respectively. Overall liking is most greatly influenced by sweetness and strawberry flavor intensity, which are undermined by environmental pressures that reduce sucrose and total volatile content. The volatile profiles among commercial strawberry varieties are complex and distinct, but a list of perceptually impactful compounds from the larger mixture is better defined. Particular esters, terpenes, and furans have the most significant fits to strawberry flavor intensity. In total, thirty-one volatile compounds are found to be significantly correlated to strawberry flavor intensity, only one of them negatively. Further analysis identifies individual volatile compounds that have an enhancing effect on perceived sweetness intensity of fruit independent of sugar content. These findings allow for consumer influence in the breeding of more desirable fruits and vegetables. Also, this approach garners insights into fruit metabolomics, flavor chemistry, and a paradigm for enhancing liking of natural or processed products.

INTRODUCTION

Modern fully ripe strawberry (*Fragaria x ananassa*) fruit is characterized by its large size [1], vibrant red color [2], reduced firmness [3], distinct aroma [4], and sweet fruity flavor [5]. The flesh of the strawberry is a swollen receptacle, a false fruit, and the seeds or achenes are the true fruit [6], which will be collectively referred to as strawberry fruit. The three stages of non-climacteric, auxin dependent strawberry fruit development; division, expansion and ripening, involve gains in diameter and fresh weight; during which color shifts from green to white to dark red in roughly forty days following anthesis [7]. Ripening of strawberry fruit results in the accumulation of multiple sugars and organic acids, culminating with peak volatile emission [8]

Flavor is the perceptual and hedonic response to the synthesis of sensory signals of taste, odor, and tactile sensation [9]. In the case of strawberry and other fruits, sensory elicitation is the result of multiple direct interactions between plant and human: sugars and acids, pigments, turgor and structure, and volatile compounds, which elicit the senses of taste, vision, tactile sensation, and olfaction, respectively, in the development of flavor [10]–[13]. A consumer based survey indicated sweetness and complex flavor as consistent favorable attributes of the "ideal" strawberry experience [14]. Much emphasis is placed on sugars, acids, and volatile compounds as these metabolites are primary sensory elicitors of taste and olfaction which attenuate the perception and hedonics of sweetness and flavor. Thus a ripe strawberry is metabolically poised to elicit the greatest sensory and hedonic responses from consumers.

During strawberry fruit development sucrose is continually imported from photosynthetic tissue. A consistently high sucrose invertase activity contributes to carbon sink strength in all developmental stages of fruit [15]. Delivered sucrose is hydrolyzed into glucose and fructose, and these three carbohydrates constitute the major soluble sugars of ripe strawberries, a result of their continual accumulation during fruit development [16]. In fact, an approximately 150% increase in their sum during ripening has been observed [8], [15]. The influx of carbon initiates a complex network of primary and secondary metabolism specific to ripening strawberry fruit [16]. For example, the metabolic activity of ripening strawberry is visualized by the late accumulation of the predominant red pigment, pelargonidin 3-glucoside [17], an anthocyanin derived from the primary metabolite phenylalanine [16].

The dynamics of fruit development are genetically driven. Microarray analysis determined nearly 15% of probed expressed genes exhibit significant differential expression (60% up, 40% down) in red compared to green fruit [18]. One up regulated gene, *Polygalacturonase 1 (FaPG1)*, contributes to fruit softening [19] by aiding in catalytic cell wall disassembly [20]. Reduction of firmness is also attributed to dissolution of middle lamella, a pectin rich cell wall layer that functions in cell-to-cell adhesion [3]. Active shifts in transcription throughout ripening result in metabolic network reconfiguration altering the chemical and physical properties.

Metabolic profiling indicates an accumulation of sugars, organic acids, and fatty acids as well as the consumption of amino acids during fruit development. Subsequently alkanes, alcohols, aldehydes, anthocyanins, ketones, esters, and furanones increase during fruit ripening [7]. Many of these chemical classes serve as precursors to volatile synthesis [21], thus facilitating a metabolic flux through biosynthetic pathways for increased and diverse volatile emissions in ripe strawberry fruit, predominantly furans, acids, esters, lactones, and terpenes [8]. Over 350 volatile compounds have been identified across *Fragaria* [22], however within a single fruit, far fewer compounds are detectable and even less contribute to aroma perception.

A cross comparison of five previous studies which analyze strawberry volatiles depicts the lack of agreement in defining chemical constituents of strawberry aroma. Each source considers a highly variable subset of volatiles, which are determined by signal intensity and/or human perception of separated compounds [4], [5], [23]–[25]. Mutual volatiles across studies include butanoic acid, methyl ester; butanoic acid, ethyl ester; hexanoic acid, methyl ester; hexanoic acid, ethyl ester; 1,6-octadien-3-ol, 3,7-dimethyl- (linalool); butanoic acid, 2-methyl-; and 3(2*H*)-furanone, 4-methoxy-2,5-dimethyl-, the current consensus of integral strawberry aroma compounds. Comparisons of consumer preference among a variety of fresh strawberries and their volatile profiles describes less preferable varieties as possessing less esters, more decalactones and hexanoic acid [4]. The breadth of volatile phenotypes previously reported highlights the diversity across strawberry genotypes and underscores the complexity of the aggregate traits of aroma and flavor.

Florida strawberry production is concentrated on ten thousand acres near the Tampa Bay. Mild winters allow for annual horticulture which requires continual harvest of ripe fruit from late November through March. Environmental effects on fruit quality are partially attributed to gradually increasing temperatures beginning in mid-January. One result is a late season decline of soluble solids content (SSC) [26], [27]. In fact, increasing temperature is known to be responsible for increasing fruit maturation rate and decreasing SSC independent of flowering date[26]. Previous work also identifies variability of SSC, as well as titratable acidity (TA) and multiple classes of volatile compounds across harvest dates [28]. The complex fruit biochemistry, which is variably affected by genetic, environmental, and developmental factors, coupled with individuals' perceptional biases has made defining strawberry flavor cumbersome.

Here we exploit the genetic and within-season variability of fruit to provide as many unique strawberry experiences as possible to a large sample of consumers. To enhance the range and diversity of flavors and chemical constituents 35 genetic backgrounds were included: public and private cultivars representing a large proportion of commercial strawberry acreage in North America, University of Florida advanced breeding selections, and European cultivars (Fig. 1). Parallel assays of ripe strawberry samples quantify fruit traits of TA, pH, and fruit firmness, as well as the content of malic acid, citric acid, glucose, fructose, sucrose, and 81 volatile compounds of diverse chemical classes. The contributions of these attributes to fruit quality is determined by simultaneously evaluating samples for perceived sensory intensities of sourness, sweetness, and strawberry flavor, as well as the hedonic responses of texture liking and overall liking by consumer panelists. Data analyses determine significant biochemical and consumer response differences between early and late season fruit, gross variation of strawberry experiences, and factors influencing hedonics and sensory perception of strawberry fruit consumption using a psychophysics approach. Ultimately, an effect of particular volatile constituents to enhance sweetness intensity independent of sugar content of fruit was found. These findings have great implications in the breeding of more desirable fruits and vegetables, as well as for the food industry as a whole.

Figure 1: Photographs of strawberry production field, plants, and harvested fruits.

METHODS

Ethics Statement

All human consumer panels are conducted at the Food Science and Human Nutrition Department at the University of Florida in Gainesville, FL. The University of Florida Institutional Review Board 2 (IRB2) chaired by Ira S. Fischler approved the protocol and written consent form (case 2003-U-0491), which participants are required to complete.

Plant Material

Thirty-five strawberry cultivars and selections were grown at or in the near vicinity of the Gulf Coast Research and Education Center (14625 County Road 672, Wimauma, FL) during the 2010–2011 (season 1) and 2011–2012 (season 2) winter seasons. Fruit are cultivated according to current commercial practices for annual strawberry plasticulture in Florida [1], [29](Fig. 1A). The cultivars are chosen to represent a large proportion of commercial strawberry acreage in North America from both public and private breeding programs. Additional breeding selections and European cultivars are added to enhance the range of diversity for flavors and chemical constituents. Weekly cultivar representation is determined by fruit availability during a particular harvest week and attempting to maximize genetic diversity, except for the highly replicated cultivar 'Festival'. Fully-ripe fruit by commercial standards, 90–100% red compared to white [30] (Fig. 1B-C), is harvested from three to five cultivars on Monday mornings, delivered to the respective laboratories, and stored at 4°C in the dark overnight for simultaneous analysis of fresh strawberry fruit volatiles, firmness, and sensory analysis on Tuesdays; as well as sample preparation for later sugar and acid measurements. Six harvests in both seasons allows for the complete analysis of 54 samples. Weather data is obtained from the Balm, FL station of the Florida Automated Weather Network (http://fawn.ifas.ufl.edu/data/reports) for date ranges January 3, 2011 through February 28, 2011 and December 26, 2011 through March 13, 2012. Daily maximum and minimum temperature recording height is 60 cm, and daily average relative humidity, rainfall, and solar radiation are recorded at 2 m.

Volatile Analysis

At least 100 g or seven berries of each sample are removed from 4°C dark overnight storage prior to volatile collection. Samples are homogenized in a blender prior to splitting into three 15 g replicates for immediate capturing of volatile emissions. The remainder is frozen in N_2 (l) and stored at –80°C for later sugar and acid quantification. A two hour collection in a dynamic headspace volatile collection system [31]allows for concentration of emitted volatiles on HaySep 80–100 porous polymer adsorbent (Hayes Separations Inc., Bandera, TX, USA). Elution from polymer is described by Schmelz [32].

Quantification of volatiles in an elution is performed on an Agilent 7890A Series gas chromatograph (GC) (carrier gas; He at 3.99 ml min^{-1}; splitless injector, temperature 220°C, injection volume 2 µl) equipped with a DB-5 column ((5%-Phenyl)-methylpolysiloxane, 30 m length ×250 µm i.d. × 1 µm film thickness; Agilent Technologies, Santa Clara, CA, USA). Oven temperature is programmed from 40°C (0.5 min hold) at 5°C min^{-1} to 250°C

(4 min hold). Signals are captured with a flame ionization detector (FID) at 280°C. Peaks from FID signal are integrated manually with Chemstation B.04.01 software (Agilent Technologies, Santa Clara, CA). Volatile emissions (ng^1 gFW^{-1} h^{-1}) are calculated based on individual peak area relative to sample elution standard peak area. GC-mass spectrometry (MS) analysis of elutions are performed on an Agilent 6890N GC in tandem with an Agilent 5975 MS (Agilent Technologies, Santa Clara, CA, USA) and retention times are compared with authentic standards (Sigma Aldrich, St Louis, MO, USA) for volatile identification [33]. Chemical Abstract Services (CAS) registry numbers were used to query SciFinder® substances database for associated chemical name and molecular formula presented in Table S1.

Sugars and Acids Quantification

Titratable acidity, pH, and soluble solids content [26] are averaged from four replicates of the supernatant of centrifuged thawed homogenates[1]. An appropriate dilution of the supernatant from a separate homogenate (centrifugation of 1.5 ml at 16,000 x g for 20 min) is analyzed using biochemical kits (per manufacturer's instructions) for quantification of citric acid, L-malic acid, D-glucose, D-fructose, and sucrose (CAT# 10-139-076-035, CAT# 10-139-068-035, and CAT# 10-716-260-035; R-Biopharm, Darmstadt, Germany) with absorbance measured at 365 nm on an Epoch Microplate Spectrophotometer (BioTek, Winooksi, VT, USA). Metabolite average concentration (mg^{1}100 gFW^{-1}) is determined from two to six technical replicates per pooled sample. Derived sucrose concentrations via D-glucose and D-fructose are mathematically pooled.

Firmness Determination

Firmness of the strawberries is determined as the resistance of the fruit to penetration at its equator with a TA.XTPlus Texture Analyzer (Texture Technologies Corp., Scarsdale, NY, USA; Stable Micro Systems, Godalming, Surrey, UK). The Texture Analyzer is equipped with a 50 kg load cell and an 8 mm diameter convex tip probe. Whole fruit is penetrated on the side to 7 mm down from the epidermis at a test speed of 2 mm^1 sec^{-1}; a flap cut off the opposite provides stability. Maximum force in kg for eight fruit is averaged and reported as a measure of firmness.

Sensory Analysis

Over the course of two annual seasons, 166 recruited strawberry consumers (58 male, 108 female) evaluate strawberry cultivars. Ages of panelist ranged from 18 to 71, with a median age of 24. Panelists self-classified themselves as 98 White or Caucasian, 11 Black or African-American, 1 Native American, Alaska Native or Aleutian, 41 Asian/Pacific Islander, and 15 Other. An average of 106 (range of 98–113) panelists evaluated between three and five cultivars per session[34]. Fresh, fully-ripe strawberry fruit is removed from overnight 4°C dark storage and allowed to warm to room temperature prior to sensory analysis. Each panelist is given one to two whole strawberries for evaluation, depending on cultivar availability. Panelists bite each sample, chew, and swallow it. Ratings for overall liking and texture liking are scaled on hedonic general labeled magnitude scale (gLMS) from −100 to +100, *i.e.* least to most pleasurable experience [34]–[37]. Perceived intensity of sweetness, sourness, and strawberry flavor are scaled in context of all sensory experiences using sensory gLMS that ranges from 0 to +100, *i.e.* none to most intense sensory stimulus[34]–[37]. Scales are employed to mediate valid comparisons across subjects and sessions.

Statistical Analysis

Means and standard errors for consumer, physical, and metabolite measurements are determined from all replicates using JMP (Version 8, SAS Institute Inc., Cary, NC, USA). One-way analysis of early and late season fruit quality and consumer response measures was subjected to mean comparison using Tukey's HSD (α=0.05). Bivariate analysis among individual measurements of samples allows for linear fit, which includes summary of fit, analysis of variance, t-test, and correlation analysis for density ellipse. Two-way Ward hierarchical cluster analysis of all quantified metabolite and strawberry samples is accomplished in JMP. Amounts of individual volatile compounds are regressed using the "enter" method in SPSS (IBM Corp., Armonk, NY, USA). This is done individually for each of the three sugars: glucose, fructose or sucrose to identify which compounds have an effect on sweetness intensity [14]independent of each of the sugars. For *p*-values ≤ 0.05, the volatile makes a contribution to perceived sweetness that is independent of the sugar tested.

Results

The inventory of 54 fully ripe (Fig. 1C) unique strawberry samples (35 cultivars, 12 harvests, two seasons) assayed for TA, pH, firmness, as well as the concentrations of malic acid, citric acid, glucose, fructose, sucrose, and quantity of 81 volatile compounds is reported (Table S2). Cluster analysis of relative chemical composition of all samples and derived hierarchy of both cultivar and metabolite relatedness is displayed (Fig. 2). The vertical dendrogram (Fig. 2) demonstrates the lack of relatedness among volatile compound quantities through large distances of initial segments, as well as the high number of clusters. Slightly more structure is observed among the samples, horizontal dendrogram (Fig. 2), due to genetic or environmental effects.

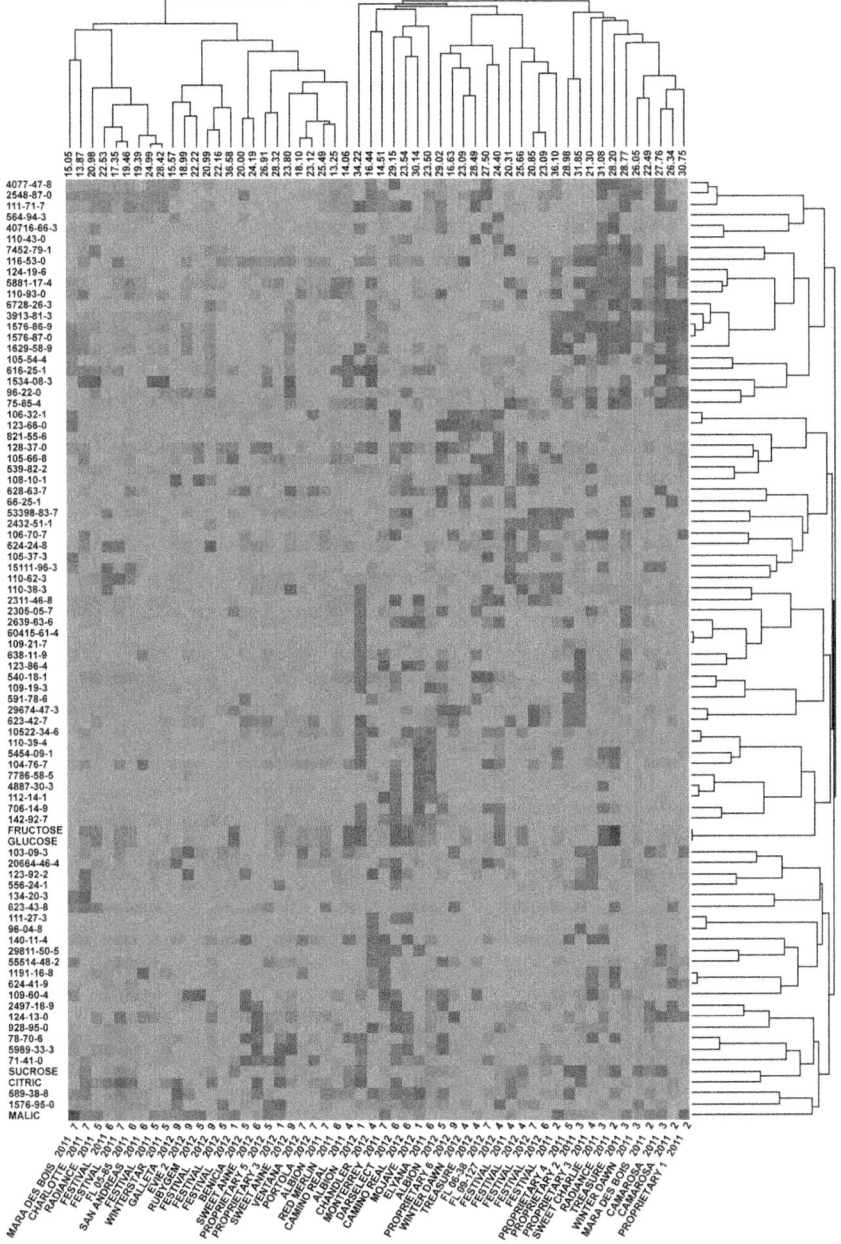

Figure 2: Cluster analysis of strawberry samples and quantified metabolites.

Progression of Harvest Season Affects Perceived Quality and Metabolite Content of Strawberry

Overall liking is a measure of pleasure derived from consuming a strawberry sample. The two samples with the greatest overall liking ratings are of cultivar 'Festival'. Fruit harvested early in the season, week 2 of season 1 and week 1 of season 2, elicit overall likings of 36.1 and 36.6, respectively (Table 1). Five weeks following both early samplings of 'Festival' the overall liking of the same cultivar decreases below the sample set median of 23.5 (Table S2) to 17.3 in season 1 week 7 and to 23.1 in season 2 week 6 (Table 1). Therefore the earlier season samples elicit a greater hedonic response than late season samples. Overall likings are determined using the hedonic general labeled magnitude scale that ranges from −100 to +100, *i.e.* least to most pleasurable experience [34]–[37]. Conversely, sweetness, sourness, and strawberry flavor are measured using the sensory intensity general labeled magnitude scale that ranges from 0 to +100, *i.e.* none to most intense sensory stimulus [34]–[37]. Consumer perception of sweetness and strawberry flavor intensity decrease significantly between the same pairs of early and late season 'Festival' fruit (Table 1). Significant biochemical differences between early and late samples include decreased content of glucose, fructose, sucrose, and total volatiles. The early 'Festival' from the first season contains 88% more total sugar and 65% more total volatiles than the late 'Festival' of the same season (Table 1), demonstrating the disparity between early and late harvest week fruit quality and its effect on consumer sensory perception and acceptability.

Table 1: Comparison of early and late season strawberry fruit

			Season 1				Season 2			
			Week 2		Week 7		Week 1		Week 6	
Mean week temperature										
	Daily maximum	°C	21.6	B	28.2	A	21.3	B	26.1	A
	Daily minimum	°C	7.4	B	13.3	A	6.7	B	13.1	A
	Daily average	°C	14.9	B	20.3	A	14.0	B	19.0	A
Consumer ratings										
	Overall liking	−100 to +100	36.1	A	17.3	B	36.6	A	23.1	B
	Texture liking	−100 to +100	35.7	A	23.8	B	34.8	A	24.3	B
	Sweetness intensity	0 to +100	30.3	A	15.9	B	34.0	A	22.2	B
	Sourness intensity	0 to +100	17.9	A	15.9	A	18.2	A	17.9	A
	Strawberry flavor intensity	0 to +100	34.3	A	20.4	B	37.5	A	25.2	B
Biochemical measures										
	Glucose	(mg^1 100 gFW^{-1})	1903	A	1127	B	2187	A	1807	B
	Fructose	(mg^1 100 gFW^{-1})	2048	A	1311	B	2327	A	1973	B
	Sucrose	(mg^1 100 gFW^{-1})	1218	A	309	B	1902	A	450	B
	Total sugar	(mg^1 100 gFW^{-1})	5169	-	2747	-	6417	-	4229	-
	Relative sucrose	-	0.37	B	0.41	A	0.34	B	0.43	A
	Relative fructose	-	0.40	B	0.48	A	0.36	B	0.47	A
	Relative sucrose	-	0.24	A	0.11	B	0.30	A	0.11	B
	Total volatiles	(ng^1 gFW^{-1} h^{-1})	19097	A	11543	B	16843	A	16001	A

Comparison of means for temperature (mean of 7 days prior to harvest), consumer ratings, and biochemical measures between early and late season strawberry fruit cultivar 'Festival' from season 1 and season 2. Mean comparison accomplished in JMP 8 using Tukey's HSD. Mean marked A is significantly greater than mean marked B (α = 0.05).
doi:10.1371/journal.pone.0088446.t001

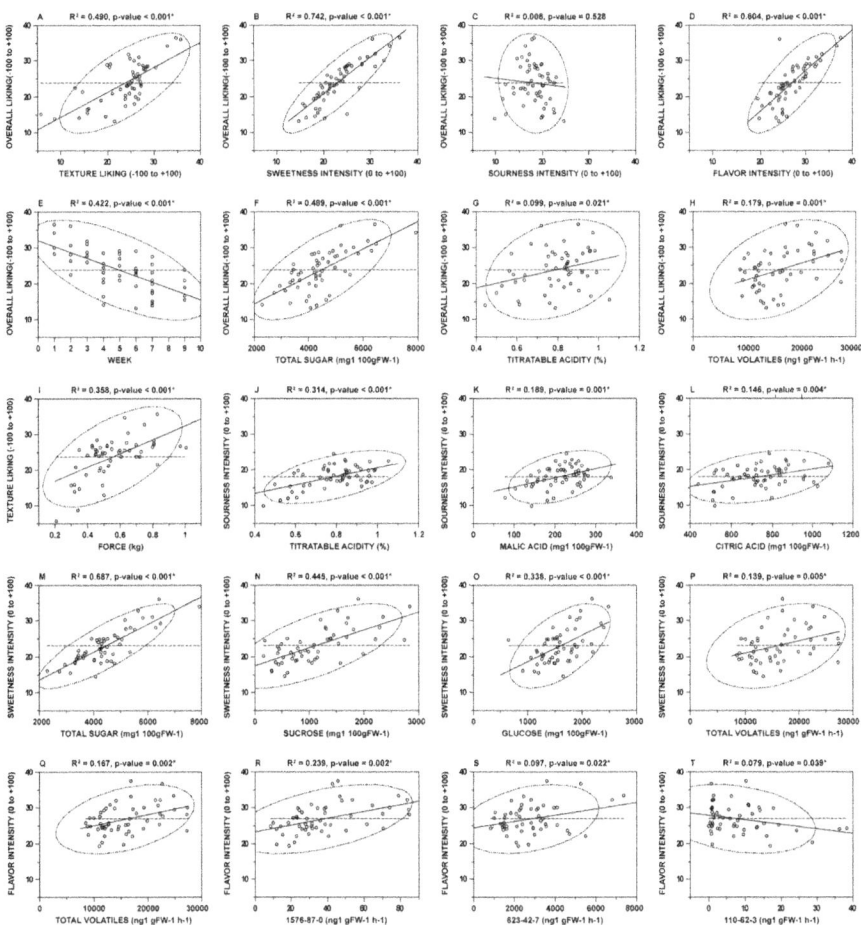

Figure 3: Regression of hedonic and sensory measures to physical and chemical fruit attributes.

Solar radiation, minimum temperature and maximum temperature increase gradually within the limits of similar ranges in season 1 and season 2 (Fig. S1 A-D). Relative humidity remains constant during and across seasons (Fig. S1 E, F). Slightly more rain fell in early season 1 than season 2 (Fig. S1 G-H) One manifestation of these environmental changes over a harvest season is the negative relationship between total sugar and harvest week (Table 1). The content of all individual sugars measured decreases between early and late season 'Festival' samples; however there is a significant decrease in the proportion of sucrose to total sugar (Table 1). The disproportionate decrease is observed for the collection of samples as well (Fig. S2A-C) (Table S3). Also, a significant correlation is observed across all 54 samples among total volatiles and sucrose (R^2=0.305*) (Fig. S2E) but not glucose (R^2=0.005) (data not shown) or fructose (R^2=0.001) (Fig. S2F). A harvest week associated decrease in total sugars, predominantly sucrose, results in a decrease in volatile content, which ultimately undermines late season overall liking (R^2=0.422*) (Fig. 3E) through sweetness and strawberry flavor intensity.

Overall Liking is Subject to Ratings of Sweetness, Flavor, and Texture but not Sourness

In order to elucidate factors contributing to a positive strawberry experience, overall liking of strawberry samples is fit against the hedonic measure of texture liking and the sensory intensities of sweetness, sourness, and strawberry flavor intensity (Fig. 3A-D). High correlation with significant fit exists for texture liking (R^2=0.490*) (Fig. 3A), sweetness intensity (R^2=0.742*) (Fig. 3B), and strawberry flavor intensity (R^2=0.604*) (Fig. 3D). However, sourness intensity shows no correlation to overall liking (R^2=0.008) (Fig. 3C). Increasing firmness contributes to greater texture liking (R^2=0.358*) (Fig. 3I), and texture liking has a significant influence on overall liking. Sweetness intensity is the strongest driver of overall liking measured in this study. The correlation between total sugar and overall liking (R^2=0.488*) (Fig. 3F) demonstrates the aggregate sugar metabolites effect on hedonic response to strawberry fruit. Total sugar concentration accounts for nearly half of the observed overall liking variation but is far from a complete measure. Sourness intensity appears to have no influence on the hedonic response to strawberry fruit, but fit of TA to overall liking is significant, even if minor (R^2=0.099*) (Fig. 3G). Total volatiles is the second aggregate metabolite measure having a significant enhancing effect on the overall liking of strawberry (R^2=0.179*) (Fig. 3H). This is not surprising, as strawberry flavor intensity exhibits the second highest correlation to overall liking (Fig. 3D).

Texture Liking Correlates to Fruit Firmness

The upper limit for hedonics of texture is comparable to that of overall liking and is observed in 'Festival' (sn 1, wk 2) with an average of 35.7, however, the low texture liking value of 5.8 for 'Mara Des Bois' (sn 1, wk 7) indicates a more drastic disliking of "off" textures than the overall liking of even the lowest rating fruit (Table S2). Firmness of samples is assayed by measuring the force required for a set penetration of the fruit, acting as a proxy for texture. The firmness of the fresh strawberry exhibited nearly a five-fold difference in force, 0.2 kg for 'Mara des Bois' (sn 1, wk 7) and 1.0 kg for 'Festival' (sn 1, wk 5) (Table S2). Increasing force of penetration, i.e. increasing firmness of berries, is positively correlated with texture liking, indicating a hedonic response to firmer fruit (Fig. 3I). However, the texture liking is less than the expected rating for the two samples with greatest firmness (Fig. 3I).

Sweetness Intensity is a Result of Sugar Content

Perceived sweetness intensity is the greatest predictor of overall liking. In fact, the same samples scoring the highest and lowest for overall liking, 'Festival' (sn 2, wk 1) and 'Red Merlin' (sn 1, wk 6), elicit the greatest (36.2) and least (14.59) intense sensations of sweetness (Table S2). The early and late harvest week samples support the observable decline in perceived sweetness intensity across harvest weeks, which is also observable for multiple sugar measures (Fig. S2A-C) (Table 1).

In the 54 samples assayed, the total sugar concentration ranged from 2.29 – 7.93%, a 3.5-fold difference (Table S2). Glucose and fructose concentrations exhibit highly similar ranges to each other, 0.66 – 2.48% and 0.75 – 2.61%, respectively (Table S2), and near-perfect correlation ($R^2=0.984*$) (data not shown) within a sample. However, the concentration of glucose or fructose is not predictive of sucrose concentration ($R^2=0.011$ and 0.004, respectively) (data not shown). Sucrose demonstrated a more dynamic state as its concentration dips as low as 0.16% and up to 2.84%, nearly a seventeen-fold difference among all samples.

Sucrose is the single metabolite with the most significant contribution to overall liking ($R^2=0.442*$) (Table S4). Individually, sucrose ($R^2=0.445*$) (Fig. 3N), glucose ($R^2=0.337*$) (Fig. 3O), and fructose ($R^2=0.300*$) (Table S4) all significantly influence the variation in sweetness intensity. However, total sugar actually only accounts for slightly more than two-thirds of sweetness intensity variation ($R^2=0.687*$) (Fig. 3M) likely a result of covariation of glucose and fructose. Interestingly, the total volatile content of a sample correlates positively with sweetness intensity, potentially accounting for up to 13.9%* of variation in sweetness intensity (Fig. 3P).

Sourness Intensity is Partially Explained by Titratable Acidity

Cultivar 'Red Merlin' (sn 1, wk 6) elicited the most intense sourness response at 24.6 (Table S2). This same sample rates as the lowest in terms of overall liking and sweetness. Acidity of strawberry fruit is assayed using measures of pH, TA, citric acid and malic acid. The pH of strawberry samples ranges from 3.35 to 4.12, while TA ranges from 0.44% to 1.05%. The range of malic acid across samples is 0.078% to 0.338% while citric acid ranged from 0.441% to 1.080% (Table S2). TA has the greatest correlation to sourness intensity ($R^2=0.314*$) (Fig. 3J), when compared to pH ($R^2=0.118*$), malic acid ($R^2=0.189*$) (Fig. 3K), or citric acid (R^2=0.146*) (Fig. 3L) concentration. Citric acid concentration is approximately three-fold greater than malic acid and has a significant effect on TA (R^2=0.49*) (data not shown). There is no correlation of malic acid to TA (R^2=0.01) (data not shown). The lack of relationship among sourness intensity and overall liking (Fig. 3C) is shadowed by the strong correlations of sweetness intensity (Fig. 3B) and flavor intensity (Fig. 3D) to overall liking. Deficiencies in perceived sweetness and flavor intensity as observed in 'Red Merlin' can result in a fruit that is negatively perceived as intensely sour.

Flavor Intensity Is Influenced by Total and Specific Volatile Content

In this study, strawberry flavor intensity accounts for the retronasal olfaction component of chemical senses, which compliments sourness and sweetness intensities' contribution to taste. The overall highest sensory intensity is 37.5 (Table S2) for strawberry flavor of 'Festival' (sn 2, wk 1), which also rates highest for overall liking and sweetness intensity. Opposite this, FL- 05-85 (sn 1, wk 6) delivers the least intense strawberry flavor experience with a score of 19.4 (Table S2). Total volatiles in 'Festival' (sn 2, wk 1) is over 50% greater than in FL 05-85 and seven more volatiles compounds are detected (Table S2). Total volatiles within a sample contribute to strawberry flavor intensity (R^2=0.167*) (Fig. 3Q), but it is not simply the sum of volatile constituents that explain the effect. For instance, the maximum total volatile content detected within a sample, 27.3 μg^1 gFW^{-1} hr^{-1} from 'Camarosa' (sn 1, wk 2), does not result in the greatest flavor intensity (30.5) and the minimum, 8.5 μg^1 gFW^{-1} hr^{-1} from 'Sweet Anne' (sn 2, wk 9), does not rate as the least flavorful (25.8) (Table S2).

The chemical diversity of the resources analyzed allows for the identification of 81 volatile compounds from fresh strawberry fruit (Fig. S3). The majority of compounds are lipid related esters, while lipid related aldehydes account for the majority of volatile mass. Terpenes, furans, and ketones are also represented in the headspace of strawberry. Forty-three of the 81 volatile compounds are not detected (<0.06 ng^1 gFW^{-1} hr^{-1}) in at least one sample. Therefore, 38 volatiles are measured in all samples; appearing to be constant in the genetic resources analyzed (Table S2). No cultivar has detectable amounts of all 81 volatiles. Samples of 'Festival', 'Camino Real', PROPRIETARY 6, and FL 06-38 are the most volatile diverse, but are lacking detectable amounts benzoic acid, 2-amino-, methyl ester (134-20-3) [5]. This methyl ester of anthranilic acid is detectable in only 'Mara des Bois' and 'Charlotte' from the final harvest (wk 7) of season 1 (Table S2). 'Chandler' (sn 2, wk 4) and 'Red Merlin' (sn 1, wk

6) are the least volatile diverse samples lacking detectable amounts of 19 and 17 compounds, respectively (Table S2).

The most abundant ester, butanoic acid, methyl ester (623-42-7) is measured at over 7 μg^1 gFW^{-1} hr^{-1} from PROPRIETARY 2 (sn 1, wk 3) and has a significant correlation to flavor ($R^2=0.097*$) (Fig. 3S). A terpene alcohol, 1,6,10-Dodecatrien-3-ol, 3,7,11-trimethyl-, (6E)- (40716-66-3) (nerolidol), with maximum content of over 600 ng^1 gFW^{-1} hr^{-1} in 'Sweet Charlie' is not detected in 'Red Merlin'. The nerolidol rich 'Sweet Charlie' garners greater flavor intensity at 32.2 than deficient 'Red Merlin' at 23.95. The impact on flavor intensity by nerolidol ($R^2=0.112*$) (Table S4) is greater than butanoic acid, methyl ester despite having maximum contents lower by one order of magnitude. Hexanal (66-25-1) is the second most abundant individual compound, an aldehyde detected in all samples, exceeds 11 μg^1 $gFW^{-1}hr^{-1}$(Table S2), and does not have a significant correlation to flavor intensity ($R^2=0.016$) (Table S4). Hexanoic acid, ethyl ester (123-66-0) exhibits over 200-fold difference across samples, and also has no bearing on sensory perception (Table S4). Conversely, two minor level aldehydes demonstrate a disparity in effect: 2-pentenal, (2E)- (1576-87-0) is enhancing toward flavor intensity ($R^2=0.239*$) (Fig. 3R), while pentanal (110-62-3) is the only compound that negatively correlates to flavor ($R^2=0.079*$) (Fig. 3T). The significant contribution of the 1,6-octadien-3-ol, 3,7-dimethyl- (78-70-6) (linalool) to flavor intensity positively correlates with increasing content (R2=0.074*) (Table S4). In 'Chandler' 3(2H)-furanone, 4-methoxy-2,5-dimethyl- (4077-47-8) is not detectable, and only has maximum content of 40 ng^1 gFW^{-1} hr^{-1} in 'Treasure' (sn 1 wk 3). The level of this characteristic strawberry furan is significantly impactful on perceived flavor intensity ($R^2=0.108*$) (Table S4). In total, thirty volatile compounds diverse in structure have a positive relationship to flavor intensity and their significance cannot be derived from content alone.

Specific Volatiles Enhance Sweetness Intensity Independent of Sugars

Multiple regression of individual volatile compounds against perceived intensity of sweetness is performed independent of glucose, fructose, or sucrose concentration (Table S5). Twenty four volatile compounds show significant correlations ($\alpha=0.05$) to perceived sweetness intensity independent of glucose or fructose concentration, twenty-two of which are mutual between the two monosaccharides. Twenty volatiles are found to enhance sweetness intensity independent of sucrose concentration; only six of these volatiles are shared with those independent of glucose and fructose: 1-penten-3-one (1629-

58-9); 2(3*H*)-furanone, dihydro-5-octyl- (2305-05-7) (γ-dodecalactone); butanoic acid, pentyl ester (540-18-1); butanoic acid, hexyl ester (2639-63-6); acetic acid, hexyl ester (142-92-7); and butanoic acid, 1-methylbutyl ester. Only three compounds are found to be negatively related to sweetness independent of at least one of the sugars: octanoic acid, ethyl ester (106-32-1) exclusively independent of glucose; 2-pentanone, 4-methyl- (108-10-1) mutually independent of glucose and fructose; and 2-buten-1-ol, 3-methyl-, 1-acetate (1191-16-8) exclusively independent of sucrose.

DISCUSSION

Exploitation of genetic diversity and environmental variation allows for a wide range of consumer hedonic and sensory responses. The cultivars in this study represent a large proportion of commercial strawberry acreage in North America, advanced breeding selections, and European cultivars. A genetic collection aimed at enhancing the diversity of physical and chemical constituents, as well as consumer experiences. Despite the perennial life cycle of strawberry much commercial production uses annual methods, which in sub-tropical Florida allows for continual harvest of ripe fruit from late November through March. A nearly three-fold difference in overall liking of strawberry is observable within all samples. The highest and lowest rating samples are 'Festival' of the first week in the second season and 'Red Merlin' of the sixth week in the first season. These two cultivars are the product of separate breeding programs, have distinct genetic backgrounds, and therefore distinct biochemical inventories. Harvested at opposite ends of the seasons the early and late season fruit are subjected to different environmental conditions, further attenuating genetic differences. The diversity of strawberries samples assayed and range of consumer liking captured (Fig. 2) indicates the chemical diversity of strawberry cultivars is not only perceivably different but certain profiles are more highly preferable.

Elevated texture liking, sweetness intensity, and strawberry flavor intensity significantly increases overall liking, while sourness intensity alone has no detectable relationship to overall liking (Fig. 3A-D). Integration and synthesis of response to sensory signals of taste, olfaction, and tactile sensation constitute an eating experience [9] and drive overall liking. The senses of taste and olfaction sample the chemicals present in food *e.g.* sugars, acids, and volatile chemical compounds. These elicitors attenuate the perception and hedonics of food [38], [39]. Ratings of strawberry fruit are correlated to specific chemical or physical attributes, especially sweetness (Fig. 3B) and flavor intensity (Fig. 3D), the two greatest drivers of overall liking.

Much work has been done to measure sugars and volatile compounds in strawberry fruit in an attempt to understand sweetness and flavor, and these aims are in line with consumer demand. A consumer survey using 36 attributes of strawberry determined "sweetness" and "complex flavor" as consistent favorable characteristics of the ideal strawberry experience [14]. Previous work in tomato [34] and this current study on strawberry surveyed participants for ideal ratings of the respective fruits. Using the same gLMS scales employed in the current study, scores for ideal strawberry and tomato overall liking, sourness intensity, and flavor intensity are highly similar. Ideal flavor evoked the highest mean response of 45 for both, exemplifying its importance to the consumer. Interestingly, a large disparity for ideal sweetness intensity is found; 42 and 33 for strawberry and tomato, respectively. Ideal sweetness intensity is much greater in strawberry, potentially due to differences in consumption. Strawberry is often consumed fresh and is a delicacy or dessert fruit, while tomato is savory and often an ingredient in complex recipes. Therefore, the desire for sweetness may be greater in strawberry.

The overall liking of strawberry fruit is significantly related to texture liking (Fig. 3A), and increasing fruit firmness accounts for more than a third of increasing texture liking (Fig. 3I). The five-fold variation in firmness can be attributed to variation in fruit development or softening (Table S2). Strawberry fruit development consists of division, expansion, and ripening [7]. Developmentally regulated, ripening associated fruit softening is multifaceted [19], including catalytic cell wall disassembly [20] and dissolution of cell-to-cell adhesion [3]. The relationship between texture liking and firmness does not appear entirely linear, because the two firmest samples are close to average texture liking (Fig. 3I). Excessively firm fruits may be perceived as under ripe while those with less firmness may be considered over ripe; affecting texture liking. Fruit can progress through ripening, from under to over ripe, in ten days [7], exemplifying the narrow window in which multiple facets of fruit quality must synchronize.

Despite a moderate range of intensity, perceived sourness has little to no bearing on overall liking (Fig. 3C). Just over 30% of sourness intensity variation can be accounted for by positive correlation with TA. The concentrations of citric acid and malic acid metabolites are likely additive toward the effect of TA on sourness intensity, and in fact both organic acids have significant correlations to sourness intensity (Fig. 3K-L). Despite a lack of influence by sourness intensity on overall liking, metabolites of sourness have a critical role in fruit biochemistry. Increased TA shows a significant minor correlation with overall liking (Table S4) and correlates significantly with total sugar (data not shown). This co-linearity may be due to accumulation of sugars and subsequent biosynthesis of organic acids during ripening of fruit [7],[8], [16]. Citric acid is the predominant organic acid in ripe fruit [40]and its concentration is fairly stable during ripening. Also, it is known to act as an intermediate between imported sucrose and fatty acid biosynthesis [16], which may facilitate enhancement of overall liking through volatile biosynthesis.

The consumer rating of sweetness intensity is the primary factor contributing to overall liking, and sweetness is the component of taste perception facilitating the detection of sugars. Sugars are simple carbohydrates, a readily available form of energy, and the degree of correlation among sweetness and overall liking is due to hedonic effect[38]. Variation in sweetness intensity is best explained by sugar content (Fig. 3L). Previously, soluble solid content (SSC) has been used as a valid indicator of sweetness in strawberry [1], [28]. However this is an aggregate measure, as previous quantification of individual sugars within a strawberry identifies sucrose, glucose, and fructose as the predominant soluble solids [1], [8], [15], [40]. Sucrose concentrations observed across samples is responsible for more variation in total sugar, sweetness intensity and overall liking than any other individual compound (Table S4). Metabolites contributing to perceived sweetness intensity have the greatest influence on the overall hedonics of strawberry. A significant decrease in sweetness intensity occurs between early and late season fruit, and unfortunately overall liking decreases as well (Table 1) (Fig. 3E).

Drastic fruit quality differences between early and late season fruit result in lower consumer response (Table 1) (Fig. 3E), which is likely due to environmental changes (Fig. S1) or plant maturity. A significant difference in the mean temperature one week prior to harvest is likely a causative factor (Table 1). Monitored development of 'Festival' fruit under elevated temperature decreases the fruit development period from 36 days at 15°C to 24 days at 22°C. Also, a simultaneous decrease in SSC is observed, both independent of flowering date i.e. plant maturity[26], [27]. The mean temperature of the week prior to harvest for early and late season 'Festival' fruit are 15°C and 20°C for the first season and 14°C and 19°C for the second season (Table 1). These differences in environment likely alter whole plant physiology and more specifically fruit biochemistry during development and ripening, affecting fruit quality. During strawberry fruit development sucrose is continually translocated from photosynthetic tissue, while a consistently high sucrose invertase activity in fruit hydrolyzes sucrose into glucose and fructose, maintaining sink strength of fruit [15] and in turn feed biosynthetic pathways [16]. Total and individual sugars decrease in ripe fruit during both seasons as the plant is subjected to increasing temperatures (Table 1). Increased maturation rate hastens fruit development, potentially decreasing cumulative period sucrose is imported to fruit, and inhibiting sucrose accumulation to affect other fruit quality attributes. These factors are likely causative of the observable decrease in sweetness and flavor intensity as the season progresses.

Although total sugar decreases between early and late fruit, a disproportionate amount of the decrease is attributed to sucrose (Table 1), which indicates sucrose as the waning constituent of sugar content (Fig. S2A-C). Glucose and fructose concentrations are tightly correlated to each other, show less seasonal influence than sucrose, and lack correlation to sucrose. These observations are indicative of tighter biochemical regulation of glucose and fructose than sucrose, which has the greatest variability in concentration among the three sugars. Total volatile content has an indirect dependence on sucrose concentration (Fig. S2E), and a decrease in total volatiles is observed between early and late season strawberry (Table 1). Influence of harvest date on headspace of fresh strawberry fruit is known [41], [42]. Increased volatile content is likely dependent on more free sucrose, *i.e.* a larger imported reserve, facilitating greater flux through primary and secondary metabolism. Generation of glucose and fructose initiates a complex network of primary and secondary metabolism specific to ripening strawberry fruit, in which sucrose is principal and limiting to the strawberry fruit biosynthetic pathways [16]. Upregulation of biosynthetic genes associated with volatile secondary metabolites [43]and the consumption of primary metabolite classes of fatty acids and amino acids, precursors of volatile compounds, happens in the final stages of ripening [16]. This sucrose dependent metabolic shift culminates in peak volatile content and diversity [8].

Strawberry flavor intensity is the second greatest determinant of overall liking (Fig. 3D) and accounts for perception of volatile compounds through retronasal olfaction. A significant positive relationship exists among total volatile content and the flavor intensity for a given sample, however, total volatile content is not entirely explanatory of flavor intensity. The maximum rating for strawberry flavor intensity by 'Festival' (sn 2, wk1) is the greatest consumer response evoked within this study (Table S2), highlighting the significance of sensory perception of aroma. However, this sample only has slightly more than 60% of total volatile mass of the greatest sample. The extent of volatile phenotype diversity is great enough across strawberry fruit to not only be discerned but be preferred.

Within the genetic resources of *Fragaria x ananassa* analyzed in this study 81 compounds are reproducibly detected, but not one cultivar has detectable amounts of all compounds. The amount of individual volatile compounds within fruit can have a significant influence on flavor intensity, but which volatiles are determinant of flavor has a lack of consensus. Previous determination of flavor relevance relied on approaches in which importance of volatiles is at least initially based on abundance. Determination of flavor descriptors or thresholds of isolated compounds were determined using human perception via orthonasal olfaction [4], [5], [23]–[25], negating the complex system of strawberry fruit or actual flavor relevant retronasal olfaction.

Of the forty-six volatile compounds cited as relevant to strawberry flavor in five studies [4], [5], [23]–[25] only seven are mutual to at least three of the studies, exemplifying the lack of agreement in defining flavor-relevant constituents. This consensus includes butanoic acid, methyl ester; butanoic acid, ethyl ester; hexanoic acid, methyl ester (106-70-7); hexanoic acid, ethyl ester; linalool; butanoic acid, 2-methyl- (116-53-0); and 3(2H)-furanone, 4-methoxy-2,5-dimethyl-, all of which are quantified in this report. These compounds exhibit adequate variability in fruit samples to discern dose dependent effect on flavor intensity. However, only linalool; butanoic acid, ethyl ester; butanoic acid, methyl ester; and 3(2H)-furanone, 4-methoxy-2,5-dimethyl- show significant positive correlation with flavor intensity (Table S4). These compounds that are found to influence flavor intensity represent diverse classes, terpene alcohol, two esters, and a furan, respectively, while the three compounds not fitting to flavor are all esters. With esters accounting for the majority of chemical compounds detected in strawberry it is possible that too much emphasis is placed on the chemical class for flavor, or that in a complex mixture less are perceivable than when smelled individually.

Over one third of volatiles in this study significantly correlate with strawberry flavor intensity, potentially enhancing perception of a complex and highly variable volatile mixture (Table S4), seventeen of which are not of previous strawberry flavor focus. Two of these unrecognized compounds, 1-hexanol (111-71-7) and butanoic acid, 3-methyl-, butyl ester (109-19-3), are present in the most flavorful strawberry sample but undetected in the least flavorful (Table S2). This pair of compounds as well as pentanoic acid, ethyl ester (539-82-2) and butanoic acid, 3-methyl-, octyl ester (7786-58-5), also present/absent in the most/least flavorful, have relatively minor amounts but show evidence of enhancing perceived sweetness intensity independent of individual sugars. Relatively low abundance volatiles are indicated as new impactful components of strawberry flavor.

Thirty-eight volatile compounds are observed to significantly enhance the perceived intensity of sweetness; twenty-two mutually independent of glucose and fructose, fourteen uniquely independent of sucrose, and six compounds mutually independent of all three sugar: 1-penten-3-one; 2(3H)-furanone, dihydro-5-octyl- (γ-dodecalactone); butanoic acid, pentyl ester; butanoic acid, hexyl ester; acetic acid, hexyl ester; and butanoic acid, 1-methylbutyl ester

(Table S5). In tomato, similar analysis of a volatile subset identifies three compounds enhancing sweetness intensity independent of fructose: geranial; 1-butanol, 3-methyl- (123-51-3); and butanal, 2-methyl- (96-17-3) [34]. These compounds are not identified in the current study; therefore the effect cannot be confirmed in a second system. Botanically, tomato is considered a true fruit and demonstrates climacteric ripening, while strawberry fruit is non-climacteric and considered an aggregate accessory fruit. The developmental origin of the flesh which is consumed is divergent, exhibiting unique biochemistries, but the observance of volatile compounds potentially enhancing perceived sweetness appears to be widespread in fruit.

Orthonasal olfaction is the result of smelling *i.e.* bringing odor in through the nose, while retronasal olfaction is elicited by odorants traveling from oral cavity or esophagus up to nasal cavity [44]. Orthonasal olfaction introduces volatile compounds to the nasal epithelium via inhalation, while retronasal olfaction is achieved during exhalation [45]. Specifically, the path of odorants distinguishes the manner of interaction between consumer and potential food, with orthonasal contributing to aroma and retronasal to flavor. Integration of sensory stimuli relies on projection of signals to various structures of the brain. Interestingly, portions of orthonasal (smell) and retronasal (flavor) olfaction project to different brain areas for processing [46], while taste activation partly overlaps that of retronasal olfaction for integration to produce flavor [47]. Co-activation of taste and retronasal olfaction, but not orthonasal, is shown to elicit responses at otherwise independently sub-threshold levels, exemplifying the ability of multiple sensory integration to intensify one another [48]. Mechanical blockage of retronasal olfaction during tasting of solutions significantly reduces the ability to correctly identify solute, including sucrose [45]. Combination of taste and retronasal olfaction results in a sensory system more adapt at analyzing the chemical content of food, but cross communication also facilitates manipulation of the system.

The food industry knows of the intensification of volatile sensations by the addition of small amounts of sweeteners to solutions containing volatiles [49]. The ability of volatiles to enhance taste is also a known phenomenon [38]. Enhancement of perceived sweetness is demonstrated by addition of volatiles amyl acetate (banana) [50] and citral [51]. Multiple studies show the ability of strawberry aroma to intensify the sweetness of a sugar solution [52], [53], as well as pineapple, raspberry, passion fruit, lychee, and peach [53], [54]. Also, sweetness enhancement has been achieved with vanilla [55], caramel[53], [56], and chocolate [45] indicating this phenomenon is not only associated with fruit volatiles. Studies to determine perceptional differences when tomato is spiked with sugars, acids, and volatiles indicates cross talk between taste and olfaction, in which volatiles impact perception of sweetness and vice versa [57]. Individual volatile compounds have been implicated in tomato to intensify perceived sweetness independent of sugar content [34], [58]. The results here narrow the previous effect of enhanced sweetness by strawberry aroma, a variable mixture, to individual compounds in the fruit. These volatiles are not present at the highest amounts in fruits and most have not been targets of flavor analysis. Also, most appear to be associated with lipid metabolism, like many other volatiles quantified in this work, yet their presence or increased content has an enhancing effect on perceived sweetness independent of sugars. Technically, sweetness is a facet of taste [38]. Therefore a means to convey sweetness via aroma can serve as an attractant to seed dispersers of wild strawberry, or perhaps it is a result of artificial selection [59] to enhance a limited sugar capacity in commercial fruit.

CONCLUSIONS

Strawberry fruit ripening culminates as the flesh softens, volatile emission peaks, and sugars accumulate. This highly coordinated process results in fruit with strong liking due primarily to texture, flavor, and sweetness. However, cultivar, environmental conditions, and their interactions influence fruit attributes, altering the composition of strawberry. This diversity allows for a spectrum of experiences such that the hedonics and intensities of these sensations can vary greatly. The importance of sucrose to sweetness intensity is evident, and the correlation of total volatiles to sucrose highlights the dependence of secondary metabolism to primary metabolism. Individual volatiles correlate to strawberry flavor intensity, helping to better define distinct, perceptually impactful compounds from the larger mixture of the fruit. The dependence of liking on sweetness and strawberry flavor is undermined by environmental pressures that reduce sucrose and total volatile content. A cultivar that exhibits minimal seasonal environmental influence presents itself as a breeding ideotype, as maintenance of sucrose concentration may alleviate loss of overall liking. Selection for increased amounts of volatile compounds that act independently of sugars to enhance sweetness can serve as an alternate approach. The volatiles described herein are sampled mainly from current commercial cultivars and are therefore feasible targets for varietal improvement in the short-term, whereas future studies will be necessary to identify sweet-enhancing volatiles not already present in elite germplasm.

ACKNOWLEDGMENTS

The authors wish to acknowledge David Moore for fruit transport, Yanina Perez for assistance with fruit chemical data collection, and Timothy Johnson for assistance with volatile collection. Portions of the results are protected by US Patent 20130280400 and International Patent WO 2013/163272 A1.

FUNDING STATEMENT

This work is supported by grants from USDA Specialty Crop Block Grant. Graduate funding is provided by USDA National Needs Fellowship. The funders had no role in study design, data collection and analysis, decision to publish, or preparation of the manuscript.

REFERENCES

1. Whitaker VM, Hasing T, Chandler CK, Plotto A, Baldwin E (2011) Historical Trends in Strawberry Fruit Quality Revealed by a Trial of University of Florida Cultivars and Advanced Selections. Hortscience46: 553–557

2. Hong V, Wrolstad RE (1990) Use of HPLC separation photodiode array detection for characterization of anthocyanins. Journal of Agricultural and Food Chemistry 38.

3. Brummell DA, Harpster MH (2001) Cell wall metabolism in fruit softening and quality and its manipulation in transgenic plants. Plant Molecular Biology 47.

4. Ulrich D, Hoberg E, Rapp A, Kecke S (1997) Analysis of strawberry flavour - discrimination of aroma types by quantification of volatile compounds. Zeitschrift Fur Lebensmittel-Untersuchung Und-Forschung a-Food Research and Technology 205.

5. Schieberle P, Hofmann T (1997) Evaluation of the character impact odorants in fresh strawberry juice by quantitative measurements and sensory studies on model mixtures. Journal of Agricultural and Food Chemistry 45.

6. Hancock JF (1999) Strawberries.

7. Zhang J, Wang X, Yu O, Tang J, Gu X, et al. . (2011) Metabolic profiling of strawberry (Fragariaxananassa Duch.) during fruit development and maturation. Journal of Experimental Botany 62.

8. Menager I, Jost M, Aubert C (2004) Changes in physicochemical characteristics and volatile constituents of strawberry (Cv. cigaline) during maturation. Journal of Agricultural and Food Chemistry 52.

9. Prescott JJ (2004) Psychological processes in flavour perception. Flavor perception.

10. Causse M, Saliba-Colombani V, Lesschaeve I, Buret M (2001)Genetic analysis of organoleptic quality in fresh market tomato. 2. Mapping QTLs for sensory attributes. Theoretical and Applied Genetics102: 273–283

11. Christensen CM (1983) Effects of color on aroma, flavor and texture judgements of foods. Journal of Food Science 48: 787–790

12. Hall RL (1968) Flavor and flavoring seeking a concensus of definition. Food Technology 22: 1496–&.

13. Stommel J, Abbott JA, Saftner RA, Camp MJ (2005) Sensory and objective quality attributes of beta-carotene and lycopene-rich tomato fruit. Journal of the American Society for Horticultural Science 130: 244–251

14. Colquhoun TA, Levin LA, Moskowitz HR, Whitaker VM, Clark DG, et al. (2012) Framing the perfect strawberry: an exercise in consumer-assisted selection of fruit crops. Journal of Berry Research 2: 45–61

15. Basson CE, Groenewald JH, Kossmann J, Cronje C, Bauer R (2010) Sugar and acid-related quality attributes and enzyme activities in strawberry fruits: Invertase is the main sucrose hydrolysing enzyme. Food Chemistry 121.

16. Fait A, Hanhineva K, Beleggia R, Dai N, Rogachev I, et al. . (2008) Reconfiguration of the achene and receptacle metabolic networks during strawberry fruit development. Plant Physiology 148.

17. Hoffmann T, Kalinowski G, Schwab W (2006) RNAi-induced silencing of gene expression in strawberry fruit (Fragaria x ananassa) by agroinfiltration: a rapid assay for gene function analysis. Plant Journal 48.

18. Aharoni A, Keizer LCP, Bouwmeester HJ, Sun ZK, Alvarez-Huerta M, et al. . (2000) Identification of the SAAT gene involved in strawberry flavor biogenesis by use of DNA microarrays. Plant Cell 12.

19. Quesada MA, Blanco-Portales R, Pose S, Garcia-Gago JA, Jimenez-Bermudez S, et al. . (2009) Antisense Down-Regulation of the FaPG1 Gene Reveals an Unexpected Central Role for Polygalacturonase in Strawberry Fruit Softening. Plant Physiology 150.

20. Trainotti L, Ferrarese L, Dalla Vecchia F, Rascio N, Casadoro G (1999) Two different endo-beta-1,4-glucanases contribute to the softening of the strawberry fruits. Journal of Plant Physiology 154.

21. Perez AG, Olias R, Luaces P, Sanz C (2002) Biosynthesis of strawberry aroma compounds through amino acid metabolism. Journal of Agricultural and Food Chemistry 50.

22. Maarse H (1991) Volatile Compounds in Foods and Beverages.

23. Hakala MA, Lapvetelainen AT, Kallio HP (2002) Volatile compounds of selected strawberry varieties analyzed by purge-and-trap headspace GC-MS. Journal of Agricultural and Food Chemistry 50.

24. Jetti RR, Yang E, Kurnianta A, Finn C, Qian MC (2007) Quantification of selected aroma-active compounds in strawberries by headspace solid-phase microextraction gas chromatography and correlation with sensory descriptive analysis. Journal of Food Science 72.

25. Olbricht K, Grafe C, Weiss K, Ulrich D (2008) Inheritance of aroma compounds in a model population of Fragaria x ananassa Duch. Plant Breeding 127.

26. MacKenzie SJ, Chandler CK (2009) The late season decline in strawberry fruit soluble solid content observed in Florida is caused by rising temperatures. Acta Horticulturae.

27. MacKenzie SJ, Chandler CK, Hasing T, Whitaker VM (2011) The Role of Temperature in the Late-season Decline in Soluble Solids Content of Strawberry Fruit in a Subtropical Production System. Hortscience 46.

28. Jouquand C, Chandler C, Plotto A, Goodner K (2008) A Sensory and Chemical Analysis of Fresh Strawberries Over Harvest Dates and Seasons Reveals Factors That Affect Eating Quality. Journal of the American Society for Horticultural Science 133: 859–867

29. Santos BM, Peres NA, Price JF, Whitaker VM, Dittmar PJ, et al. . (2012) Strawberry Production in Florida. Vegetable Production Handbook. Gainesville, FL: University of Florida Institute of Food and Agricultural Sciences.

30. Strand L (2008) Integrated Pest Management for Strawberries: ANR Publications.

31. Underwood BA, Tieman DM, Shibuya K, Dexter RJ, Loucas HM, et al. (2005) Ethylene-regulated floral volatile synthesis in petunia corollas. Plant Physiology 138: 255–266

32. Schmelz EA, Alborn HT, Banchio E, Tumlinson JH (2003)Quantitative relationships between induced jasmonic acid levels and volatile emission in Zea mays during Spodoptera exigua herbivory.Planta 216: 665–673

33. Schmelz EA, Alborn HT, Tumlinson JH (2001) The influence of intact-plant and excised-leaf bioassay designs on volicitin- and jasmonic acid-induced sesquiterpene volatile release in Zea mays.Planta 214: 171–179

34. Tieman D, Bliss P, McIntyre LM, Blandon-Ubeda A, Bies D, et al. (2012) The Chemical Interactions Underlying Tomato Flavor Preferences. Current Biology 22: 1035–1039

35. Bartoshuk LM, Duffy V, Green BG, Hoffman HJ, Ko CW, et al. (2004) Valid across-group comparisons with labeled scales: the gLMS versus magnitude matching. Physiology & Behavior 82: 109–114

36. Bartoshuk LM, Duffy VB, Fast K, Green BG, Prutkin J, et al. (2003) Labeled scales (eg, category, Likert, VAS) and invalid across-group comparisons: what we have learned from genetic variation in taste. Food Quality and Preference 14: 125–138

37. Bartoshuk LM, Fast K, Snyder DJ (2005) Differences in our sensory - Invalid comparisons with labeled scales. Current Directions in Psychological Science 14: 122–125

38. Lindemann B (2001) Receptors and transduction in taste. Nature413: 219–225

39. Fujimaru T, Lim J (2013) Effects of Stimulus Intensity on Odor Enhancement by Taste. Chemosensory Perception 6: 1–7

40. Mikulic-Petkovsek M, Schmitzer V, Slatnar A, Stampar F, Veberic R (2012) Composition of Sugars, Organic Acids, and Total Phenolics in 25 Wild or Cultivated Berry Species. Journal of Food Science 77: C1064–C1070

41. Watson R, Wright CJ, McBurney T, Taylor AJ, Linforth RST (2002) Influence of harvest date and light integral on the development of strawberry flavour compounds. Journal of Experimental Botany 53.

42. Pelayo-Zaldivar C, Ebeler SE, Kader AA (2005) Cultivar and harvest date effects on flavor and other quality attributes of California strawberries. Journal of Food Quality 28.

43. Cumplido-Laso G, Medina-Puche L, Moyano E, Hoffmann T, Sinz Q, et al. . (2012) The fruit ripening-related gene FaAAT2 encodes an acyl transferase involved in strawberry aroma biogenesis. Journal of Experimental Botany 63.

44. Pierce J, Halpern BP (1996) Orthonasal and retronasal odorant identification based upon vapor phase input from common substances. Chemical Senses 21: 529–543

45. Masaoka Y, Satoh H, Akai L, Homma I (2010) Expiration: The moment we experience retronasal olfaction in flavor. Neuroscience Letters 473: 92–96

46. Small DM, Jones-Gotman M (2001) Neural substrates of taste/smell interactions and flavour in the human brain. Chemical Senses. pp. 1034.

47. Small DM, Voss J, Mak YE, Simmons KB, Parrish T, et al. . (2004) Experience-dependent neural integration of taste and smell in the human brain. Journal of Neurophysiology 92.

48. Veldhuizen MG, Shepard TG, Wang MF, Marks LE (2010)Coactivation of Gustatory and Olfactory Signals in Flavor Perception.Chemical Senses 35: 121–133

49. SjÖStrÖM Loren B, Cairncross Stanley E (1955) Role of Sweeteners in Food Flavor. Use of Sugars and Other Carbohydrates in the Food Industry: American Chemical Society. pp. 108–113.

50. Burdach KJ, Kroeze JHA, Koster EP (1984) Nasal, Retronasal, and Gustatory Perception - An Experimental Comparison. Perception & Psychophysics 36.

51. Murphy C, Cain WS (1980) Taste and Olfaction - Independence vs Interaction. Physiology & Behavior 24.

52. Frank RA, Byram J (1988) Taste smell interactions are tastant and odorant dependent. Chemical Senses 13.

53. Stevenson RJ, Prescott J, Boakes RA (1999) Confusing tastes and smells: How odours can influence the perception of sweet and sour tastes. Chemical Senses 24.

54. Cliff M, Noble AC (1990) Time-intensity evaluation of sweetness and fruitiness and their interaction in a model solution. Journal of Food Science 55.

55. Lavin JG, Lawless HT (1998) Effects of color and odor on judgments of sweetness among children and adults. Food Quality and Preference 9.

56. Prescott J (1999) Flavour as a psychological construct: implications for perceiving and measuring the sensory qualities of foods. Food Quality and Preference 10.

57. Baldwin EA, Goodner K, Plotto A (2008) Interaction of volatiles, sugars, and acids on perception of tomato aroma and flavor descriptors. Journal of Food Science 73.

58. Bartoshuk LM, Klee HJ (2013) Better Fruits and Vegetables through Sensory Analysis. Current Biology 23: R374–R378

59. Aharoni A, Giri AP, Verstappen FWA, Bertea CM, Sevenier R, et al. . (2004) Gain and loss of fruit flavor compounds produced by wild and cultivated strawberry species. Plant Cell 16.

Chapter 5

PHOTOSYNTHETIC PIGMENTS IN DIATOMS

Paulina Kuczynska [1], Malgorzata Jemiola-Rzeminska [1,2] and Kazimierz Strzalka [1,2]

[1]Faculty of Biochemistry, Biophysics and Biotechnology, Department of Plant Physiology and Biochemistry, Jagiellonian University, Gronostajowa 7, Krakow 30-387, Poland

[2]Małopolska Centre of Biotechnology, Gronostajowa 7A, Krakow 30-387, Poland

ABSTRACT

Photosynthetic pigments are bioactive compounds of great importance for the food, cosmetic, and pharmaceutical industries. They are not only responsible for capturing solar energy to carry out photosynthesis, but also play a role in photoprotective processes and display antioxidant activity, all of which contribute to effective biomass and oxygen production. Diatoms are organisms of a distinct pigment composition, substantially different from that present in plants. Apart from light-harvesting pigments such as chlorophyll a, chlorophyll c, and fucoxanthin, there is a group of photoprotective carotenoids which includes β-carotene and the xanthophylls, diatoxanthin, diadinoxanthin, violaxanthin, antheraxanthin, and zeaxanthin, which are engaged in the xanthophyll cycle. Additionally, some intermediate products of biosynthetic pathways have been identified in diatoms as well as unusual pigments, e.g., marennine. Marine algae have become widely recognized as a source of unique bioactive compounds for potential industrial, pharmaceutical, and medical applications. In this review, we summarize current knowledge on diatom photosynthetic pigments complemented by some new insights regarding their physico-chemical properties, biological role, and biosynthetic pathways, as well as the regulation of pigment level in the cell, methods of purification, and significance in industries.

INTRODUCTION

Diatoms are becoming more and more prominent microalgae. More advanced knowledge about them has also enhanced their importance and usefulness in commercial and industrial applications such as biofuels, pharmaceuticals, health foods, biomolecules, materials relevant to nanotechnology, and as bioremediators of contaminated water [1]. The singularity of these organisms is physiological in nature, involving, for example, novel metabolic pathways and compounds, but is also due to their importance in the evolutionary history of eukaryotes as well as their ecological success. They are mainly associated with the silicon metabolism engaged in the biogeochemical cycling of Si in the sea. The siliceous structures in their cell wall create the unique morphotypes that are used as taxonomic keys [2]. Another interesting process is the ornithine-urea cycle which is absent in green algae and plants and is essential for diatom growth and their contribution to marine productivity [3]. Furthermore, these organisms are rich in bioactive compounds capable of antiviral activity, including naviculan [4] and neuroexcitatory amino acid derivative domoic acid [5], not to mention the cytotoxic and blood platelet inhibitory activity caused by adenosine [6]. There is a high diversity of beneficial diatom cell components in lipids and pigments, whose amount in the cell may be partially regulated by certain abiotic stresses or genetic modifications of metabolic pathways. For example, the overexpression of endogenous Δ5 desaturase leads to an accumulation of eicosapentaenoic acid, a fatty acid known to have a variety of health benefits including anti-inflammatory effects and better neuronal functioning [7]. On the other hand, enhanced biosynthesis of strong antioxidant fucoxanthin was obtained through low light and nitrogen treatment [8]. Although not all the compounds in diatoms are known, due to the importance of these organisms, there is continual research to find, identify, and examine their properties. In the marine diatom *Haslea ostrearia*, a water soluble blue pigment named marennine was identified and further experiments have shown its allelopathic, antioxidant, antibacterial, antiviral, and growth-inhibiting properties [9]. Another diatom, *Haslea karadagensis*, also has a blue pigment. Although it has quite different absorption maxima from those of marennine, its similar bioactivity has come to be called marennine-like [10]. While these unusual pigments occur only in selected species, there are some which are included in the more general group of photosynthetic pigments and are common among diatoms. These also have great benefits in medicine, pharmacy, cosmetics, food, and supplements. However, the pigment profile of diatoms is quite different than that found in plants and some algae. Chlorophyll a (Chl *a*) and chlorophyll c (Chl *c*) together with fucoxanthin (Fx) are components of fucoxanthin-chlorophyll protein (FCP) complexes which replace plants light harvesting complexes (LHC) in performing the light-harvesting function [11]. Furthermore, the three carotenoids—β-carotene (β-car), diadinoxanthin (Ddx), and diatoxanthin (Dtx)—are known to play an important role in photoprotection and, additionally, violaxanthin (Vx), antheraxanthin (Ax), and zeaxanthin (Zx) may be engaged in this process. This article is focused on the above-described photosynthetic pigments which are essential for diatom life and which are commonly used in various industries. Despite the fact that studies in this field have been carried out for many years, a lot of aspects that require further analysis still remain. What follows below represents the current knowledge of their physical and chemical properties, their biosynthetic pathways, their regulation of pigment level, as well as their localization in the cell, their role in photosynthesis and photoprotection, methods for their identification and purification, and their significance in industries.

PHYSICAL AND CHEMICAL PROPERTIES OF PHOTOSYNTHETIC PIGMENTS OF DIATOMS

Diatoms contain two types of pigments involved in light harvesting and photoprotection: chlorophylls and carotenoids. Chlorophylls trap light energy—blue and red portions of the electromagnetic spectrum, in particular, which are used in photosynthesis. Generally, chlorophylls can be defined as a magnesium coordination complex of cyclic tetrapyrroles containing a fifth isocyclic ring, referred to as porphyrin, with a long-chain isoprenoid alcohol ester group. Being the highly conservative structural motif of Chls, however, the phytyl chain is not present in the majority of Chl c pigments found in diatoms. The second group of pigments, carotenoids, are engaged mainly in photoprotection; however, Fx also participates in light harvesting. They are comprised of carotenes (hydrocarbons) and their oxygenated derivatives, xanthophylls. Extensive data compiled for 47 of the most important chlorophylls and carotenoids found in marine algae were published by Jeffrey and Vesk [12].

Chlorophylls

Several kinds of chlorophylls are found in photosynthetic organisms; however, only two forms occur in diatoms: Chl a and, identified in various algae, Chl c. The predominant Chl a plays a central role in the photochemical energy conversion of the majority of photosynthesizing organisms, while Chl c participates effectively in photosynthesis as an accessory pigment, similar in its functional activity to the Chl b of higher plants. From among different forms of Chls c which were described in diatoms, the most abundant are Chl c_1 and c_2 (Figure 1). The distinct structure of a Chl c brings changes in the absorption spectrum to produce a strong Soret (blue) absorption band in comparison with a weak band in the red region. The ratios of band I (at ~630 nm) to band II (at ~580 nm) are >1 for Chl c_1-like chromophores, ~1 for Chl c_2-like chromophores, and <1 for Chl c_3-like chromophores [13].

Among 51 species (71 isolates) of tropical and sub-tropical diatoms from 13 out of 22 families examined by Stauber and co-workers [14], Chl c_2 was present in all the diatoms tested and occurred together with Chl c_1 in 88% of them. Where Chl c_1 was absent or occurred in trace amounts only, it was usually replaced by a Chl c_3, identified by Fookes and Jeffrey as (7-methoxycarbonyl)-Chl c_2 [15]. The presence of a methoxycarbonyl group ($-COOCH_3$) at C-7 (ring B) explains the difference in molecular weight (653 m/z) compared to Chl c_2 (609 m/z), as well as the significant decrease in absorption band I (Qy) intensity compared to the Chl c_1 and c_2 values [16]. Exceptions are *Nitzschia closterium* (CS-114), with merely Chl c_2, and *Nitzschia bilobata* (CS-47), which contains all three Chls (c_1, c_2, and c_3) in approximately equal amounts. Five species that have Chls c_1 and c_2 also contain Chlc_3 in trace quantities. Additionally, diatoms may possess the Chl c_2-P.gyrans-type found, for example, in *Pseudo-nitzschia multiseries* [16]. Finally, traces of (DV)-PChlide with a propionic acid side chain at C-17 instead of acrylic acid are commonly present in diatoms. An overview of the Chl c distribution in microalgae classes is given by Jeffrey and Vesk [12].

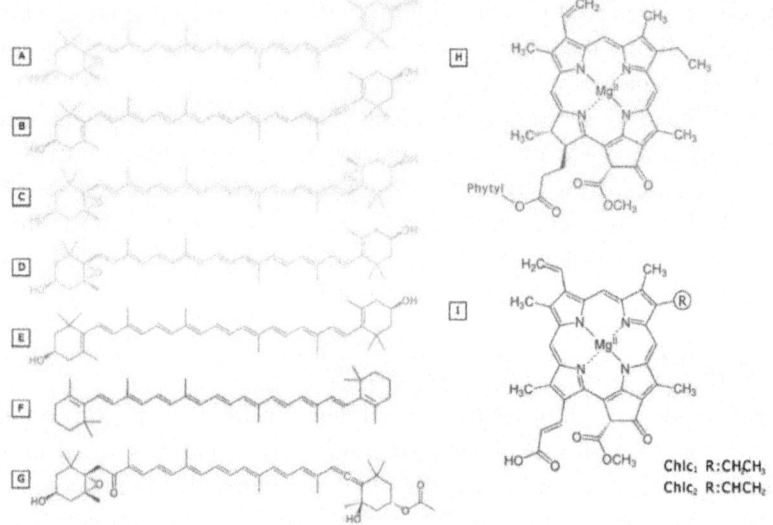

Figure 1: Structural formula of photosynthetic pigments in diatoms including *all-trans* carotenoids: (A) diadinoxanthin; (B) diatoxanthin; (C) violaxanthin; (D) antheraxanthin; (E) zeaxanthin; (F) β-carotene; (G) fucoxanthin; and chlorophylls: (H) chlorophyll *a*; (I) chlorophyll *c*.

In the aquatic environment, pigments may undergo degradation in response to chemical, photochemical, and biological processes. Studies of marine organic matter from the central equatorial Pacific were able to weigh the overall reactivity of various biochemical classes and found that pigments, especially chlorophylls, were the most labile compounds [17]. However, possessing nitrogen, chlorophylls are more prone to being salvaged during senescence and biological breakdown than carotenoids. Although the photooxidation of chlorophyll was studied almost exclusively in terms of the porphyrin moiety of the molecule, its unsaturated chain is also susceptible to a reaction with singlet oxygen or the hydroxy and peroxy radicals which are generated during Chl photodegradation. It has been demonstrated by Rontani and co-workers [18] that several free and esterified oxidized isoprenoid compounds are produced during the photodegradation of Chl *a* in seawater. Moreover, they suggested that in the dead cells of *Phaeodactylum tricornutum*, the Chl phytyl chain will be reduced to up to one half of its initial concentration after prolonged exposure to high light.

Carotenoids

More than 700 types of carotenoids were identified in nature [19]. They are commonly synthesized by plants, algae, and some micro-organisms. Seven kinds of carotenoids were found in diatoms with β-car as an example of carotenes, as well as Fx, Dtx, Ddx, Zx, Ax, and Vx, which represent xanthophylls (Figure 1). Although derivatives of all mentioned pigments, including isomers and products of degradation, may occur in the cell, the all-*trans*-isomers are the most abundant and functionally active forms. The presence of a conjugated polyene chain in carotenoids may be the cause of carotenoid instability, which is related to their susceptibility to oxidation, E/Z isomerization by heat, light, and chemicals. As a result of the *cis*-isomerization of a chromophore's double bond there is a slight loss of color, as well as a small hypsochromic shift and a hypochromic effect, accompanied by the appearance of a *cis* peak about 142 nm below the longest wavelength absorption maximum of the *trans*-carotenoid when measured in hexane [16]. Moreover, the geometrical structures of *cis* and *trans* carotenoids, which are differentially oriented in the thylakoid membrane, have an impact on membrane physical properties. Intercalation of carotenoids changes permeability for the oxygen and other small molecules, which is associated with their protective activity [20,21]. The activation energy for allenic (R/S) isomerization in carotenoids is higher than that associated with geometrical (E/Z) isomerization. Consequently, E/Z stereomutation of Fx can occur in sunlight in the absence of a catalyst, whereas allenic isomerization was shown to occur only to a very low extent. Recently, data has been published on the iodine or diphenyl diselenide-mediated photoisomerization of Fx, which, in view of the increased reaction rate with UVA radiation, support the radical mechanism of R/S isomerization [22]. Additionally, some carotenoids are highly unstable in the presence of acid. Under weak acidic conditions, 5,6-epoxide is readily rearranged into furanoid 5,8-epoxide. On the other hand, the lability of Fx towards alkali has been established, which precludes the use of saponification in the isolation procedure of carotenoid mixture containing Fx. First, chromophoric changes upon treatment of this allenic carotenoid with a weak base (K_2CO_3) were reported and, then, two products were identified: (i) isofucoxathinol as the kinetically controlled product and (ii) fucoxanthinol hemiketal, with a shorter chromophore, as the thermodynamically controlled product [23]. Unlike chlorophylls, carotenoids are often broken down into a colorless compound by the destruction of the long chain of alternating double bonds and cannot be detected by regular pigment analysis [24].

Carotenoids exhibit intense absorption between 400 and 500 nm. The conjugation length and type of the functional groups that are attached to the ionone rings terminating the polyene chain largely determine the absorption properties of the carotenoids [17,25]. The absorption spectra of carotenoids are markedly solvent-dependent, which should be considered when analyzing pigment extracts by high performance liquid chromatography with a photodiode array detector (HPLC-DAD), because in most cases spectra are taken in mixed solvents. The λ_{max} values of carotenoids in hexane or petroleum ether are practically the same as in diethyl ether, methanol, ethanol, and acetonitrile, and are higher by 2–6 nm in acetone, 10–20 nm in chloroform, 10–20 nm in dichloromethane, and 18–24 nm in toluene; see [16] for ultraviolet and visible absorption data for common carotenoids. To give an idea of the spectral fine structure, the values of %III/II are also given along with the λ_{max} values.

The main light-harvesting carotenoid in diatoms is Fx, which is also abundant in brown algae. Unusually, small amounts of a 19′-butanoyloxyfucoxanthin-like pigment, in addition to Fx, were found in one diatom species, *Thalassiothrix heteromorpha*, as reported in [26]. Fx has an allenic bond, a conjugated carbonyl, a 5,6-monoepoxide, and acetyl groups that contribute to the unique structure and spectral properties of the molecule (Figure 1). In contrast to other carotenoids, its broad absorption band (between 460 and 570 nm) covers much of the gap left by chlorophyll in the green region. Diatoms also possess the β-car of carotenes as well as two xanthophylls, Ddx and Dtx, which are also asymmetric molecules, containing an acetylenic group at one of the ionone rings. Moreover, three other xanthophylls, characteristic of higher plants, Vx, Ax, and Zx, may also occur (Figure 1) [24,27]. However, these carotenoids accumulate only under specific conditions, e.g., during long-time illumination with strong light (see Section 8).

BIOSYNTHESIS PATHWAYS

The biosynthetic pathways of both chlorophylls and carotenoids in plants have been extensively studied and complete information in this field is available. This opens up many opportunities for studies on genetically modified organisms and for *in vitro* approaches using recombinant proteins. In diatoms, some pathway points remain unclear and the production of cell lines with an enhanced amount of a selected photosynthetic pigment is currently difficult. Most of the genes which encode enzymes in these steps of the pathway were sought out by genome alignment, but they have not been identified. The whole genomes of two diatom species, *Phaeodactylum tricornutum* [28] and*Thalassiosira pseudonana* [29], were sequenced, but few analogues of the genes which occur in plants or algae were found in these diatoms.

Two pathways, methylerythritol phosphate (MEP) and mevalonate (MEV), of the early steps of carotenoid biosynthesis have been described. Their occurrence is not clear but a few studies show that it depends on the taxon or the growth rate. However, the products of both are dimethylallyl diphosphate (DMAPP) and its isomer, isopentenyl pyrophosphate (IPP) [14,16,30]. The next steps on the pathway to lycopene synthesis are the conversion of DMAPP to geranylgeranyl pyrophosphate (GGPP), which is catalyzed by GGPP synthase, then to phytoene by phytoene synthase (PSY), afterwards to ζ-carotene, which is catalyzed by phytoene desaturase (PDS), and, finally, the product of ζ-carotene desaturase (ZDS) is lycopene [30]. The biosynthesis pathway from lycopene to xanthophylls is presented in Figure 2. Firstly, lycopene as a long and straight molecule is cyclized by lycopene β-cyclase (LCYB) to β-car, having two β-ionone rings at both ends of the yield. In the next step, xanthophyll is first formed, and this reaction requires hydroxylation. However, a gene encoding β-carotene hydroxylase (BCH) was not found in the diatom genome and another one that is similar to LUT1 has been proposed as a putative enzyme to make the formation of Zx from β-car possible [31]. Two further light-dependent and reversible reactions lead to Vx formation via the intermediate product Ax. Both are catalyzed by Vx de-epoxidases (VDEs) in high light conditions, but reverse reactions are catalyzed by Zx epoxidases (ZEPs) in low light or in the dark [32]. In *T. pseudonana*, two VDEs and two ZEPs were identified [33], while in *P. tricornutum*, three isoforms of ZEPs (ZEP1, ZEP2, ZEP3) and VDEs (VDE, VDL1, VDL2) were found [31]. Alternatively, Vx might be formed from Zx through β-cryptoxanthin (Cx) and β-cryptoxanthin-epoxide (CxE) [27]. Further steps which lead to the formation of Fx, Ddx, and Dtx are still unclear because of missing data about the enzymes engaged in the process. However, to date, two models of possible conversions from Vx to Fx have been presented. The first model was described by Lohr and

Wilhelm [27], who proposed Vx as a precursor of Ddx, Dtx, and Fx, with the reaction from Vx to Dtx as well as to Fx proceeding via Ddx. This hypothesis was confirmed experimentally using norflurazon, which inhibits the *de novo* synthesis of carotenoids and which was used after the accumulation of Vx. In low light, an increase in the Fx level was detected. The other model is based on speculation about the chemical properties of these xanthophylls and neoxanthin (Nx) was regarded as a precursor of both Ddx and Fx [34]. Nx is the most likely candidate for enabling the formation of the acetylenic bond in Ddx or the allenic double bond in Fx. However, the formation of Fx requires two modification steps: the ketolation of Nx and the acetylation of an intermediate, probably fucoxanthinol. To support one of these hypotheses, the identification of the enzymes is necessary. The most powerful approach in this field is to look for genes encoding the proteins of interest on databases. Unfortunately, the gene encoding Nx synthase (NXS), which catalyzes the conversion of Vx to Nx in *Arabidopsis thaliana*, was not found in brown seaweeds. Moreover, Nx has not been detected either. However, LCYB shares a 64% amino acid identity with NXS and is proposed to be engaged in Nx production, although no LCYB-like NXS in brown seaweeds was identified [35]. It is important to reveal the whole xanthophyll biosynthetic pathway because of the many opportunities to further studies and also to prepare transgenic organisms with an increased xanthophyll level.

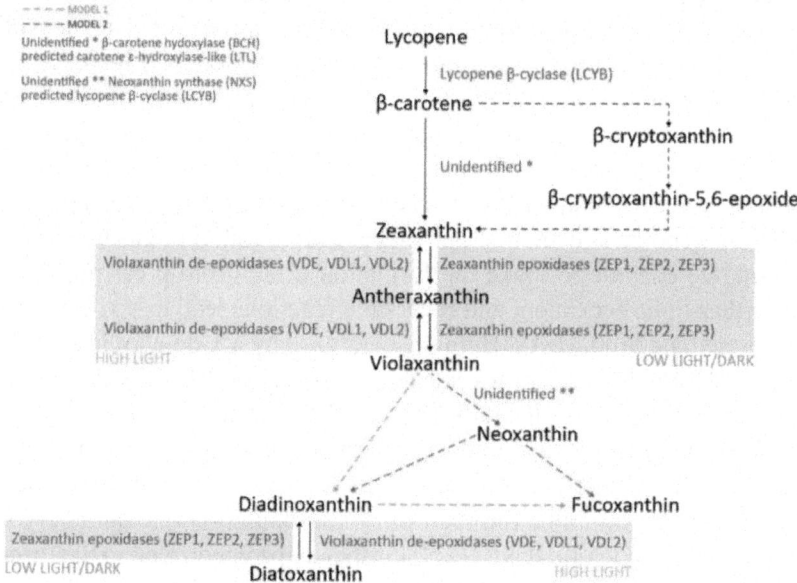

Figure 2: Biosynthetic pathway of photosynthetic carotenoids in the diatom *Phaeodactylum tricornutum* from lycopene to fucoxanthin and diatoxanthin.

The chlorophyll biosynthetic pathway has been extensively studied in higher plants and also in some groups of algae, but in the case of diatoms, it has been poorly investigated. Nevertheless, the main aspects are similar in all photosynthetic organisms. Three general steps are needed for Chl synthesis: aminolevulinic acid formation, its transformation into Mg-porphyrins, and protochlorophyllide conversion to Chl [36]. The initial steps rely on tetrapyrrole cyclization, the insertion of Mg leading to diviny-PChlide *a* formation, and its reduction to PChlide *a*. Afterwards, light-independent PChlide oxidoreductase (DPOR) and a light-dependent enzyme (LPOR) catalyze PChlide hydrogenation to Chlide and further steps lead to Chl *a* formation. In four diatom species, more than one isoform of LPOR was found [37]. The final step is the introduction of phytol residue, which is associated with the MEP pathway also used in carotenoid formation [38]. In higher plants, the chlorophyll cycle relies on conversion between Chl *a* and Chl *b*, allowing the adjustment of their ratio to light conditions [39]; however, in diatoms, instead of Chl *b*, Chl *c* (Chl c_1, Chl c_2, and rarely Chl c_3) was identified [13,14] (see also Section 2). The molecular structure of Chl *c* may suggest that PChlide is its precursor in the biosynthetic pathway where oxidation and dehydration are required, but no enzyme carrying out these steps has been described [40]. In general, data about Chl biosynthesis and the enzymes which catalyze each step are still poorly understood.

REGULATION OF PIGMENT LEVEL IN THE CELL

Changes in photosynthetic pigment levels are usually a fast response to environmental conditions because they are engaged in basic processes such as photosynthesis and photoprotection, which are essential for cell life (Table 1). Diatom cell physiology displays some significant differences that are observed during growth, including an exponential, stationary, and declining phase. In the exponential phase of growth, the metabolism is the highest, especially that of amino acids, whereas in the declining phase, catabolism is predominant and is related to an increase in metabolites like terpens and putrescine [41]. Experiments which concern cell responses to stress factors and pigment measurements should be performed during the exponential phase of growth.

Pigment level is mostly regulated by light conditions, resulting in fast conversions of them, usually without any changes in gene expression, although long-term acclimation causes changes in the transcript level of the genes which encode proteins in biosynthetic pathways. The effect of white and blue-green light on pigment content during the exponential and stationary growth phases of three diatom species was studied [42] and it was reported that Chl *a* content decreased by nearly 50% in the stationary phase and carotenoid content increased in blue-green light. In high light, the amount of Chl *a*, Fx, and β-car decreased, but with unrelated variations in Chl *c*, and it was also reported that the Chl *c* content in FCP complexes might be regulated by light, but not that of Chl *a* and Fx [43]. Under low light, FCP trimers contain mainly Lhcf5 proteins with a high Fx:Chl *c* ratio, whereas trimers in high light contain mainly Lhcf4 proteins with a low Fx:Chl *c* ratio. A different variation of Chl *a*, β-car, and Fx is correlated with a decreasing number of PSII units in high light. This photoacclimation strategy described in species growing in variable light conditions enables the efficient regulation of photosystem structures to the amount of absorbed energy [43,44]. The spectral composition of light plays a crucial role in growth rate, photoprotective mechanisms, and photosynthetic efficiency, and thereby in pigment content. In blue light-acclimated *P. tricornutum* cells, the pool of xanthophyll cycle pigments was higher than those growing in red light, but in neither case was Dtx detected [45]. Studies on the long-term dark incubation of diatoms in the sediment from a tidal mud flat showed that after one year of microscope observations, fluorescence was detected. Pigment analysis including Fx, Chl *c*, Chl *a*, Dtx, and Ddx showed a rapid decrease in the first weeks and, then, slower changes, and although Ddx was not detected after two months, Dtx remained steady over time [46]. It is not only light conditions that have an impact on pigment levels, but also nutrient limitation along with the heavy metals which are more and more abundant in the environment [47]. Iron limitation results in a wide range of changes in diatoms, including in gene expression in factors which regulate pigment content. As many as 20 genes of iron-responsive regulation were described in *P. tricornutum* [48]. It has been reported that under iron limitation, there was an increase in the transcript level of 16 genes involved in Chl biosynthetic pathways, although the Chl content decreased [49]. With an increased cadmium concentration, a rapid decrease in Chl *a* as well as carotenoids was observed in epiphytic diatoms of*Myriophyllum triphyllum* [50]. Although the pigment composition is similar in each diatom species, the ratio between them is different and highly variable. Unfortunately, a comparison of pigment content in different diatom species growing under optimal or stress conditions is very complicated because of the variety of ways in which it is calculated. This is very often done in relation to Chl *a* content, but the absolute content in dry or wet weight is also measured, or the percentage of each pigment is measured.

Table 1: Changes in pigment content (Chl *a*: chlorophyll *a*; Chl *c*: chlorophyll *c*; β-car: β-carotene; Fx: fucoxanthin; Ddx: diadinoxanthin; Dtx: diatoxanthin; Vx: violaxanthin; Ax: antheraxanthin; Zx: zeaxanthin) in diatoms in response to selected stress conditions. The down arrow represents a decrease of pigment content, the up arrow is the opposite, and *const* means no changes

Conditions	Species	Changes in Pigment Content								
		Chl *a*	Chl *c*	β-Car	Fx	Ddx	Dtx	Vx	Ax	Zx
HL (140 μmol photons m⁻² s⁻¹) in comparison to LL (40 μmol photons m⁻² s⁻¹), 16 h light/8 h dark photoperiod [51]	*Cyclotella meneghiniana*	N/A	const	const	const	↑	↑	N/A	N/A	N/A
Iron-replete medium (12 μM) compared to iron-reduced medium (1 μM) in HL (140 μmol photons m⁻² s⁻¹) [52]	*Cyclotella meneghiniana*	↑	N/A	N/A	N/A	↓		N/A	N/A	N/A
Iron-replete medium (12 μM) compared to iron-reduced medium (1 μM) in LL (40 μmol photons m⁻² s⁻¹) [52]	*Cyclotella meneghiniana*	↑	N/A	N/A	N/A		const	N/A	N/A	N/A
HL (300 μmol photons m⁻² s⁻¹) compared to LL (50 μmol photons m⁻² s⁻¹), 14 h light/10 h dark photoperiod [53]	*Phaeodactylum tricornutum*	N/A	↓	↑	↓	↑	↑	N/A	N/A	N/A
B-HL (450 PFD) compared to BR-HL (450 PFD in R:B ratio 0.25) [43]	*Pseudonitzschia multistriata*	↓	↑	const	↓	↓	↓	N/A	N/A	N/A
B-LL (250 PFD) compared to BR-LL (250 PFD in R:B ratio 0.25) [43]	*Pseudonitzschia multistriata*	↓	↑	const	↓	↓	↓	N/A	N/A	N/A
B-LL (24 (10 absorbed) μmol photons m⁻² s⁻¹) compared to W-LL (40 (10 absorbed) μmol photons m⁻² s⁻¹) [54]	*Phaeodactylum tricornutum*	↑	const	const	const	const	N/A	↓	N/A	N/A
R-LL (41 (10 absorbed) μmol photons m⁻² s⁻¹) compared to W-LL (40 (10 absorbed) μmol photons m⁻² s⁻¹) [54]	*Phaeodactylum tricornutum*	↑	↓	↑	↓	↓	N/A	↓	N/A	N/A
HL (1250 μmol photons m⁻² s⁻¹) in comparison to LL (40 μmol photons m⁻² s⁻¹), 12 h light/12 h dark photoperiod [55]	*Phaeodactylum tricornutum*	↓	const	const	const	↓	↑	N/A	N/A	N/A
6 days acclimated to shift from BL (24 (10 absorbed) μmol photons m⁻² s⁻¹) to RL (40 (10 absorbed) μmol photons m⁻² s⁻¹) [45]	*Phaeodactylum tricornutum*	↑	N/A	N/A	N/A	↓	N/A	N/A	N/A	N/A
6 days acclimated to shift from RL (40 (10 absorbed) μmol photons m⁻² s⁻¹) to BL (24 (10 absorbed) μmol photons m⁻² s⁻¹) [45]	*Phaeodactylum tricornutum*	const	N/A	N/A	N/A	↑	N/A	N/A	N/A	N/A
14 days dark storage culture [56]	*Thalassiosira weissflogii*	↓	↓	↓	↓		↓	N/A	N/A	N/A
HL (700 μmol photons m⁻² s⁻¹) in comparison to LL (40 μmol photons m⁻² s⁻¹), 16 h light/8 h dark photoperiod [24]	*Cyclotella meneghiniana*	N/A	N/A	N/A	N/A	↓	↑	const	↑	↑
high nitrogen culture (18 mM) compared to low nitrogen culture (6 mM) in LL (100 μmol photons m⁻² s⁻¹) [8]	*Odontella aurita*	N/A	N/A	N/A	↑	N/A	N/A	N/A	N/A	N/A

B-HL: high blue light; B-LL: low blue light; BR-HL: high blue/red light; BR-LL: low blue/red light; HL: high light; LL: low light; PFD: photon flux density; R-LL: low red light; W-LL: low white light; N/A: not available.

LOCALIZATION IN THE CELL

Diatom cells contain either a few small chloroplasts or one large chloroplast [57]. In diatoms, the thylakoid membranes, where pigments responsible for the absorption of light for photosynthesis are located, are not differentiated into granal and stromal lamellae, *i.e.*, granal stacking is absent [58]. Instead, the diatom thylakoid membranes are arranged into groups of three loosely stacked lamellae which span through the whole length of the chloroplast (Figure 3) [59]. Consequently, no lateral heterogeneity in the distribution of photosystems (PS) I and PS II has been detected so far. The organization of the LHC proteins, recently reviewed in detail by Gundermann and Büchel [60], also exhibit differences when compared to the LHCs of higher plants; in diatoms, FCP complexes are the main light-harvesting antennae. Moreover, the attribution of the different FCPs to the two photosystems and/or their supramolecular structure remains unknown. However, when comparing plant LHCs with FCPs the most obvious difference is in pigmentation.

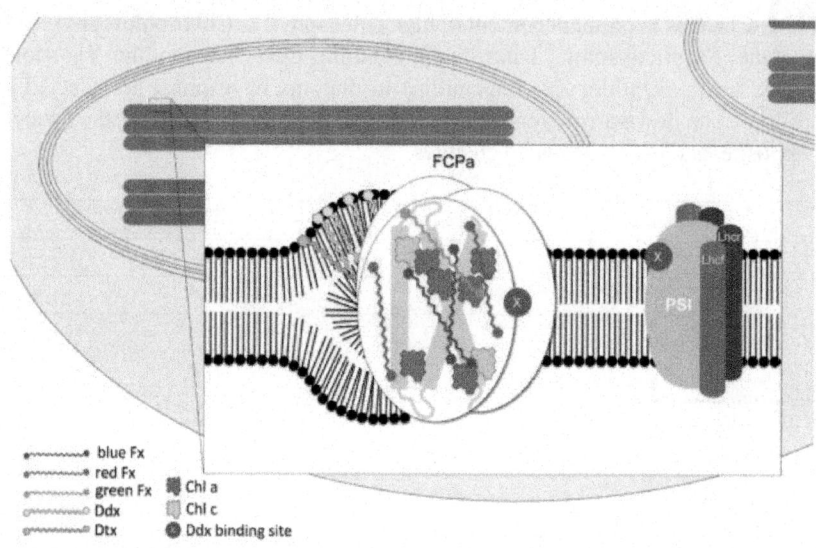

Figure 3|: Simplified model of diatom thylakoid membrane showing the localization of photosynthetic pigments within FCP, PS I, and those localized within an monogalactosyldiacylglycerol (MGDG) shield surrounding the FCP. See the text for more information. Based on Gundermann and Büchel [60].

Fx is found in much higher amounts in FCPs than carotenoids are in LHCII: the molar Chl/carotenoid ratio is almost 1:1 and 14:4, respectively [52,61]. Upon binding to the protein, Fx undergoes extreme bathochromic shifts, and since it depends strongly on the polarity of the protein environment, several populations can be distinguished, *i.e.*, Fx red, Fx green, and Fx blue (Figure 3) [62,63,64]. In diatoms, the Ddx pool is heterogeneous. Recently, three different pools of diadinoxanthin cycle pigments were proposed. Two of them are bound to special antenna proteins within PS I and FCP, respectively, and since their turnover is very low, they play no direct role in the Ddx cycle [27]. The largest pool which would be localized within an MGDG shield surrounding the FCP (Figure 3) [65,66] is convertible to Dtx during the Ddx cycle. This spatial separation has been explained in terms of the probable functional heterogeneity of these pigments. The protein-bound diadinoxanthin cycle pigments would participate in the non-photochemical quenching (NPQ) mechanism, while the lipid-associated ones would essentially play an antioxidant function, scavenging $^1O_2$ and peroxylipids. Interestingly, Alexandre and co-workers [51] reported that the additional Ddx molecules observed when cells are grown in high light conditions adopt a more twisted conformation than the lower levels of Ddx present when the cells are grown in low light conditions. They conclude that this pool of Ddx is more tightly bound to a protein-binding site, which must differ from the one occupied by the Ddx present in low light conditions.

In the thylakoid membranes of diatoms, other xanthophylls like Vx, Ax, and Zx could also occur [24,27]. However, these carotenoids accumulate only under specific conditions, e.g., during long-term illumination with strong light. Moreover, it has been shown that Vx can be either a direct or an indirect (through the formation of Ddx) precursor of Dtx (see Section 3).

A structural homology model based on LHCII structure and spectroscopic analyses was recently published by Gundermann and Büchel [60]. It is based on the previously postulated one by Eppard and Rhiel [67] with five conserved Chl *a* binding sites (a602, a603, a610, a612, and a613; nomenclature according to Liu *et al.* [68]) and modified after Premvardhan *et al.* [62], who identified two further conserved binding sites: a614 and b609. The model accounts for some major requirements listed in [60] as follows: (i) the pigment content of FCPa and FCPb is based on 2 Chl *c* per monomer, *i.e.*, the FCPs contain 6–8 Chl *a*: 2 Chl *c*: 5–6 Fx in total; (ii) the Chl *a* molecules cannot be arranged in a way as to favor excitonic interactions; (iii) Fx does not transfer energy to Chl *c* and Fx and Chl *c* should thus be at an appropriate distance and/or orientation; (iv) Chl *c* to Chl *a* transfer is extremely fast,*i.e.*, Chl *c* has to be in close proximity to a Chl *a* molecule. For further details the reader is referred to the text [60].

PIGMENTS INVOLVED IN PHOTOSYNTHESIS

Diatoms are responsible for about 40% of marine productivity and, because the oceans cover about 70% of the Earth's surface, they make a great contribution to global productivity [69]. However, the marine environment, especially for planktonic species, is changing continuously because of water turbulence. The cells are exposed to varied light intensities as well as light spectrums, depending on the depth (Figure 4). The cell morphology and physiology are strongly connected to the types of habitats [70,71,72], which have a great importance, especially in ecological studies based on organisms in natural environments. The most common diatom species with their typical habitats and lifestyles are described in Table 2. Consequently, diatoms are well adapted to these conditions through efficient photon accumulation and CO_2 uptake and also through fast response to strong light to prevent photodamage. Moreover, according to Falkowski and Knoll [73], the ecological success of diatoms as one of the most important groups of planktonic species is related to their pigment profile, including Chl*a*, Chl *c*, β-car, Ddx, Dtx, and Fx, which enables them to harvest light more efficiently than do green [74,75] and red algae [74]. Chlorophylls play a light-harvesting role and are also known for their electron transfer function. This applies to Chl c, which serves exclusively as an antenna pigment. However, free monomeric Chls can easily transmute into an excited state, inducing free radicals. Therefore, the closely localized carotenoids Ddx and Dtx in diatoms are able to quench it efficiently [36].

Table 2: Ecological specification of the most common diatom taxa, cell morphology, colony lifestyle, habitats. The presence of species in ecological region is variable and dependent on, e.g., the season, nutrient availability, salinity, and conductivity; however, the most frequent occurrence is specified. One group is benthic species including epiphytic (attached to plants), epilithic (attached to rock surfaces), epipelic (on mud), and epipsammic (on sand) species and the second is pelagic diatoms (free living in the water column) [76,77]

Species	Morphology	Colony-Forming	Lifestyle	Habitat
Actinella punctata	eunotioid	yes	benthic	acidic, humic lakes, and ponds
Actinocyclus normanii	centric	no	planktonic	coasts, brackish waters, sediment core
Amphora minutissima	asymmetrical biraphid	no	benthic	marine habitats, often epiphytic
Bacillaria paradoxa	nitzschioid	yes	benthic	marine, brackish, and freshwaters
Campylodiscus hibernicus	surirelloid	no	benthic	epipelon in fresh, brackish, marine waters
Cocconeis pediculus	monoraphid	no	benthic	planktonic, epiphytic, epilithic habitats
Cyclotella distinguenda	centric	no	planktonic	preferentially alkaline waters
Cymbella amplificata	asymmetrical biraphid	yes	benthic	oligotrophic waters
Diatoma vulgaris	araphid	yes	benthic	fresh and brackish water
Discostella stelligera	centric	no	planktonic	primarily in lakes and large rivers
Distrionella incognita	araphid	yes	benthic	alkaline lakes and streams
Epithemia turgida	epithemioid	no	benthic	epiphyte on coarse filamentous algae
Eucocconeis alpestris	monoraphid	no	benthic	the littoral zone of oligotrophic lakes
Eunotia exigua	eunotioid	N/A	benthic	moist soils, wet walls, streams, waterfalls
Fragilaria crotonensis	araphid	yes	planktonic	mesotrophic lakes, water column
Gyrosigma acuminatum	symmetrical biraphid	no	benthic	primarily an epipelic species
Navicula reinhardtii	symmetrical biraphid	no	benthic	fresh water, slightly brackish
Nitzschia regula	nitzschioid	no	benthic	cold-water, ponds, and streams
Phaeodactylum tricornutum	fusiform, triradiate, oval	no	planktonic	marine coastal waters
Pinnularia rabenhorstii	symmetrical biraphid	no	benthic	cold oligotrophic waters in the mountains
Pleurosira laevis	centric	yes	benthic	naturally saline or polluted waters
Thalassiosira weissflogii	centric	N/A	planktonic	primarily in marine waters

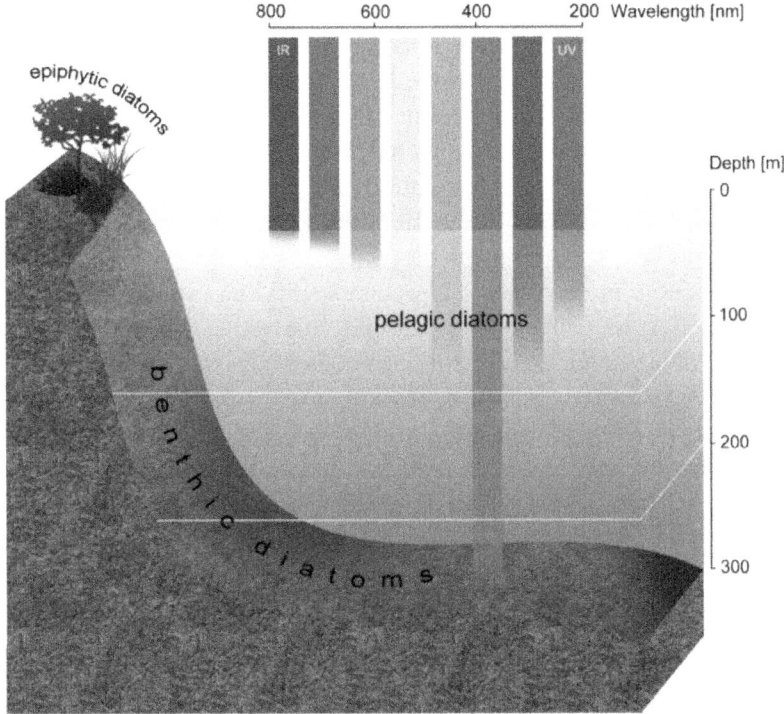

Figure 4: Spectra composition of light depending on the depth.

Studies on acclimation to different light conditions, not only in terms of the quantity but also the quality of light, have been conducted. It was observed that during daily insolation the Chl content increased and it was accompanied by a decrease in the Fx content [78]. Both pigments are present in FCP complexes and have different absorption maxima, which results in efficient photon accumulation. Although the ratio between light harvesting and photoprotective pigments is balanced and, therefore, photosynthetic efficiency is lowered under high irradiance, because the amount of alternative electrons is kept to a minimum under fluctuating light, diatoms can convert energy into biomass efficiently [78]. Additionally, there is valuable data about the adaptation of diatom cells to their environment, achieved through experiments with selected light spectra. Blue light has a significant impact on diatoms, resulting in the production of more cells with greater photosynthetic activity in comparison to white light [79]. Moreover, planktonic diatoms which float in the water column are exposed to changes in the ratio between blue and red light. Recently, Jungandreas and co-workers [45] found that a shift from blue to red light results in a slightly diminished growth rate within 48 h, whereas a shift from red to blue light completely abolished growth within 24

h. Furthermore, metabolic reorganization based on the carbon allocation in diatom cells is the first response in acclimation to new light conditions [45]. Beyond light stress, another group of factors such as nutrient availability have a substantial impact on photosynthetic efficiency. Silicon is a special inorganic compound in diatoms, occurring abundantly in their cell wall. The energy for silicification is more linked to respiration than to photosynthesis and, therefore, it has no effect on oxygen production but leads to decreased utilization of carbohydrates during biomineralization [2]. It has been reported that silicon limitation results in increased Chl $a$content [80,81], but its addition had no significant effect on Chl accumulation [82]. Nevertheless, silicate importance for photosynthetic efficiency is similar to other nutrients [83], including iron, which plays an important role in marine primary productivity. A decreased Chl content, lowered growth rate, up-regulated expression of genes engaged in early steps of carotenoid biosynthesis, and enhanced NPQ were observed under iron deficiency [49], which altogether resulted in a down-regulation of photosynthesis, nitrate assimilation, and mitochondrial electron transport.

Although photosynthetic pigments are related to processes depending on light, dark phase reactions are equally important. Diatoms might perform both C_3 as well as C_4 photosynthesis and, indeed, both types of initial products were found in one species, and it was reported that the CO_2 concentration has no effect on the transcripts of genes engaged in this process [84]. As in other photosynthetic organisms, the enzyme responsible for carbon fixation is RubisCO, which also has the ability to fix oxygen. Additionally, diatoms' cytoplasmic membrane is highly permeable and a lot of CO_2 is lost by diffusion. However, diatoms are well adapted to low CO_2 concentrations through an efficient CO_2concentration mechanism based on the active transport of carbon from the cytoplasm into the chloroplast [85]. As described above, despite environmental fluctuations, diatoms are well-adapted marine organisms with great importance to global productivity.

PIGMENTS INVOLVED IN PHOTOPROTECTION

Light conditions change significantly depending on the daily cycle, season, habitat, and environment. These conditions determine the life of photosynthetic organisms that have to react to differing situations, accumulate photons efficiently in low light, and dissipate excesses of energy in high light. Excess light might lead to photoinhibition, inactivation of photosystems, and induce the formation of reactive oxygen species (ROS), resulting in photodamage and wide-ranging changes in the cells caused by oxidative stress. To avoid this, phototrophs have developed photoprotective mechanisms such as NPQ and the xanthophyll cycle.

The NPQ relies on the dissipation of excessive excitation energy as heat and might be detected as a quenching of Chl *a* fluorescence. Various mechanisms have been postulated to describe this process; however, all of them require structural changes in photosynthetic antenna that allow a transition from light harvesting to an energy dissipation state [86]. The role of the proton gradient between thylakoid lumen and chloroplast stroma, as well as the presence of xanthophylls and the LHCII structure, have all been thoroughly investigated in higher plants (for a review see [86]), but there has also been significant progress in this field when it comes to diatoms. Studies carried out so far indicate the importance of Dtx in NPQ that is observed only in the presence of this pigment. Contrary to plants and green algae, the proton gradient across the thylakoid membrane in diatoms is not sufficient to induce NPQ [87]. Furthermore, the decrease in the proton gradient does not result in NPQ relaxation, which is correlated with Dtx epoxidation [88].

Six types of xanthophyll cycles have been described in photosynthetic organisms and two of them occur in diatoms [89]. Typical of this group is the diadinoxanthin cycle (DD cycle, see Figure 5), which exists in algae such as diatoms, phaeophytes, dinophytes, and haptophytes. The two pigments engaged in this process, Ddx and Dtx, are normally present in their cells, although their amount depends on the light intensity (see Section 8). In high light, Ddx, which is a monoepoxide xanthophyll, is converted to epoxy-free Dtx by Ddx de-epoxidase (DDE, also called VDEs, three isoforms were even identified); in low light or dark conditions there is the reverse reaction, leading to Ddx formation which is catalyzed by Dtx epoxidase (DEP, also called ZEPs, including three isoforms). Overlong light stress leads to the formation of Zx, one of three xanthophylls involved in the violaxanthin cycle (VAZ cycle), as shown in Figure 6, which occurs mostly in higher plants, mosses, and lichens, but in several conditions also observed in diatoms [24]. It relies on a cyclic conversion of di-epoxy Vx and epoxy-free Zx via an intermediate product, mono-epoxy Ax. De-epoxidation of Vx is catalyzed by VDEs and epoxidation by ZEPs. Nevertheless, it is still not known which of the three isoforms of both enzymes identified in *P. tricornutum* are engaged in the DD or VAZ cycle.

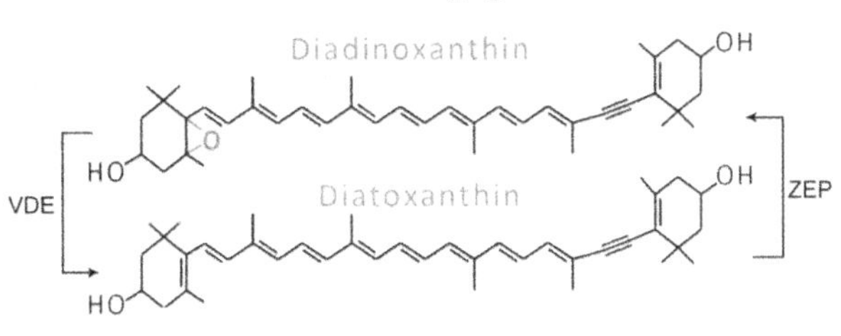

Figure 5: The diadinoxanthin cycle: in high light, diadinoxanthin with one epoxy group is converted to epoxy-free diatoxanthin by violaxanthin de-epoxidase (VDE);

the reverse reaction is observed in low light and dark and is catalyzed by zeaxanthin epoxidase (ZEP).

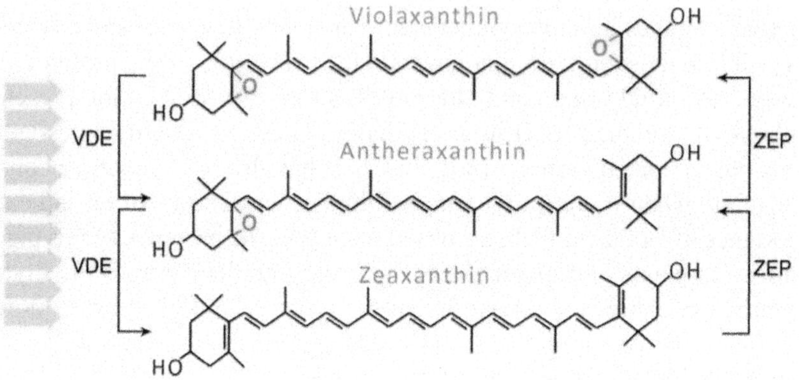

Figure 6: The violaxanthin cycle: under high light, violaxanthin (which is normally a precursor of fucoxanthin) is converted to zeaxanthin via the intermediate antheraxanthin and this reaction is catalyzed by violaxanthin de-epoxidase (VDE), whereas in low light and dark, two single steps of oxygenation catalyzed by zeaxanthin epoxidase (ZEP) lead to violaxanthin formation.

Studies on ZEPs and VDEs in this species have shown that their transcript levels vary (the lowest for ZEP3 and VDL1) and are stimulated by white and blue light [31], but there is no data on their correlation with the type of xanthophyll cycle. Although there are similarities of these enzymes in diatoms and higher plants based on structure, substrate, and cosubstrate requirements, the properties and the mechanism of light-dependent activation/deactivation are not yet understood. Epoxidation in diatoms under low light is faster than those in higher plants and green algae [88]. Although it is known that this reaction in higher plants occurs in both light and dark conditions, in diatoms, the presence of a proton gradient between thylakoid lumen and chloroplast stroma in high light almost completely inhibits epoxidation [88,90]. In *P. tricornutum*, the rate of Ddx de-epoxidation is higher than that of Vx, whereas Dtx and Zx epoxidation rates are similar [24]. Moreover, VDEs in diatoms are active at higher pH values than those in plants and need a lower ascorbate concentration as a cosubstrate [91,92]. Although the photoprotective role of both xanthophyll cycles as well as the ability of Dtx and Zx to quench excited Chl and free radicals are known, the reason why they exist together in diatom cells and the activation mechanism of each remain unclear. Indeed, Zx might be formed from both Vx and β-car and de-epoxidation, and *de novo* carotenoid synthesis could play a substantial role in its accumulation under high light [24]. It is interesting that in *Chlamydomonas reinhardtii* with no functional xanthophyll cycle, the accumulation of Zx was observed, which indicates the crucial role of β-car as a substrate [93] and diminishes the importance of the VAZ cycle. Moreover, the possibility that Zx may not be an obligate precursor of Vx was put forward when Cx and CxE were detected in *P. tricornutum*, although conversions between β-car, Cx, CxE, Vx, Ax, and Zx require detailed studies [27]. A comparative analysis of the photoprotective properties of Zx and Dtx is needed. The quenching efficiency of Dtx has been tested and it was shown that this pigment is a better quencher in low light because additional molecules of Dtx and Ddx bind to FCP, and while not participating in NPQ, they are precursors of Fx [94]. This may suggest that Zx is necessary to avoid photodamage in high light and could explain its accumulation.

METHODS OF PIGMENT ANALYSIS AND IDENTIFICATION

Photosynthetic pigments are used as indicators of the relative abundance of major microalgal taxonomic groups and offer a means for assessing changes or differences in the relative abundance of phototropic functional groups in mixed assemblages [95]. This approach provides measurements of the quantitative changes in the relative abundance of major microalgal groups under similar environmental conditions and habitats. As reported by Schagerl and Kunzl [96], within the year 2005, in nine international journals, 15 different concentrations and six solvents were used, whereas pigment quantification was done by measuring pigment extracts by means of fluorometry (35%) and spectrophotometry (35%), followed by HPLC (25%). Given the fact that Chl *a* is extensively used as an indirect measure of overall algal biomass, and other pigments like Chl *c* or carotenoids have been successfully applied as markers for specific taxonomic groups within algal communities [96] and the references therein, the varieties of methods make comparisons very difficult. Although the determination of chlorophylls is routinely conducted by spectrophotometric [97] or spectrofluorometric [98] methods, the results may be erroneous due to the fact that the absorption and emission bands of Chl *a* overlap with those of other Chls and the degradation products of chlorophylls are neither detected nor determined along with their parent chlorophylls. Moreover, the determination of individual carotenoids is difficult to achieve by these methods. Thus, pigment analysis by HPLC has become the favorite tool for marine researchers.

Extraction of Pigments

The best extraction technique is still a matter for debate. The efficiency of pigment extraction may vary depending on the properties of the solvent, the duration of the extraction, cell concentration, species of algae, and whether mechanical disruption is used. Physical methods of sample disruption such as grinding, bath sonication, high power sonication, or soaking are in use and the most suitable solvents currently applied include acetone, methanol, and non-volatile *N,N*-dimethylformamide (DMF) (see [96] and the references therein). The need for the complete extraction of all pigments present in the sample, compatibility with chromatographic technique, and stability of the pigments are to be considered before the choice is made.

The application of various extraction solvents has been thoroughly discussed by Wright and co-workers [99], of which both acetone and methanol are widely applied in the extraction of algal pigments. Although the highest extraction efficiency is generally achieved by using methanol as an extraction solvent in combination with mechanical disruption, the stability of pigments in methanol is low. It has often been claimed to promote allomerization of Chl *a* [100,101] and, accordingly, Mantoura and Llewellyn [102] found that methanol led to the formation of Chl *a*-derivative products. On the other hand, there is evidence that certain chlorophylls and carotenoids are more thoroughly extracted with methanol [103] or dimethyl sulfoxide [104]. Altogether, acetone seems to be the best choice when there is no information on the specific composition of the sample. While acetone and methanol have the same polarity index, acetone has greater eluotrophic strength than methanol for carbon-rich substrates [105]. Moreover, it has been proved to cause fewer artifacts than methanol and DMF as the derivatives of Chl *a* (mainly Chlide *a*, Chl *a* allomer and epimer) were found at only up to 5% of measured Chl *a* [96]. The extraction efficiency of acetone increases considerably when, prior to extraction, freeze-drying is applied, which breaks up the protein matrix of membranes, creating accessibility for the extraction solvent even with more recalcitrant algae [100,106]. Additionally, freeze-drying inhibits the activity of chlorophyllase, the enzyme found in many marine taxa [101,107], which catalyzes the hydrolysis of the bond linking the phytol chain to the propionic acid residue of Chls*a* and *b*, thereby converting Chl to its corresponding chlorophyllide [108,109]. Since chlorophyllase is mainly active in an aqueous environment [108], it appears that dehydration by freeze-drying creates conditions that inhibit enzyme activity. To conclude, it is worth remembering that algal pigments are extracted differentially by various solvents, and there is no single combination of solvent and extraction time that is best for all species [103].

HPLC Analysis of Pigments

HPLC was applied to phytoplankton studies in the early 1980s [110]. All current methods use reverse phase separation where compounds are resolved primarily on the basis of their polarity. Up until now, different columns have been tested, including octadecylsilica (ODS) C18 and octylsilica (OS) C8, as well as C30 stationary phase, with different diameters, lengths, and particle sizes, which provide different degrees of effectiveness in the resolution of key pigments. Pigment separation and resolution can be also tuned by the column temperature, mobile phase composition, and the gradient program. For the resolution of acidic chlorophylls, buffering, ion-pairing, or an ion suppression reagent is required, with ammonium acetate as the one of choices.

A comprehensive monograph where both modern analytical methods and their applications to biological oceanography were reviewed is that of Jeffery and co-workers from 1997 [111] and 1999 [112]. Nineteen HPLC methods published between 1983 and 1998 have been reported. In short, the work has been started by Mantoura and Llewellyn [102], who developed a reverse-phase HPLC technique for the rapid separation and quantification of 17 carotenoids and 14 chlorophylls together with their breakdown products from acetone extracts of algal culture and natural waters. They also suggested that the presence of an ion-pairing reagent during the chromatographic separation was essential to obtain good resolution for chloropigments (chlorophylls, chlorophyllides, and phaeophorbides). A modified technique was described by Wright and Shearer [113] in which photosynthetic pigments from phytoplankton were separated by HPLC using a lineal elution gradient from 90% acetonitrile to ethyl acetate. Although a better resolution was obtained by the latter method for carotenoid pigments, it does not resolve the polar chlorophylls c_1, c_2, c_3, and Mg-2,4 divinyl phaeoporphyrin a_5 monomethyl ester (Mg2,4D), which coelute as a single peak, or adequately resolve the chlorophyllides a and b. A more comprehensive method was published in 1991 [114], and has been used thereafter for routine work. Further improvement has been achieved by replacing ammonium acetate with either a pyridine or tetrabutylammonium (TBAA) modifier. In particular, Zapata and co-workers [115], using a gradient from aqueous methanol/acetonitrile to methanol/acetonitrile/acetone with a pyridine modifier (as the acetate, pH 5.0), succeeded in resolving, on a C8 column, seven polar Chl c derivatives, the Chl a/DV Chl a pair, and, partially, Chl$b$/DV Chl b. The method of Van Heukelem and Thomas [116] is based on an aqueous methanol to methanol gradient at 60 °C with a TBAA modifier (pH 6.5). The effectiveness of two HPLC methods (a monomeric C18 column with a high ion strength mobile phase composed of methanol, acetonitryl, and ethyl acetate by Kraay and co-workers [117] (Figure 7 and Figure 8) and a monomeric

C8 column with a pyridine-containing mobile phase [116]) in the separation, identification, and quantification of pigments were compared by Mendes and co-workers [118], using phytoplankton, microphytobenthos, and algal cultures. Although the C8 method allows the separation of chlorophylls c_1, c_2, and Mg-2, 4-divinyl phaeoporphyrin a_5 monomethyl ester, as well as the pair Chl a/DV Chl a, the C18 method had a shorter elution program and a lower solvent flow rate, making it cheaper and faster than the C8 method. The excessive separations feasible with C30 columns have proven useful for unique applications such as pigment isolation [116], although the long analysis times are a deterrent if routine separations of complex mixtures are required.

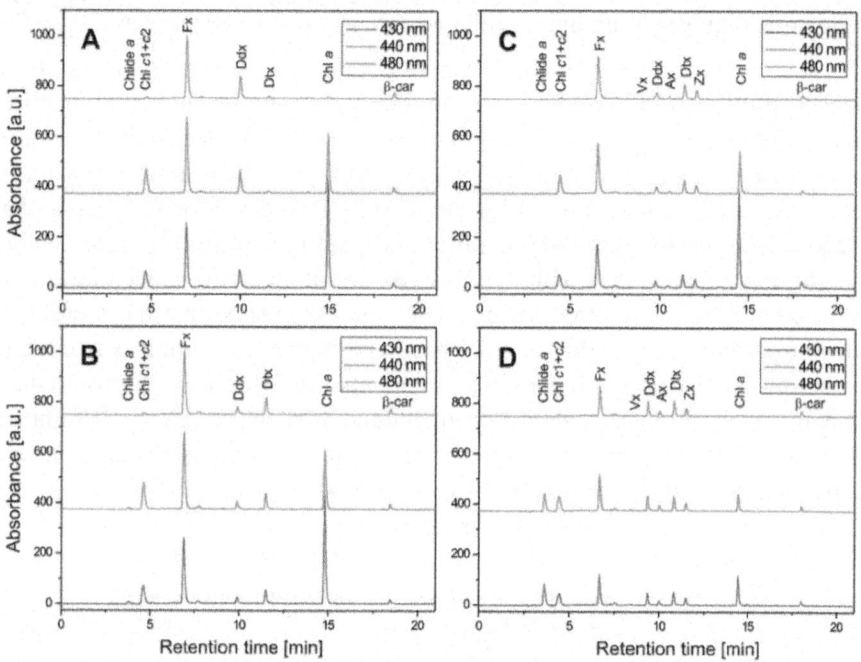

Figure 7: Examples of HPLC chromatograms of absorbance recorded at 430, 440, and 480 nm obtained with the method by Kraay and co-workers [117] for pigment extracts from the diatom *P. tricornutum*: (**A**) 1 h dark incubated cells growing in LL; (**B**) 1 h HL illuminated cells growing in LL; (**C**) 1 h HL illuminated cells growing in ML; (**D**) 48 h HL illuminated cells growing in LL. LL: white light with the intensity of 100 μmol $m^{-2} \cdot s^{-1}$ in a 16 h light/8 h dark photoperiod; ML: white light with the intensity of 700 μmol $m^{-2} \cdot s^{-1}$ in a 6 h light/18 h dark photoperiod; HL: white light with the intensity of 1250 μmol $m^{-2} \cdot s^{-1}$.

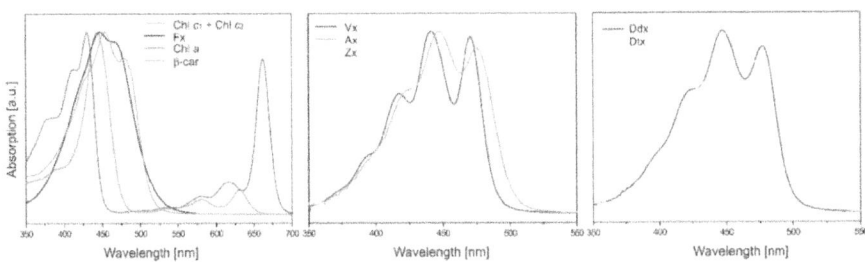

Figure 8: Absorption spectra of photosynthetic pigments recorded during HPLC-DAD analysis performed on extracts from the diatom *P. tricornutum* with the method by Kraay and co-workers [118]. The spectra were normalized at λ_{max}.

For the detection and quantitative analysis of pheophytin *a*, pheophorbide *a*, and their derivatives, the wavelengths of 420–430 nm are useful, while the region of 450–480 nm detects all carotenoids and Chl *c* without interference from Chl *a* derivatives. Peaks can be quantified either using internal or external standards. The first approach ensures higher accuracy as it accounts for any changes resulting from evaporation or dilution as well as providing a check on the injection status. Commercially available internal standards in common use are ethyl 8'-apo-β-carotenoate, 8'-apo-β-carotenoal, vitamin E, and canthaxanthin, the first of which attracts special interest due to its stability and non-occurrence in natural systems. Table 3 shows the elution order of photosynthetic pigments from the diatom *P. tricornutum*, analyzed with the method by Kraay and co-workers [117], and their visible absorption characteristics. The absorbance spectra of more algal pigments and their elution order are available in the literature [53,102,114,118,119].

Pigment identification can be confirmed by performing LC/MS and comparing the resulting mass spectra with published fragment ion abundance and/or mass spectral libraries of pigment standards.

Table 3: Elution order of photosynthetic pigments from the diatom *P. tricornutum* and their visible absorption characteristics

Pigment	Literature Data				HPLC Data *	
	Solvent	λ_{max} (nm)	E (L g^{-1}·cm^{-1})	Reference	λ_{max} (nm)	Retention Time (min)
Chlorophyll c_1	acetone (90%)	443	318	[120]		
Chlorophyll c_2	acetone (90%)	444	374	[120]		
Chlorophyll $c_1 + c_2$	N/A	N/A	N/A	N/A	443	4.7
Fucoxanthin	petrol ether	449	165	[121]	447	7.0
	ethanol	450	114	[122]		
	acetone	443	165	[123]		
Violaxanthin	N/A	443	N/A	[114]	441	8.4
Diadinoxanthin	acetone	448	224	[124]	447	9.9
	methanol	445	225	[124]		
	hexane	446	211	[124]		
Anteraxanthin	ethanol	446	235	[125]	447	10.8
Diatoxanthin	acetone	454	N/A	[114]	454	11.3
Zeaxanthin	ethanol	452	254	[126]	453	11.9
Chlorophyll *a*	acetone	662	88.15	[127]	429	14.9
β-carotene	etanol	453	262	[128]	454	18.7
	hexane	453	259.2	[128]		
	acetone	454	250	[129]		

*: data obtained with the method by Kraay and co-workers [117] for pigment extracts from the diatom *P. tricornutum*.

SIGNIFICANCE OF DIATOM PIGMENTS

Marine algae are rich sources of structurally diverse bioactive compounds with various biological activities. Among them, natural pigments have received particular attention as they exhibit beneficial activities such as antioxidant, anticancer, anti-inflammatory, anti-obesity, anti-angiogenic, and neuroprotective effects important in the fields of food, cosmetics, and pharmacology [130]. Although chemically synthesized pigments are widely available, natural products are more attractive to consumers as they exhibit higher bioavailability and bioactivity. Since bioreactors allow successful cultivation of different species of marine microorganisms, including diatoms, the latter are considered a good source of natural pigments. A relatively fast growth rate along with with low costs is by far the best advantage; however, it has to be taken into account that pigments for food or pharmaceutical applications need to be of a consistently high quality that is safe, traceable, measurable, and readily extractable, which makes the technology more demanding. Nevertheless, the great health benefits and nutraceutical properties represented by diatom pigments make them interesting for pharmaceutical companies.

Carotenoids, especially β-car as the most efficient natural quencher, are involved in the scavenging of ROS, singlet molecular oxygen, and peroxyl radicals, so they may protect against lipid peroxidation [131]. β-car is also an effective stimulator of gap junctional communication, which is associated with cell growth, differentiation, and apoptosis and, together with α-tocopherol, inhibits the lipid peroxidation better than on its own [132]. The anticancer activity of β-car was confirmed, e.g., for lung cancer, although the risk of cancer increased when high doses of a dietary supplement rich in this carotenoid were consumed by smokers [133]. Lutein and Zx are responsible for the coloration of the macula lutea, which is a part of the retina, the area of maximal visual acuity; thus, both pigments are extremely important in ophthalmology and are also used in protection against age-related macular degeneration [134]. This photoprotection arises from their function as filters for damaging blue light and as quenchers of excited oxygen. Strong antiproliferative activity of Vx on breast cancer cells was demonstrated, which makes this pigment a potent compound of drugs for the induction of apoptosis [135]. Although diatoms contain a few photosynthetic pigments including chlorophylls and carotenoids, Fx has the greatest significance in these organisms. This xanthophyll is present in large amounts in diatoms and brown algae, whereas Zx, Ax, Vx, and β-car are sourced mainly from plants where they are present abundantly. Nevertheless, two other xanthophylls, Dtx and Ddx, are also specific to diatoms and might be obtained from them. Unfortunately, the biological activity of their extracts, health benefits, and importance in the diet are still unknown and should be a major motivation for future research.

Fx has great commercial importance because of its varied and numerous bioactivities, such as antioxidant, anticancer, anti-inflammatory, anti-obesity, anti-angiogenic, neuroprotective, photoprotective, and prevention of osteoporosis [130]. The extracts of a few seaweed species such as *Hyaleucerea fusiformis*, *Cladosiphon okamuranus*, *Undaria pinnatifida*, and *Sargassum fulvellum* showed strong DPPH radical scavenging activity [136]. Fx may inhibit intracellular ROS formation, DNA damage, and apoptosis induced by H_2O_2 and its antioxidant activity is comparable to that of α-tocopherol [136]. The anti-inflammatory activity of Fx is associated with its ability to reduce the nitric oxide level, prostaglandin E2, tumor necrosis factor-α, interleukin-1β, interleukin-6, and histamine [137,138]. Strong antiproliferative activity is exhibited by Fx which may induce the apoptosis of human leukemia cell line HL-60, and reduce the viability of PC-3 human prostate cancer cells and human colon cancer cell lines Caco-2, HT-29, and DLD-1 [139,140] (for a review see [136]). Moreover, Fx is a promising compound that might be used in obesity prevention through the upregulation of UCP1 which promotes energy expenditure by thermogenesis [141]. It was reported that Fx mixed with pomegranate seed oil reduced body weight, body fat, and liver fat content [142].

METHODS OF PURIFICATION OF DIATOM PIGMENTS

From among photosynthetic pigments occurring in diatoms, neither chlorophylls nor β-car are of interest to be extracted and purified on a preparative scale since they are more abundant in plants. On the contrary, the broad application potential of Fx has prompted researchers to explore the commercial production of this photosynthetic pigment from algae. To find an efficient source of Fx, brown macroalgae, such as *Laminaria japonica*, *Eisenia bicyclis*, and *Undaria pinnatifida* [143,144], were studied; they are traditional foods in Southeast Asia and some European countries. However, the Fx concentration from these macroalgae ranged from 0.02 to 0.58 mg·g^{-1} in fresh samples and 0.01 to 1.01 mg·g^{-1} in dried samples [26], which makes the production of Fx from brown macroalgae not commercially viable. Since the reported Fx concentration in microalgae is one to three orders of magnitude greater than that found in macroalgae, ranging from 2.24 to 18.23 mg·g^{-1} [26], diatoms are considered a promising source of Fx for various commercial applications. In particular, *P. tricornutum* was found to contain Fx as a main carotenoid. Studies by the Rebolloso-Fuentes research group [145] showed that 60% of the total carotenoids in the acetone extract was Fx, which gives Fx a concentration of 1.81 mg/g dw. Xia and co-workers [8] investigated the marine diatom *Odontella aurita* and determined that it can accumulate high concentrations of Fx (>20 mg·g^{-1} of dry weight). The Fx concentration of 3.33–21.67 mg·g^{-1} and the volumetric concentration of 18.15–79.56 mg·L^{-1} obtained in this study have set the records for algae-based Fx production. A dataset containing the Fx concentrations in samples of different macroalgae and microalgae is available in [8,26].

There are numerous methods used to recover and purify Fx from brown algae, including: (i) centrifugal partition chromatography; (ii) pressurized liquid extraction; (iii) enzymatic pre-treatment followed by co-solvent extraction; (iv) supercritical carbon dioxide with ethanol as co-solvent; and (v) traditional solvent extraction followed by chromatographic methods, as reviewed by Gomez-Loredo in [146]. However, the costs associated with new technologies and equipment are major drawbacks to their implementation. Thus, there is a need to develop low cost, simple, and scalable strategies for the recovery of Fx as one of the high value compounds occurring naturally in abundance in diatoms. The results show that Fx extraction efficiency, as with other phytoplankton pigments, is highly dependent on the extraction conditions—the solvent type, in particular (see Section 8). Kim and co-workers [26] have assessed a number of extraction procedures to investigate the effect of solvent type, extraction time, temperature, and extraction method (maceration, ultrasound-assisted extraction, Soxhlet extraction, and pressurized liquid extraction) on Fx extraction efficiency. Among the investigated solvents, ethanol provided the best Fx extraction yield (15.71 mg g^{-1} freeze-dried sample weight), whereas its content in the extracts produced by the different methods was quite constant (15.42–16.51 mg·g^{-1} freeze-dried sample weight). The pigment extracts obtained from freeze-dried powder of *P. tricornutum* extracted with 100% ethanol at room temperature for 3 h were then subjected to silica-gel adsorption chromatography with an eluent of n-hexane:acetone (7:3, v/v). These results are in accordance with findings reported by Xia and co-workers [8], who additionally pointed to the fact that selecting a proper ratio of solvent to dry algal biomass (v/w) is important, as it may affect the quantity and quality of Fx. In particular, they showed that an ethanol/dry biomass ratio of 20:1 is sufficient for the effective extraction of Fx from this microalga. In this work, the crude pigment extracts were subjected to open silica gel column chromatography with n-hexane:acetone (6:4, v/v) as the eluting solvent system. An orange-red colored Fx-rich fraction containing Fx, *cis*-Fx, Ddx, and Dtx was separated by the column. Further purification of Fx, whose purity was estimated as 86.7% in the mixture, was carried out by a prep-HPLC yielding a final Fx purity >97%.

Alternatively, the two-phase solvent system of n-hexane:ethanol:water with a volume ratio of 10:9:1 was determined to be the best system for the separation of Fx and lipids from extracts of *Isochrysis galbana*. Under these conditions, Fx was fractionated in the hydroalcohol phase apart from the hexane phase containing lipids [147]. Recently, Gómez-Loredo and co-workers [146] proposed a different potential strategy for the recovery and partial purification of Fx. In their study, the partition behavior of Fx in aqueous two-phase systems (ATPS) composed of ethanol and potassium phosphate was evaluated. The optimal extraction conditions were found to be un-acidified methanol (2.5% w/v for *P. tricornutum*) or ethanol (5% w/v for *P. tricornutum*) as a solvent with orbital agitation at 250 rpm for 1 h at 20 °C. The best results were observed in systems with a tie-line length, TLL 50% w/w, and a volume ratio of 1 for *P. tricornutum* methanolic extracts (95.36% recovery and 66.01% purity).

To the best of our knowledge, there is no method published so far to obtain Ddx and Dtx on a preparative scale, except for a semi-preparative HPLC approach. Thus, the method based on column chromatography recently elaborated in our laboratory (see Acknowledgments) seems to be promising for the potential application of these xanthophylls naturally found in diatoms.

CONCLUSIONS

Apart from chlorophylls *a* and *c*, and fucoxanthin, which play a light-harvesting function, a group of photosynthetic pigments in diatoms comprises those engaged in photoprotection, including β-carotene, diatoxanthin, diadinoxanthin, violaxanthin, antheraxanthin and zeaxanthin. Physical and chemical properties of diatom pigments have been extensively studied, which contributed to a more complete understanding of the function they play in cells as well as enabled explanation of their health promoting activity and beneficial effects. There are, however, still further studies required to gain the knowledge on Chls c to a level comparable with the well-studied Chls a and b. Moreover, it is of great importance to explore biosynthetic pathways of Chls c and xanthophylls by identifying the enzymes involved and the specificity of their reactions. There is a growing body of studies on diatom pigments, which refers to their photosynthetic and photoprotective functions and explains the mechanisms of these processes. Nevertheless, there are still many questions to be answered in the field of xanthophyll cycle and non-photochemical quenching research. Photoacclimation and pigment level variation are also commonly studied, which is related to the fact that marine environment is highly changeable. Furthermore, analysis of diatom pigments is widely used in global oceanography to assess phytoplankton biomass, productivity, community structure and ecological processes. Natural pigments, including those derived from algae have received particular attention as they exhibit beneficial activities important in terms of their commercial and industrial applications. In this respects, diatoms seem to be a promising source of unique bioactive compounds, with fucoxanthin, diadinoxanthin and diatoxanthis as representatives of photosynthetic pigments.

ACKNOWLEDGMENTS

This work was supported by project No. 2011/01/M/NZ1/01170. Faculty of Biochemistry, Biophysics and Biotechnology of Jagiellonian University is a partner of the Leading National Research Center (KNOW) supported by the Ministry of Science and Higher Education.

A method of diadinoxanthin and diatoxanthin purification has been reported as (1) Patent application No. P. 412178, priority date 29.04.2015, authors: Paulina Kuczynska and Malgorzata Jemiola-Rzeminska; (2) Patent application No. P. 412177, priority date 29.04.2015, authors: Paulina Kuczynska and Malgorzata Jemiola-Rzeminska.

Malgorzata Jemiola-Rzeminska would like to express her sincere gratitude to Susann Schaller-Laudel and Reimundt Goss for the cooperation and all the scientific support she has received from the Institute of Biology I, Plant Physiology, University of Leipzig, Germany.

AUTHOR CONTRIBUTIONS

Paulina Kuczynska made a substantial contribution to conception and design of paper; analyzed data and wrote parts relating to biological aspects of this review including physiology, ecology, biosynthesis and genetics; prepared Figure 1, Figure 2, Figure 3, Figure 4, Figure 5 and Figure 6, Table 1 and Table 2; participated in revising the manuscript in response to the reviewers' suggestions. Malgorzata Jemiola-Rzeminska analyzed information and wrote parts concerning physical, chemical, structural and methodical aspects of this review; prepared Figure 7 and Figure 8 and Table 3; contributed to revising critically for important intellectual content and made necessary modifications in response to the reviewers' comments. Kazimierz Strzalka contributed to proofreading and final approval of the version to be published.

REFERENCES

1.   Bozarth, A.; Maier, U.G.; Zauner, S. Diatoms in biotechnology: Modern tools and applications.Appl. Microbiol. Biotechnol. 2009, 82, 195–201.

2.   Martin-Jézéquel, V.; Hildebrand, M.; Brzezinski, M.A. Silicon metabolisim in diatoms: Implications for growth. J. Phycol. 2000, 36, 821–840.

3.   Allen, A.E.; Dupont, C.L.; Oborník, M.; Horák, A.; Nunes-Nesi, A.; McCrow, J.P.; Zheng, H.; Johnson, D.A.; Hu, H.; Fernie, A.R.; et al. Evolution and metabolic significance of the urea cycle in photosynthetic diatoms. Nature 2011, 473, 203–207.

4.   Lee, J.-B.; Hayashi, K.; Hirata, M.; Kuroda, E.; Suzuki, E.; Kubo, Y.; Hayashi, T. Antiviral sulfated polysaccharide from Navicula directa, a diatom collected from deep-sea water in Toyama Bay. Biol. Pharm. Bull. 2006, 29, 2135–2139.

5.   Perl, T.M.; Bédard, L.; Kosatsky, T.; Hockin, J.C.; Todd, E.C.; Remis, R.S. An outbreak of toxic encephalopathy caused by eating mussels contaminated with domoic acid. N. Engl. J. Med.1990, 322, 1775–1780.

6.   Prestegard, S.K.; Oftedal, L.; Nygaard, G.; Skjaerven, K.H.; Knutsen, G.; Døskeland, S.O.; Coyne, R.T.; Herfindal, L. Marine benthic diatoms contain compounds able to induce leukemia cell death and modulate blood platelet activity. Mar. Drugs 2009, 7, 605–623.

7.   Peng, K.-T. Delta 5 Fatty Acid Desaturase Upregulates the Synthesis of Polyunsaturated Fatty Acids in the Marine Diatom Phaeodactylum tricornutum. J. Agric. Food Chem. 2014, 62, 8773–8776.

8. Xia, S.; Wang, K.; Wan, L.; Li, A.; Hu, Q.; Zhang, C. Production, characterization, and antioxidant activity of fucoxanthin from the marine diatom odontella aurita. Mar. Drugs 2013,11, 2667–2681.

9. Gastineau, R.; Turcotte, F.; Pouvreau, J.B.; Morançais, M.; Fleurence, J.; Windarto, E.; Arsad, S.; Prasetiya, F.S.; Jaouen, P.; Babin, M.; et al. Marennine, promising blue pigments from a widespread Haslea diatom species complex. Mar. Drugs 2014, 12, 3161–3189.

10. Gastineau, R.; Davidovich, N.A.; Bardeau, J.F.; Caruso, A.; Leignel, V.; Hardivillier, Y.; Rince, Y.; Jacquette, B.; Davidovich, O.I.; Gaudin, P.; et al. Haslea karadagensis (Bacillariophyta): A second blue diatom, recorded from the Black Sea and producing a novel blue pigment. Eur. J. Phycol. 2012, 47, 469–479.

11. Gelzinis, A.; Butkus, V.; Songaila, E.; Augulis, R.; Gall, A.; Büchel, C.; Robert, B.; Abramavicius, D.; Zigmantas, D.; Valkunas, L. Mapping energy transfer channels in fucoxanthin-chlorophyll protein complex. Biochim. Biophys. Acta Bioenerg. 2015, 1847, 241–247.

12. Jeffrey, S.; Vesk, M. Introduction to marine phytoplankton and their pigment signatures. InPhytoplankton Pigments in Oceanography: Guidelines to Modern Methods; Jeffrey, S., Mantoura, R., Wright, S., Eds.; UNESCO Publishing: Paris, France, 1997; pp. 37–84.

13. Zapata, M.; Garrido, J.; Jeffrey, S. Chlorophyll c Pigments: Current Status. In Chlorophylls and Bacteriochlorophylls; Grimm, B., Porra, R., Rüdiger, W., Scheer, H., Eds.; Springer Netherlands: Dordrecht, The Netherlands, 2006; pp. 39–53.

14. Stauber, J.L.; Jeffrey, S.W. Photosynthetic pigments in fifty-one species of marine diatoms. J. Phycol. 1988, 24, 158–172.

15. Fookes, C.; Jeffrey, S. The structure of chlorophyll c3, a novel marine photosynthetic pigment. J. Chem. Soc. Chem. Commun. 1989, 23, 1827–1828.

16. Britton, G. UV/visible spectroscopy. In Carotenoids: Spectroscopy, Vol 1B; Liaaen-Jensen, S., Britton, G., Pfander, H., Eds.; Birkhäuser Verlag: Basel, Switzerland, 1995; pp. 13–63.

17. Zigmantas, D.; Hiller, R.G.; Sharples, F.P.; Frank, H.A.; Sundstrom, V.; Polivka, T. Effect of a conjugated carbonyl group on the photophysical properties of carotenoids. Phys. Chem. Chem. Phys. 2004, 6, 3009–3016.

18. Rontani, J.-F.; Beker, B.; Raphel, D.; Baillet, G. Photodegradation of chlorophyll phytyl chain in dead phytoplanktonic cells. J. Photochem. Photobiol. A Chem. 1995, 85, 137–142.

19. Britton, G.; Liaaen-Jensen, S.; Pfander, H. Carotenoids, 1st ed.; Britton, G., Liaaen-Jensen, S., Pfander, H., Eds.; Birkhäuser Verlag: Basel, Switzerland, 2004.

20. Subczynski, W.K.; Markowska, E.; Sielewiesiuk, J. Effect of polar carotenoids on the oxygen diffusion-concentration product in lipid bilayers. An EPR spin label study. Biochim. Biophys. Acta Biomembr. 1991, 1068, 68–72.

21. Berglund, A.H.; Nilsson, R.; Liljenberg, C. Permeability of large unilamellar digalactosyldiacylglycerol vesicles for protons and glucose— Influence of α-tocopherol, β-carotene, zeaxanthin and cholesterol. Plant Physiol. Biochem. 1999, 37, 179–186.

22. Refvem, T.; Strand, A.; Kjeldstad, B.; Haugan, J.A.; Liaaen-Jensen, S. Stereoisomerization of Allenic Carotenoids—Kinetic, Thermodynamic and Mechanistic Aspects. Acta Chem. Scand. 1999, 53, 114–123.

23. Haugan, J.A.; Englert, G.; Liaaen-Jensen, S.; Frimpong-Manso, S.; Springborg, J.; Wang, D.-N.; Christensen, S.B. Algal Carotenoids. 50. Alkali Lability of Fucoxanthin—Reactions and Products. Acta Chem. Scand. 1992, 46, 614–624.

24. Lohr, M.; Wilhelm, C. Algae displaying the diadinoxanthin cycle also possess the violaxanthin cycle. Proc. Natl. Acad. Sci. USA 1999, 96, 8784–8789.

25. Frank, H.A.; Cogdell, R.J. Carotenoids in photosynthesis. Photochem. Photobiol. 1996, 63, 257–264.

26. Kim, S.M.; Jung, Y.J.; Kwon, O.N.; Cha, K.H.; Um, B.H.; Chung, D.; Pan, C.H. A potential commercial source of fucoxanthin extracted from the microalga Phaeodactylum tricornutum. Appl. Biochem. Biotechnol. 2012, 166, 1843–1855.

27. Lohr, M.; Wilhelm, C. Xanthophyll synthesis in diatoms: Quantification of putative intermediates and comparison of pigment conversion kinetics with rate constants derived from a model. Planta 2001, 212, 382–391.

28. Bowler, C.; Allen, A.E.; Badger, J.H.; Grimwood, J.; Jabbari, K.; Kuo, A.; Maheswari, U.; Martens, C.; Maumus, F.; Otillar, R.P.; et al. The Phaeodactylum genome reveals the evolutionary history of diatom genomes. Nature 2008, 456, 239–244.

29. Armbrust, E.V.; Berges, J.A.; Bowler, C.; Green, B.R.; Martinez, D.; Putnam, N.H.; Zhou, S.; Allen, A.E.; Apt, K.E.; Bechner, M.; et al. The genome of the diatom Thalassiosira pseudonana: Ecology, evolution, and metabolism. Science 2004, 306, 79–86.

30. Bertrand, M. Carotenoid biosynthesis in diatoms. Photosynth. Res. 2010, 106, 89–102.

31. Coesel, S.; Oborník, M.; Varela, J.; Falciatore, A.; Bowler, C. Evolutionary origins and functions of the carotenoid biosynthetic pathway in marine diatoms. PLoS ONE 2008, 3, 1–16.

32. Stransky, H.; Hager, A. Das Carotinoidmuster und die Verbreitung des lichtinduzierten Xanthophyllcyclus in verschiedenen Algenklassen. Arch. Mikrobiol. 1970, 73, 315–323.

33. Montsant, A.; Allen, A.E.; Coesel, S.; Mangogna, M.; Siaut, M.; Heijde, M.; Jabbari, K.; Maheswari, U.; Rayko, E.; Vardi, A.; et al. Identification and comparative genomic analysis of signaling and regulatory components in the diatom Thalassiosira pseudonana. J. Phycol. 2007, 43, 585–604.

34. Dambek, M.; Eilers, U.; Bretenbach, J.; Steiger, S.; Büchel, C.; Sandmann, G. Biosynthesis of fucoxanthin and diadinoxanthin and function of initial pathway genes in Phaeodactylum tricornutum. J. Exp. Bot. 2012, 63, 5607–5612.

35. Mikami, K.; Hosokawa, M. Biosynthetic pathway and health benefits of fucoxanthin, an algae-specific xanthophyll in brown seaweeds. Int. J. Mol. Sci. 2013, 14, 13763–13781.

36. Grimm, B.; Porra, R.J.; Rüdiger, W.; Scheer, H. Chlorophylls and Bacteriochlorophylls: Biochemistry, Biophysics, Functions and Applications, 1st ed.; Grimm, B., Porra, R.J., Rüdiger, W., Scheer, H., Eds.; Springer Netherlands: Dordrecht, The Netherlands, 2006; Volume 25.

37. Hunsperger, H.M.; Randhawa, T.; Cattolico, R.A. Extensive horizontal gene transfer, duplication, and loss of chlorophyll synthesis genes in the algae. BMC Evol. Biol. 2015, 15, 1–19.

38. Disch, A.; Schwender, J.; Muller, C.; Lichtenthaler, H.K.; Rohmer, M. Distribution of the mevalonate and glyceraldehyde phosphate/pyruvate pathways for isoprenoid biosynthesis in unicellular algae and the cyanobacterium Synechocystis PCC 6714. Biochem. J. 1998, 333, 381–388.

39. Ito, H.; Ohtsuka, T.; Tanaka, A. Conversion of chlorophyll b to chlorophyll a via 7-hydroxymethyl chlorophyll. J. Biol. Chem. 1996, 271, 1475–1479.

40. Porra, R.J. Recent Progress in Porphyrin and Chlorophyll Biosynthesis. Photochem. Photobiol.1997, 65, 492–516.

41. Vidoudez, C.; Pohnert, G. Comparative metabolomics of the diatom Skeletonema marinoi in different growth phases. Metabolomics 2012, 8, 654–669.

42. Sánchez-Saavedra, M.P.; Voltolina, D. Effect of photon fluence rates of white and blue-green light on growth efficiency and pigment content of three diatom species in batch cultures. Cienc. Mar. 2002, 28, 273–279.

43. Brunet, C.; Chandrasekaran, R.; Barra, L.; Giovagnetti, V.; Corato, F.; Ruban, A.V. Spectral radiation dependent photoprotective mechanism in the diatom Pseudo-nitzschia multistriata.PLoS ONE 2014, 9, 1–10.

44. Lepetit, B.; Goss, R.; Jakob, T.; Wilhelm, C. Molecular dynamics of the diatom thylakoid membrane under different light conditions. Photosynth. Res. 2012, 111, 245–257.

45. Jungandreas, A.; Costa, B.S.; Jakob, T.; von Bergen, M.; Baumann, S.; Wilhelm, C. The acclimation of Phaeodactylum tricornutum to blue and red light does not influence the photosynthetic light reaction but strongly disturbs the carbon allocation pattern. PLoS ONE2014, 9, 1–14.

46. Veuger, B.; van Oevelen, D. Long-term pigment dynamics and diatom survival in dark sediment.Limnol. Oceanogr. 2011, 56, 1065–1074.

47. Järup, L. Hazards of heavy metal contamination. Br. Med. Bull. 2003, 68, 167–182.

48. Yoshinaga, R.; Niwa-Kubota, M.; Matsui, H.; Matsuda, Y. Characterization of iron-responsive promoters in the marine diatom Phaeodactylum tricornutum. Mar. Genom. 2014, 16, 55–62.

49. Allen, A.E.; Laroche, J.; Maheswari, U.; Lommer, M.; Schauer, N.; Lopez, P.J.; Finazzi, G.; Fernie, A.R.; Bowler, C. Whole-cell response of the pennate diatom Phaeodactylum tricornutumto iron starvation. Proc. Natl. Acad. Sci. USA 2008, 105, 10438–10443.

50. Sivaci, R.E.; Sivaci, A.; Eroglu, S. Changes in photosynthetic pigments and species diversity of epiphytic diatoms on Myriophyllum triphyllum exposed to cadmium. ScienceAsia 2013, 39, 100–104.

51. Alexandre, M.; Gundermann, K.; Pascal, A.; van Grondelle, R.; Büchel, C.; Robert, B. Probing the carotenoid content of intact Cyclotella cells by resonance Raman spectroscopy. Photosynth. Res. 2014, 119, 273–281.

52. Beer, A.; Juhas, M.; Büchel, C. Influence of different light intensities and different iron nutrition on the photosynthetic apparatus in the diatom Cyclotella meneghiniana (bacillariophyceae). J. Phycol. 2011, 47, 1266–1273.

53. Brotas, V.; Plante-Cuny, M.R. The use of HPLC pigment analysis to study microphytobenthos communities. Acta Oecol. 2003, 24, 109–115.

54. Costa, B.S.; Jungandreas, A.; Jakob, T.; Weisheit, W.; Mittag, M.; Wilhelm, C. Blue light is essential for high light acclimation and methylation and

chromatin patterning photoprotection in the diatom Phaeodactylum tricornutum. J. Exp. Bot. 2013, 64, 483–493.

55. Domingues, N.; Matos, A.R.; da Silva, J.M.; Cartaxana, P. Response of the DiatomPhaeodactylum tricornutum to photooxidative stress resulting from high light exposure. PLoS ONE 2012, 7, 1–6.

56. Katayama, T.; Murata, A.; Taguchi, S. Responses of pigment composition of the marine diatomThalassiosira weissflogii to silicate availability during. Plankt. Benthos Res. 2011, 6, 1–11.

57. Lavaud, J. Fast regulation of photosynthesis in diatoms: Mechanisms, evolution and ecophysiology. Funct. Plant Sci. Biotechnol. 2007, 1, 267–287.

58. Gibbs, S. The comparative ultrastructure of the algal chloroplast. Ann. N. Y. Acad. Sci. 1970,175, 454–473.

59. Pyszniak, A.; Gibbs, S. Immunocytochemical localization of photosystem I and the fucoxanthinchlorophyll a/c light-harvesting complex in the diatom Phaeodactylum tricornutum.Protoplasma 1992, 166, 208–217.

60. Gundermann, K.; Büchel, C. The Structural Basis of Biological Energy Generation, 1st ed.; Hohmann-Marriott, M., Ed.; Springer Netherlands: Dordrecht, The Netherlands, 2014.

61. Papagiannakis, E.; van Stokkum, I.H.M.; Fey, H.; Büchel, C.; van Grondelle, R. Spectroscopic characterization of the excitation energy transfer in the fucoxanthinchlorophyll protein of diatoms. Photosynth. Res. 2005, 86, 241–250.

62. Premvardhan, L.; Sandberg, D.J.; Fey, H.; Birge, R.R.; Büchel, C.; van Grondelle, R. The charge-transfer properties of the S2 state of fucoxanthin in solution and in fucoxanthin chlorophyll-a/c2 protein (FCP) based on stark spectroscopy and molecular-orbital theory. J. Phys. Chem. B 2008, 112, 11838–11853.

63. Premvardhan, L.; Bordes, L.; Beer, A.; Büchel, C.; Robert, B. Carotenoid structures and environments in trimeric and oligomeric fucoxanthin chlorophyll a/c2 proteins from resonance raman spectroscopy. J. Phys. Chem. B 2009, 113, 12565–12574.

64. Premvardhan, L.; Réfrégiers, M.; Büchel, C. Pigment organization effects on energy transfer and Chl a emission imaged in the diatoms C. meneghiniana and P. tricornutum in vivo: A confocal laser scanning fluorescence (CLSF) microscopy and spectroscopy study. J. Phys. Chem. B 2013,117, 11272–11281.

65. Gundermann, K.; Schmidt, M.; Weisheit, W.; Mittag, M.; Büchel, C. Identification of several sub-populations in the pool of light harvesting

proteins in the pennate diatom Phaeodactylum tricornutum. Biochim. Biophys. Acta Bioenerg. 2013, 1827, 303–310.

66. Lepetit, B.; Volke, D.; Gilbert, M.; Wilhelm, C.; Goss, R. Evidence for the existence of one antenna-associated, lipid-dissolved and two protein-bound pools of diadinoxanthin cycle pigments in diatoms. Plant Physiol. 2010, 154, 1905–1920.

67. Eppard, M.; Rhiel, E. The genes encoding light-harvesting subunits of Cyclotella cryptica(Bacillariophyceae) constitute a complex and heterogeneous family. Mol. Gen. Genet. 1998, 260, 335–345.

68. Liu, Z.; Yan, H.; Wang, K.; Kuang, T.; Zhang, J.; Gui, L.; An, X.; Chang, W. Crystal structure of spinach major light-harvesting complex at 2.72 A resolution. Nature 2004, 428, 287–292.

69. Falkowski, P.G. Biogeochemical Controls and Feedbacks on Ocean Primary Production. Science1998, 281, 200–206.

70. Stevenson, R.J.; Peterson, C.G. Variation in benthic diatom (Bacillariophyceae) immigration with habitat characteristics and cell morphology. J. Phycol. 1989, 25, 120–129.

71. Padisák, J.; Soróczki-Pintér, É.; Rezner, Z. Sinking properties of some phytoplankton shapes and the relation of form resistance to morphological diversity of plankton—An experimental study.Hydrobiologia 2003, 500, 243–257.

72. Verleyen, E.; Hodgson, D.A.; Leavitt, P.R.; Sabbe, K.; Vyverman, W. Quantifying habitat-specific diatom production: A critical assessment using morphological and biogeochemical markers in Antarctic marine and lake sediments. Limnol. Oceanogr. 2004, 49, 1528–1539.

73. Falkowski, P.G.; Knoll, A.H. Evolution of Primary Producers in the Sea. In Evolution of Primary Producers in the Sea; Falkowski, P., Knoll, A.H., Eds.; Academic Press: New York, NY, USA, 2007; p. 456.

74. Haxo, F.T.; Blinks, L.R. Photosynthetic action spectra of marine algae. J. Gen. Physiol. 1950,33, 389–422.

75. Nicklisch, A. Growth and light absorption of some planktonic cyanobacteria, diatoms and Chlorophyceae under simulated natural light fluctuations. J. Plankton Res. 1998, 20, 105–119.

76. Spaulding, S.A.; Lubinski, D.J.; Potapova, M. Diatoms of the United States. Available online: http://westerndiatoms.colorado.edu (accessed on 6 July 2015).

77. Johansen, J.R. Morphological variability and cell wall composition of Phaeodactylum tricornutum(Bacillariophyceae). Great Basin Nat. 1991, 51, 310–315.

78. Wagner, A.H.; Jakob, T.; Wilhelm, C.; Wagner, H. Balancing the energy captured under fluctuating light conditions to. New Biotechnol. 2005, 169, 95–108.

79. Holdsworth, E.S. Effect of growth factors and light quality on the growth, pigmentation and photosynthesis of two diatoms, Thalassiosira gravida and Phaeodactylum tricornutum. Mar. Biol.1985, 86, 253–262.

80. Darley, W.; Volcani, B. Role of silicon in diatom metabolism. A silicon requirement for deoxyribonucleic acid synthesis in the diatom Cylindrotheca fusiformis Reimann and Lewin. Exp. Cell Res. 1969, 58, 334–342.

81. Lombardi, A.; Wangersky, P. Influence of phosphorus and silicon on lipia class production by the marine diatom Chaetoceros gracilis grown in turbidostat cage cultures. Mar. Ecol. Prog. Ser.1991, 77, 39–47.

82. Brzezinski, M.A.; Baines, S.B.; Balch, W.M.; Beucher, C.P.; Chai, F.; Dugdale, R.C.; Krause, J.W.; Landry, M.R.; Marchi, A.; Measures, C.I.; et al. Co-limitation of diatoms by iron and silicic acid in the equatorial Pacific. Deep. Res. II Top. Stud. Oceanogr. 2011, 58, 493–511.

83. Lippemeier, S.; Hartig, P.; Colijn, F. Direct impact of silicate on the photosynthetic performance of the diatom Thalassiosira weissflogii assessed by on- and off-line PAM fluorescence measurements. J. Plankton Res. 1999, 21, 269–283.

84. Roberts, K.; Granum, E.; Leegood, R.C.; Raven, J.A. C3 and C4 pathways of photosynthetic carbon assimilation in marine diatoms are under genetic, not environmental, control. Plant Physiol. 2007, 145, 230–235.

85. Hopkinson, B.M.; Dupont, C.L.; Allen, A.E.; Morel, F.M.M. Efficiency of the CO_2-concentrating mechanism of diatoms. Proc. Natl. Acad. Sci. USA 2011, 108, 3830–3837.

86. Goss, R.; Lepetit, B. Biodiversity of NPQ. J. Plant Physiol. 2014, 172, 13–32.

87. Lavaud, J.; Rousseau, B.; Etienne, A.L. In diatoms, a transthylakoid proton gradient alone is not sufficient to induce a non-photochemical fluorescence quenching. FEBS Lett. 2002, 523, 163–166.

88. Goss, R.; Ann Pinto, E.; Wilhelm, C.; Richter, M. The importance of a highly active and ΔpH-regulated diatoxanthin epoxidase for the regulation of the PS II antenna function in diadinoxanthin cycle containing algae. J. Plant Physiol. 2006, 163, 1008–1021.

89. García-Plazaola, J.I.; Matsubara, S.; Osmond, C.B. The lutein epoxide cycle in higher plants: Its relationships to other xanthophyll cycles and possible functions. Funct. Plant Biol. 2007, 34, 759–773.

90. Mewes, H.; Richter, M. Supplementary ultraviolet-B radiation induces a rapid reversal of the diadinoxanthin cycle in the strong light-exposed diatom Phaeodactylum tricornutum. Plant Physiol. 2002, 130, 1527–1535.

91. Jakob, T.; Goss, R.; Wilhelm, C. Unusual pH-dependence of diadinoxanthin de-epoxidase activation causes chlororespiratory induced accumulation of diatoxanthin in the diatomPhaeodactylum tricornutum. J. Plant Physiol. 2001, 158, 383–390.

92. Grouneva, I.; Jakob, T.; Wilhelm, C.; Goss, R. Influence of ascorbate and pH on the activity of the diatom xanthophyll cycle-enzyme diadinoxanthin de-epoxidase. Physiol. Plant. 2006, 126, 205–211.

93. Baroli, I.; Do, A.D.; Yamane, T.; Niyogi, K.K. Zeaxanthin accumulation in the absence of a functional xanthophyll cycle protects Chlamydomonas reinhardtii from photooxidative stress.Plant Cell 2003, 15, 992–1008.

94. Schumann, A.; Goss, R.; Jakob, T.; Wilhelm, C. Investigation of the quenching efficiency of diatoxanthin in cells of Phaeodactylum tricornutum (Bacillariophyceae) with different pool sizes of xanthophyll cycle pigments. Phycologia 2007, 46, 113–117.

95. Pinckney, M.; Micheli, F. Microalgae on seagrass mimics: Does epiphyte community structure differ from live seagrasses. J. Exp. Mar. Biol. Ecol. 1998, 221, 59–70.

96. Schagerl, M.; Künzl, G. Chlorophyll a extraction from freshwater algae—A reevaluation. Biol. Bratisl 2007, 62, 270–275.

97. Strickland, J.; Parsons, T. A Practical Handbook of Seawater Analysis, 2nd ed.; Stevenson, J.C., Ed.; Fisheries Research Board of Canada: Ottawa, ON, Canada, 1972; Volume 167.

98. Holm-Hansen, O.; Lorenzen, C.; Holmes, R.; Strickland, J. Fluorometric determination of chlorophyll. J. Cons. Perm. Int. Explor. Mer 1965, 30, 3–15.

99. Wright, S.W.; Jeffrey, S.W.; Mantoura, R.F.C. Evaluation of methods and solvents for pigment extraction. In Phytoplankton Pigments in Oceanography: Guidelines to Modern Methods; Wright, S.W., Jeffrey, S.W., Mantoura, R.F.C., Eds.; UNESCO: Paris, France, 1997; pp. 261–282.

100. Chen, N.; Bianchi, T.; Bland, J. Novel decomposition products of chlorophyll-a in continental shelf (Louisiana shelf) sediments: Formation and transformation of carotenol chlorine esters.Geochim. Cosmochim. Acta 2003, 67, 2027–2042.

101. Jeffrey, S. Profiles of photosynthetic pigments in the ocean using thin-layer chromatography.Mar. Biol. 1974, 26, 101–110.

102. Mantoura, R.; Llewellyn, C. The rapid determination of algal chlorophyll and carotenoid pigments and their breakdown products in natural waters by reverse-phase high-performance liquid chromatography. Anal. Chim. Acta 1983, 151, 297–314.

103. Bowles, N.; Paerl, H.; Tucker, J. Effective solvents and extraction periods employed in phytoplankton carotenoi d and chlorophyll determination. Can. J. Fish. Aquat. Sci. 1985, 42, 1127–1131.

104. Shoaf, W.T.; Lium, B.W. Improved extraction of chlorophyll a and b from algae using dimethyl sulfoxide. Limnol. Oceanogr. 1976, 21, 926–928.

105. Stock, R.; Rice, C.B.F. Chromatographic Methods, 1st ed.; Springer US: New York, NY, USA, 1974.

106. Buffan-Dubau, E.; Carman, K.R. Extraction of benthic microalgal pigments for HPLC analyses.Mar. Ecol. Prog. Ser. 2000, 204, 293–297.

107. Jeffrey, S.; Hallegraeff, G. Chlorophyllase distribution in ten classes of phytoplankton: A problem for chlorophyll analysis. Mar. Ecol. Prog. Ser. 1987, 35, 293–304.

108. Barrett, J.; Jeffrey, S.W. Chlorophyllase and formation of an atypical chlorophyllide in marine algae. Plant Physiol. 1964, 39, 44–47.

109. Draziewicz, M. Chlorophyllase: Occurrence, functions, mechanisms of action, effects of external and internal factor. Photosynthesis 1994, 30, 321–331.

110. Rowan, K. Photosynthetic Pigments of Algae; Cambridge University Press: Cambridge, UK, 1989.

111. Jeffrey, S.W.; Mantoura, R.F.C.; Wright, S.W. Phytoplankton Pigments in Oceanography: Guidelines to Modern Methods; Jeffrey, S., Mantoura, R., Wright, S., Eds.; UNESCO Publishing: Paris, France, 1997.

112. Jeffrey, S.; Wright, S.; Zapata, M. Recent advances in HPLC pigment analysis of phytoplankton.Mar. Freshw. Res. 1999, 50, 879–896.

113. Wright, S.; Shearer, J. Rapid extraction and HPLC of chlorophylls and carotenoids from marine phytoplankton. J. Chrom. 1984, 294, 281–295.

114. Wright, S.W. Improved HPLC method for the analysis of chlorophylls and carotenoids from marine phytoplankton. Mar. Ecol. Prog. Ser. 1991, 77, 183–196.

115. Zapata, M.; Rodríguez, F.; Garrido, J.L. Separation of chlorophylls and carotenoids from marine phytoplankton: A new HPLC method using a

reversed phase C8 column and pyridine-containing mobile phases. Mar. Ecol. Prog. Ser. 2000, 195, 29–45.

116. Van Heukelem, L.; Thomas, C. Computer-assisted high-performance liquid chromatography method development with applications to the isolation and analysis of phytoplankton pigments. J. Chromatogr. A 2001, 910, 31–49.

117. Kraay, G.W.; Zapata, M.; Veldhuis, M.J.W. Separation of chlorophylls c1c2, and c3 of marine phytoplankton by reversed-phase-C18-high-performance liquid chromatography. J. Phycol.1992, 28, 708–712.

118. Mendes, C.R.; Cartaxana, P.; Brotas, V. HPLC determination of phytoplankton and microphytobenthos pigments: Comparing resolution and sensitivity of a C18 and a C8 method.Limnol. Oceanogr. Methods 2007, 5, 363–370.

119. Jeffrey, S.; Mantoura, R.; Bjørnland, T. Data for the identification of 47 key phytoplankton pigments. In Phytoplankton Pigments in Oceanography: Guidelines to Modern Methods; Jeffrey, S., Mantour, R., Wright, S., Eds.; UNESCO Publishing: Paris, France, 1997; pp. 447–559.

120. Jeffrey, S.W. Preparation and some properties of crystalline chlorophyll c1 and c2 from marine algae. Biochim. Biophys. Acta 1972, 279, 15–33.

121. Jensen, A. Algal carotenoids, I. Fucoxanthin monoacetate. Acta Chem. Scand. 1961, 15, 1604–1628.

122. Antia, N.J. The optical activity of fucoxanthin. Can. J. Chem. 1965, 43, 302–303.

123. Haugan, J.A.; Liaaen-Jensen, S. Improved isolation procedure for fucoxanthin. Phytochemistry1989, 28, 2797–2798.

124. Johansen, J.E.; Svec, W.A.; Liaaen-Jensen, S. Carotenoids of the Dinophyceae. Phytochemistry1974, 13, 2261–2271.

125. Hager, A.; Meyer-Bertenrath, T. Die Isolierung und quantitative Bestimmung der Carotinoide und Chlorophylle von Blättern, Algen und isolierten Chloroplasten mit Hilfe dünnschichtchromatographischer Methoden. Planta 1966, 69, 198–217.

126. Davies, B.H. Carotenoids. In Chemistry and Biochemistry of Plant Pigments Vol. II; Goodwin, T., Ed.; Academic Press: London, UK, 1976; pp. 38–165.

127. Jeffrey, S.W.; Humphrey, G.F. New spectrophotometric equations for determining chlorophylls a, b, c1 and c2 in higher plants, algae and natural phytoplankton. Biochem. Physiol. Pflanz.1975, 167, 191–194.

128. Isler, O.; Lindlar, H.; Montavon, M.; Rüegg, R.; Zeller, P. Die technische Synthese von beta-Carotin. Helv. Chim. Acta 1956, 39, 249–259.

129. Hiyama, T.; Nishimura, M.; Chance, B. Determination of carotenes by thin-layer chromatography. Anal. Biochem. 1969, 29, 339–342.

130. Pangestuti, R.; Kim, S.K. Biological activities and health benefit effects of natural pigments derived from marine algae. J. Funct. Foods 2011, 3, 255–266.

131. Stahl, W.; Sies, H. Bioactivity and protective effects of natural carotenoids. Biochim. Biophys. Acta Mol. Basis Dis. 2005, 1740, 101–107.

132. Palozza, P.; Krinsky, N.I. β-Carotene and α-tocopherol are synergistic antioxidants. Arch. Biochem. Biophys. 1992, 297, 184–187.

133. Omenn, G.S.; Goodman, G.E.; Thornquist, M.D.; Balmes, J.; Cullen, M.R.; Glass, A.; Keogh, J.P.; Meyskens, F.L.; Valanis, B.; Williams, J.H.; et al. Risk factors for lung cancer and for intervention effects in CARET, the Beta-Carotene and Retinol Efficacy trial. J. Natl. Cancer Inst.1997, 89, 325–326.

134. Krinsky, N.I.; Landrum, J.T.; Bone, R.A. Biologic mechanisms of the protective role of lutein and zeaxanthin in the eye. Annu. Rev. Nutr. 2003, 23, 171–201.

135. Pasquet, V.; Morisset, P.; Ihammouine, S.; Chepied, A.; Aumailley, L.; Berard, J.B.; Serive, B.; Kaas, R.; Lanneluc, I.; Thiery, V.; et al. Antiproliferative activity of violaxanthin isolated from bioguided fractionation of Dunaliella tertiolecta extracts. Mar. Drugs 2011, 9, 819–831.

136. Peng, J.; Yuan, J.P.; Wu, C.F.; Wang, J.H. Fucoxanthin, a marine carotenoid present in brown seaweeds and diatoms: Metabolism and bioactivities relevant to human health. Mar. Drugs2011, 9, 1806–1828.

137. Heo, S.-J.; Ko, S.-C.; Kang, S.-M.; Kang, H.-S.; Kim, J.-P.; Kim, S.-H.; Lee, K.-W.; Cho, M.-G.; Jeon, Y.-J. Cytoprotective effect of fucoxanthin isolated from brown algae Sargassum siliquastrum against H2O2-induced cell damage. Eur. Food Res. Technol. 2008, 228, 145–151.

138. Kim, K.-N.; Heo, S.-J.; Yoon, W.-J.; Kang, S.-M.; Ahn, G.; Yi, T.-H.; Jeon, Y.-J. Fucoxanthin inhibits the inflammatory response by suppressing the activation of NF-κB and MAPKs in lipopolysaccharide-induced RAW 264.7 macrophages. Eur. J. Pharmacol. 2010, 649, 369–375.

139. Hosokawa, M.; Wanezaki, S.; Miyauchi, K.; Kurihara, H.; Kohno, H.; Kawabata, J.; Odashima, S.; Takahashi, K. Apoptosis-Inducing Effect of

Fucoxanthin on Human Leukemia Cell Line HL-60.Food Sci. Technol. Res. 1999, 5, 243–246.

140. Asai, A.; Sugawara, T.; Ono, H.; Nagao, A. Biotransformation of fucoxanthinol into amarouciaxanthin a in mice and HepG2 cells: Formation and cytotoxicity of fucoxanthin metabolites. Drug Metab. Dispos. 2004, 32, 205–211.

141. Maeda, H.; Hosokawa, M.; Sashima, T.; Funayama, K.; Miyashita, K. Fucoxanthin from edible seaweed, Undaria pinnatifida, shows antiobesity effect through UCP1 expression in white adipose tissues. Biochem. Biophys. Res. Commun. 2005, 332, 392–397.

142. Abidov, M.; Ramazanov, Z.; Seifulla, R.; Grachev, S. The effects of XanthigenTM in the weight management of obese premenopausal women with non-alcoholic fatty liver disease and normal liver fat. Diabetes Obes. Metab. 2010, 12, 72–81.

143. Jaswir, I.; Dedi, N.; Salleh, H.M.; Taher, M.; Miyashita, K.; Ramli, N. Analysis of fucoxanthin content and purification of all-trans-fucoxanthin from Turbinaria turbinata and Sargassum plagyophyllum by SiO2 open column chromatography and reversed phase HPLC. J. Liq. Chromatogr. 2013, 36, 1340–1354.

144. Kim, S.M.; Shang, Y.F.; Um, B.H. A preparative method for isolation of fucoxanthin from Eisenia bicyclis by centrifugal partition chromatography. Phytochem. Anal. 2011, 22, 322–329.

145. Rebolloso-Fuentes, M.M.; Navarro-Pérez, A.; García-Camacho, F.; Ramos-Miras, J.J.; Guil-Guerrero, J.L. Biomass nutrient profiles of the microalga Nannochloropsis. J. Agric. Food Chem.2001, 49, 2966–2972.

146. Gómez-loredo, A.; Benavides, J.; Rito-palomares, M. Partition behavior of fucoxanthin in ethanol-potassium phosphate two-phase systems. J. Chem. Technol. Biotechnol. 2014, 89, 1637–1645.

147. Kim, S.; Kang, S.-W.; Kwon, O.-N.; Chung, D.; Pan, C.-H. Fucoxanthin as a major carotenoid in Isochrysis aff. galbana: Characterization of extraction for commercial application. J. Korean Soc. Appl. Biol. Chem. 2012, 55, 477–483.

Chapter 6

META-ANALYSIS OF THE DETECTION OF PLANT PIGMENT CONCENTRATIONS USING HYPERSPECTRAL REMOTELY SENSED DATA

Jingfeng Huang[1] , Chen Wei[1,2], Yao Zhang[1] , George Alan Blackburn[3] , Xiuzhen Wang[4], Chuanwen Wei[1] , Jing Wang[1]

[1] Institute of Agricultural Remote Sensing & Information Application, Zijingang Campus, Zhejiang University, Hangzhou, China

[2] Zhejiang Meteorological Service Center, Hangzhou, China

[3] Lancaster Environment Centre, Lancaster University, Lancaster, United Kingdom

[4] Institute of Remote Sensing and Earth Sciences, Hangzhou Normal University, Hangzhou, China

ABSTRACT

Passive optical hyperspectral remote sensing of plant pigments offers potential for understanding plant ecophysiological processes across a range of spatial scales. Following a number of decades of research in this field, this paper undertakes a systematic meta-analysis of 85 articles to determine whether passive optical hyperspectral remote sensing techniques are sufficiently well developed to quantify individual plant pigments, which operational solutions are available for wider plant science and the areas which now require greater focus. The findings indicate that predictive relationships are strong for all pigments at the leaf scale but these decrease and become more variable across pigment types at the canopy and landscape scales. At leaf scale it is clear that specific sets of optimal wavelengths can be recommended for operational methodologies: total chlorophyll and chlorophyll a quantification is based on reflectance in the green (550–560nm) and red edge (680–750nm) regions; chlorophyll b on the red, (630–660nm), red edge (670–710nm) and the near-infrared (800–810nm); carotenoids on the 500–580nm region; and anthocyanins on the green (550–560nm), red edge (700–710nm) and near-infrared (780–790nm). For total chlorophyll the optimal wavelengths are valid across canopy and landscape scales and there is some evidence that the same applies for chlorophyll

INTRODUCTION

A pigment is a material that changes the spectral distribution of reflected or transmitted light as the result of wavelength-selective absorption which is determined by the physical properties of the pigment itself. Plant pigments play an important role in light capture, photosystem protection, and in various growth and development functions. The photosynthetic pigments control the amount of solar radiation absorbed by a leaf and thus determine photosynthetic potential and primary production [1,2]. Pigment concentrations are also related to plant stress (excess direct sunlight, UV–B irradiation, low temperature, water stress, nitrogen deficiencies and so on) and senescence (e.g., [3–9]). Therefore, accurate measurements of the temporal dynamics and spatial variations of pigment concentration using remotely sensed data can provide a basis for monitoring physiological and ecological processes [10,11].

The spectral absorbance properties of pigments offer the possibility of using measurements of reflected radiation as a non-destructive method for quantifying pigments. Different approaches have arisen recently to remotely estimate pigment concentrations from a wide variety of wavelengths and sensor types. These studies produced variable results, and none have been demonstrated to have satisfactory performance under all growth and environmental conditions. These inconsistencies may stem from the fact that the experimental results are influenced by a number of factors including different species, experimental conditions and analytical methods used [11].

Recent review articles have attempted to assimilate knowledge in this field of passive optical hyperspectral remote sensing with the sun as energy source. Blackburn [10] reviewed the developing technologies and analytical methods for quantitative estimation of pigment across a range of spatial scales using passive optical hyperspectral remote sensing. Ustin et al. [11] appraised the most widely used methodologies for retrieving pigment information with hyperspectral data at the leaf scale. However, it has been demonstrated that traditional qualitative reviewers may subjectively select their preferred studies when faced with conflicting results on a single question [12]. In contrast, it has been argued that meta-analysis can take the results from primary research articles and quantitatively analyze and synthesize these data in an attempt to arrive at more robust conclusions. As such, meta-analysis review papers make the shift from a narrative-driven to a data-driven approach [13,14].

Glass [15] published the first article to lay out the essential rationale of meta-analysis. As a fully general set of methods, meta-analysis has been widely applied to the integration of literatures in many areas of empirical science, including ecology [14]. This form of analysis has, for example, been used to determine the response of biodiversity to intensive biomass production, the effects of elevated CO_2 on plant–arthropod interactions, the influence of plant invasion on carbon and nitrogen cycles and the causes and consequences of variations in leaf mass per area [16–19]. Today, many findings and advances are being made not only by those who do primary research studies, but also by those who use meta-analysis to discover the latent meaning of existing research literatures [13]. Recently, meta-analysis has been employed in remote sensing research. Garbulsky et al. [20] performed a meta-analysis to assess the use of the photochemical reflectance index (PRI) as an indicator of radiation use efficiencies at the leaf, canopy and ecosystem scales for different time scales and vegetation types. Zolkos et al. [21] conducted a meta-analysis of publications on LiDAR remote sensing estimation of terrestrial aboveground biomass. These investigations show that meta-analysis can be used to systematically integrate the results from a collection of studies, and through statistical comparison, assess the relationships between remotely sensed measurements and variables of interest.

Here, a meta-analysis of data from a wide selection of studies reporting the passive optical hyperspectral remote sensing of pigments was used to quantify the development of this scientific field, identify optimal wavelengths for retrieval of individual pigments and evaluate the strength of the relationships between pigment concentration and remotely sensed data across pigment types and scales.

MATERIALS AND METHODS

Study Selection and Data Extraction

Databases of Elsevier, Springer and Web of Science, licensed to Zhejiang University, were used for source data from inception to August 2014. The following key words were used: pigment, chlorophyll, carotenoids, carotene, xanthophyll, anthocyanins, anthoxanthin in combination with the terms reflectance, estimation, quantification, retrieval, prediction and remote sensing. More than 4500 citations were collected as a result of this initial search.

Then the abstracts of these articles were reviewed and considered for inclusion in the meta-analysis. The following criteria were applied to ensure homogeneity in methodology. First, the studies had to include a chemical measurement of pigment concentration (total chlorophyll, chlorophyll a, chlorophyll b, carotenoids, xanthophyll, carotene or anthocyanins). Second, the article had to report the quantification of pigments using remotely sensed data. Third, the authors must have provided the following statistical information: (1) coefficient of determination for the relationships between pigment concentration and remotely sensed measurements; (2) the wavelength(s) used to estimate pigment concentration; and (3) training sample sizes.

Based on the first two decision rules, 135 articles were selected. According to the final criterion, 50 studies were excluded because of insufficient statistical information. Finally, 85 articles were used in the meta-analysis, which reported results at different spatial and temporal scales and from a wide range of vegetation types between 1977 and 2014. The number of studies selected at various stages is shown in the flow diagram in Fig 1. Some studies reported multiple results for different pigment types or vegetation types. Different types of sensors were used in these studies, from spectrophotometers and hand-held spectroradiometers to satellite sensors. All the sensors were working in reflectance mode. Within the selected articles 44 were working at the leaf scale, 21 at the canopy scale, 15 at the landscape scale, 2 at the leaf and canopy scales, 1 at the leaf and landscape scales, and 2 covered the leaf, canopy and landscape scales. The term "canopy" refers to either a single plant or a monospecific stand where the experimental results are influenced by a number of controlling factors, such as orientation of leaves (leaf angle distribution; LAD), variations in number of leaf layers (LAI), presence of non-leaf elements, multiple scattering and areas of shadow [10,22], the term "landscape" refers to a mixed-species stand where the reflectance spectrum from airborne and spaceborne sensors is subject to even more controlling factors, such as atmospheric conditions, instrucment sensitivity (signal-to-noise ratio) and spatial resolution. In total, the sample size from all the selected studies is 16100. The Preferred Reporting Items for Meta-Analyses is shown in S1 PRISMA Checklist.

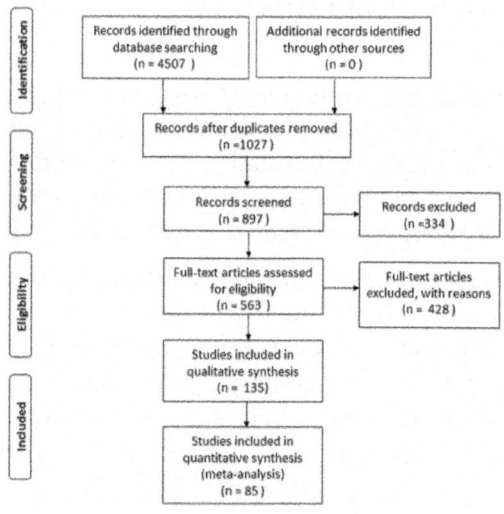

Figure 1: Selection of studies for inclusion in the meta-analysis.

Relevant information was extracted from each study in the final set: ①
scales (leaf, canopy, landscape), ② pigment types, ③ species, ④ wavelengths,
⑤ coefficient of determination, ⑥ sample sizes, ⑦ sensors, ⑧ authors and ⑨
year of publication. In order to reduce human error in data extraction and coding,
two sets of reviewers independently screened articles in accordance with those
inclusion criteria discussed above, evaluated the quality and extracted the data
from the eligible studies. The results from one group were cross-checked by the
other group. Divergences of opinion about article selection and data extraction
were settled by discussion. Table 1 is a summary of the studies contained in
this research. This list is not exhaustive but it does cover most papers published
related to quantification of pigments using remotely sensed data that met the
selection criteria. Table 2 provides a statistical summary of the data extracted
from the studies included in the meta-analysis.

Table 1: A summary of the studies contained in this research that linked remotely
sensed data with pigment

Scale	Pigment Type	Year	Species	Sensor	Reference
leaves	Chl tot	1992	Amaranthus tricolor	Specpho	[23]
leaves	Chl tot	1995	Slash pine	Specrad	[24]
leaves	Chl tot	1995	Bigleaf maple	Specrad	[25]
leaves	Chl tot	1996	Horse Chestnut, Norway maple,Cotoneaster, Tobacco	Specpho	[26]
leaves	Chl tot	1996	Norway Maple, Horse Chestnut	Specpho	[27]
leaves	Chl tot	1997	Norway Maple, Horse Chestnut, Fig, Cotoneaster, Tobacco,Oleander, Hibiscus, Vine, Rose	Specpho	[28]
leaves	Chl tot	1998	Tobacco, Horse Chestnut, Cotoneaster	Specpho	[29]
leaves	Chl tot	1999	Beech tree, Elm tree,Wild vine shurb	Specpho	[30]
leaves	Chl tot	1999	Bragg Soybean	Specrad	[31]
leaves	Chl tot	2002	53 species	Specrad	[32]
leaves	Chl tot	2002	Paper birch	Specrad	[33]
leaves	Chl tot	2003	Bigleaf Maple, Horse Chestnut, Wild vine, Beech	Specpho	[34]
leaves	Chl tot	2005	Cotton	Specrad	[35]
leaves	Chl tot	2007	Winter wheat	Specpho	[36]
leaves	Chl tot	2012	15 different species(Beech, Fraxinus lanuginosa, Acer Japonicum, Magnolia obovata and so on)	Specrad	[37]
leaves	Chl tot	2014	Douglas fir	Specrad	[38]
leaves	Chl a	1994	Norway Maple, Horse Chestnut	Specpho	[39]
leaves	Chl a	1994	Norway Maple, Horse Chestnut	Specpho	[40]
leaves	Chl a	1996	Norway Maple, Horse Chestnut	Specpho	[41]
leaves	Cars/Chl tot	1977	Cantaloupe, Corn, Spinach Cotton, Cucumber, tobacco, Head lettuce, Grain sorghum	Specpho	[42]
leaves	Cars/Chl tot	1992	Sunflower	Specrad	[43]
leaves	Cars/Chl tot	1999	Norway Maple, Potato, Lemon, Apple, Coleus	Specpho	[7]
leaves	Cars/Chl tot	2006	24 species of woody trees and shrubs	Specpho	[44]
leaves	Anths/Cars/Chl tot	1999	Quercus agrifolia, Pseudotsuga menziesii	Specpho	[45]
leaves	Anths/Cars/Chl tot	2003	Apple	Specpho	[46]
leaves	Anths/Cars/Chl tot	2004	Norway maple, Maize, Dogwood,Horse chestnut, Second-flush beech, Wild vine shrub, Cotoneaster, Pelargonium zonale	Specpho	[47]
leaves	Chl tot/Anths	2014	Chilean strawberry	Specrad	[48]
leaves	Cars/Chl a/ Chl b	1992	Soybean	Specrad	[49]
leaves	Cars/Chl a/Chl b	1998	Beech, Oak, Maple, Sweet chestnut	Specrad	[50]
leaves	Cars/Chl a/Chl b	2005	Rice	Specrad	[51]
leaves	Chl tot/Chl a/Chl b	1999	Norway Maple, Horse Chestnut, Beech, Oak	Specrad	[52]
leaves	Chl tot/Chl a/Chl b	2001	Croton, Elaeagnus, Japanese pittosporum,Benjamin fig	Specrad	[53]
leaves	Chl tot/Chl a/Chl b	2010	Flowering cherry	Specrad	[54]
leaves	Chl tot/Chl a	1996	Tobacco	Specpho	[55]
leaves	Chl tot/Chl a	1999	Eucalyptus	Specrad	[56]
leaves	Cars	2002	Norway maple, Horse chestnut,Second-flush beech	Specpho	[57]
leaves	Cars	2009	Scot pine	Specpho	[58]
leaves	Cars	2011	Bur oak, Sugar maple, LOPEX database	Specrad	[59]
Scale	Pigment Type	Year	Species	Sensor	Reference

leaves	Anths	2001	Norway maple, Cotoneaster, Dogwood	Specpho	[60]
leaves	Anths	2009	Grapevine	Specrad	[61]
leaves	Anths	2009	European hazel, Siberian dogwood, Norway maple, Virginia creeper	Specpho	[62]
leaves	Anths	2011	Grapevine	Specrad	[63]
leaves	Anths	2011	Sweet cherries	Specpho	[64]
leaves	Anths	2011	Norway maple, Horse chestnut, Beech, Virginia creeper, Dogwood	Specpho&specrad	[65]
Leaves/canopy	Chl tot	2009	Maize	Specpho	[66]
Leaves/canopy	Chl tot	2013	Irrigated maize	Specrad	[67]
Leaves/landscape	Chl tot	2014	Black Spruce, Sugar maple	Specrad&MERIS	[68]
Leaves/canopy/ landscape	Chl tot	2010	Winter Wheat, Winter Rapeseed	Specrad	[69]
Leaves/canopy/ landscape	Cars/Chl tot	2000	Sugar maple	Specrad	[70]
canopy	Chl tot	1990	Slash pine	Airborne spectro	[1]
canopy	Chl tot	1994	pepper	Specrad	[71]
canopy	Chl tot	2005	Maize, Soybean	Specrad	[72]
canopy	Chl tot	2006	Rice	Specrad	[73]
canopy	Chl tot	2007	Cotton	Specrad	[74]
canopy	Chl tot	2008	Winter wheat, Corns	Specrad	[75]
canopy	Chl tot	2008	Heterogeneous grassland	Specrad	[76]
canopy	Chl tot	2008	Heterogeneous grassland	Specrad	[77]
canopy	Chl tot	2008	Corn, Cotton	Specrad	[78]
canopy	Chl tot	2010	Rice	Specrad	[79]
canopy	Chl tot	2011	Rice	Specrad	[80]
canopy	Chl tot	2012	Potato, Grassland	Specrad	[81]
canopy	Chl tot	2013	Irrigated maize	Specrad	[82]
canopy	Chl tot	2014	Winter wheat	Specrad	[83]
canopy	Chl a	2003	Rice	Specrad	[84]
canopy	Chl a	2007	Winter Wheat	Specrad	[85]
canopy	Chl a/Chl b	2004	Winter wheat	Specrad	[86]
canopy	Chl tot/Chl a	2006	Wheat	Specrad	[87]
canopy	Cars/Chl tot	2010	Tall fescue	Specrad	[88]
canopy	Cars	2008	Kermes oak	Specrad	[89]
canopy	Cars	2008	Douglas fir	Specrad	[90]
landscape	Chl tot	2002	Corn	CASI	[91]
landscape	Chl tot	2003	Eucalypt	CASI-2	[92]
landscape	Chl tot	2004	Jack pine	CASI	[93]
landscape	Chl tot	2004	Douglas fir	MERIS	[94]
landscape	Chl tot	2007	Corn, Wheat	CASI	[95]
landscape	Chl tot	2008	Rice, Cotton	EO-1	[96]
landscape	Chl tot	2008	Garlic, Alfalfa, Onion, Sunflower, Corn, Potato, Wheat, Vineyard, Sugar beet	PROBA/CHRIS	[97]
landscape	Chl tot	2010	Flax, Tea, Chestnut, Corn, Potato, Pine, Bamboo	EO-1	[98]
landscape	Chl tot	2010	Garlic, Onion, Corn, Alfalfa, Sugar beet, Sunflower, Potato, Vineyard, Wheat	PROBA/CHRIS	[99]
landscape	Chl tot	2014	London plane, Canary Island date palm, European nettle tree, White mulberry	CASI	[100]
landscape	Chl a	2004	Winter Wheat	AVIS	[101]
landscape	Cars/Chl tot	2002	Quercus petrea, Pinus sylvestris	CASI	[102]
landscape	Chl a/Cars	2005	Rice	PHI	[103]
landscape	Cars/Chl tot/Chl a/Chl b	2008	Aspen, Birch, Spruce, Balsam fir	CASI	[104]
landscape	Anths	2009	Austrocedrus chilensis forest	Hyperion	[105]

doi:10.1371/journal.pone.0137029.t001

Specrad = spectroradiometer; Specpho = spectrophotometer; Chl tot = total chlorophyll; Chl a = chlorophyll a; Chl b = chlorophyll b; Cars = carotenoids; Anths = anthocyanins.

doi:10.1371/journal.pone.0137029.t001

Table 2: Summary statistics for the selected studies and extracted data for different pigment types at leaf, canopy and landscape scales

Scale	Pigment type	Number of studies	Number of effect sizes	Total sample size	Number of wavelengths
leaves	Chl tot	34	53	6431	131
	Chl a	11	23	1595	53
	Chl b	6	10	860	24
	Cars	14	15	1381	40
	Anths	10	17	1752	43
canopy	Chl tot	20	23	1146	55
	Chl a	4	4	162	6
	Chl b	1	1	35	0
	Cars	3	2	45	7
	Anths	0	0	0	0
landscape	Chl tot	15	17	1883	46
	Chl a	3	3	153	6
	Chl b	1	1	24	2
	Cars	3	3	573	4
	Anths	1	1	60	2

doi:10.1371/journal.pone.0137029.t002

Statistical Analysis of Effect Size

The calculation of effect size for each study.

The coefficient of determination (R^2) was used to evaluate the strength of relationships between spectral reflectance and pigment concentration in each article we selected. The value of R^2, however, is affected by the number of selected wavelengths. The more wavelengths included in the model, be they relevant or not, the larger would be the R^2 [106]. The increase of R^2 is not without cost. The increasing number of selected wavelengths reduces the degrees of freedom, which reduces model robustness. The adjusted coefficient of determination was applied to correct for the degrees of freedom:

$$R_A^2 = 1 - (1 - R^2)\frac{n - 1}{n - k}$$

(1)

where n is the sample size for each study, k is the number of independent variables in the linear or nonlinear model. Eq (1) shows that R_A^2 is always smaller than R^2 when k > 1, which means the growth rate of R_A^2 is lower than that of R^2 as the number of parameters increase. This result is straightforward and it has been shown that when the added parameter explains a significant amount of the behavior of the dependent variable, R_A^2 will increase; otherwise, R_A^2 will decrease [107]. So R_A^2 was chosen as the effect size statistic, the variance of effect size is calculated as [108]:

$$V_i = \frac{(1 - R_A^2)^2}{n - 1},$$

(2)

The resulting data set was categorized by pigment type at the scales of leaf, canopy and landscape to allow comparison.

Test of heterogeneity for effect sizes.

It is important to assess the heterogeneity among the results from a collection of studies before computing the mean effect size [109]. Basically, there are two possible sources of heterogeneity in meta-analysis: methodological heterogeneity and statistical heterogeneity. To ensure homogeneity in methodology, we applied a series of criteria to identify the studies to be used in the meta-analysis (as described in section 2.1). Here the I^2 statistic was used to test for the statistical heterogeneity. The I^2 statistic measures the extent of true heterogeneity dividing the difference between the result of the Q test and its degrees of freedom by the Q value itself [110]:

$$I^2 = 100\% \times \frac{Q_{tot} - df}{Q_{tot}}$$

(3)

where df = $N_{tot} - 1$, N_{tot} is the total number of effect sizes from all the selected studies, Q_{tot} is computed as [111]:

$$Q_{tot} = \sum_{i=1}^{N_{tot}} W_i E_i^2 - \frac{\left(\sum_{i=1}^{N_{tot}} W_i E_i\right)^2}{\sum_{i=1}^{N_{tot}} W_i}$$

(4)

where $W_i = 1/v_i$, E_i is adjusted coefficient of determination (R_A^2).

The I^2 statistic can be interpreted as the percentage of heterogeneous component in the total variability of effect size (Q_{tot}), so the larger the I^2 statistic is, the stronger the heterogeneity is. If I^2 exceeds 50%, the null hypothesis of homogeneity is rejected. The I^2 statistic for different pigments at different scales were calculated, all the results were lower than 50%, the null hypothesis of homogeneity for this study was accepted.

The calculation of mean effect size for different pigments at different scales.

In contrast to studies based on original data, the unit of meta-analysis is the individual research study. Distinctive aspects of data analysis follow from this difference. The first complication is that the studies incorporated into the meta-analysis generally use different sample sizes and this controls the statistical properties of effect sizes [112]. From a statistical perspective, larger sample studies have less sampling error than smaller sample studies, thus more weight should be assigned to larger sample studies in the computation of the mean effect size. The other complication is inter-study variability, which is caused by the influence of an indeterminate number of characteristics that vary among the studies.

Considering the two sources of variability discussed above, a random effects model was used to compute the weighted mean of R_A^2 for different pigment types. In contrast to a fixed effects model, the weight applied to each effect size in a random effects model must represent both subject-level sampling error and the additional random variance component [112]. As such, the mean effect size becomes a reasonable estimate of the true strength of the effect in the population. Because of the generality of the random effects model, it is the preferred strategy in meta-analysis [113]. The mean effect size is computed as:

$$M_{rand} = \frac{\sum_{i=1}^{N_p} W_{i(rand)} E_i}{\sum_{i=1}^{N_p} W_{i(rand)}}$$ (5)

The variance is:

$$V_{rand} = \frac{1}{\sum_{i=1}^{N_p} W_{i(rand)}}$$ (6)

where

$$W_{i(rand)} = \frac{1}{V_i + \sigma^2},$$

$$\sigma^2 = \frac{Q_p - (N_p - 1)}{\sum_{i=1}^{N_p} W_i - \frac{\sum_{i=1}^{N_p} W_i^2}{\sum_{i=1}^{N_p} W_i}}, \quad W_i = 1 \Big/ V_i,$$

$$Q_p = \sum_{i=1}^{N_p} W_i E_i^2 - \frac{\left(\sum_{i=1}^{N_p} W_i E_i\right)^2}{\sum_{i=1}^{N_p} W_i} E_i$$

is adjusted coefficient of determination (R_A^2) and N_p is the total number of effect sizes for a specific type of pigment at each different scales (Table 2). Using this approach the mean effect size of Chl tot, Chl a, Chl b, Cars and Anths at the scales of leaf, canopy and landscape were calculated.

A confidence interval gives the range of values within which the mean effect size is likely to be, it is useful in indicating the degree of precision of the estimate of the mean effect size. A 95% confidence interval is subsequently calculated as follows:

$$Conf_{95} = M_{rand} \pm 1.96 SE_{rand} \tag{7}$$

where

$$SE_{rand} = \sqrt{V_{rand}}$$

If the confidence intervals of multiple mean effect sizes donot overlap, then there are significant differences between these mean effect sizes.

Optimal Wavelengths for Pigment Quantification

A large number of narrow-band indices were proposed to measure plant pigments in the selected articles. These narrow band indices include difference vegetation index (NBDVI), ratio vegetation index (NDRVI), normalized difference vegetation index (NBNDVI), anthocyanin reflectance index (ARI), soil-adjusted vegetation index (SAVI), perpendicular vegetation index (PVI) and so on. The wavelengths used in these studies are different and there is lack of agreement on optimal wavelengths for pigment quantification.

Histograms and quantile plots were used to identify the optimal wavelengths for individual pigment quantification at different scales. The histogram partitions the data distribution of wavelengths into subsets of 10nm width. This enabled us to provide an overview of suitable wavelengths, which is difficult to achieve if the analysis is performed at higher spectral resolutions. Also this approach avoided inaccuracies of spectral calibration associated with the use of many different instruments across the studies incorporated into the meta-analysis. In the histogram each subset is represented by a rectangle whose height is equal to the count of observations that fall into the wavelength interval. A quantile plot is a simple and effective way to compare different wavelength distributions. Let λ_i ($i = 1$ to G) be the wavelengths sorted in increasing order so that λ_1 is the smallest wavelength and λ_G is the largest. Each wavelength, λ_i, is paired with a percentage, f_i, which indicates that approximately $100 f_i\%$ of the data are below or equal to the value, λ_i.

$$f_i = \frac{i - 0.5}{G} \, (i = 1, \ldots, G) \tag{8}$$

In a quantile plot, λ_i is graphed against f_i. This allows us to compare different wavelength distributions based on their quantiles [114].

RESULTS

Quantifying the Development of Remote Sensing Of Plant Pigment Concentrations

The number of studies used in the meta-analysis published over the period from 1977 to 2014 are shown in Fig 2, along with the 5-year running mean which summarises the overall trajectory of development in this scientific field. After the first two studies were published in 1977 there were no other publications for 11 years, but then there was fast rate of growth from 1990 to 1999. The number of publications reached top in 1999 after which the publication rate stopped increasing, indicating that research in passive optical hyperspectral remote sensing of plant pigment concentrations is within a mature phase. The overall trajectory of publications shows three periods covering the origins, development and proliferation of research in this field. This trajectory corresponds to the developmental phases of hyperspectral instruments, which started with spectrophotometers and hand-held spectroradiometers enabling leaf and canopy-scale work. With the more recent advent of airborne and spaceborne imaging spectrometers, more landscape scale analyses have become possible.

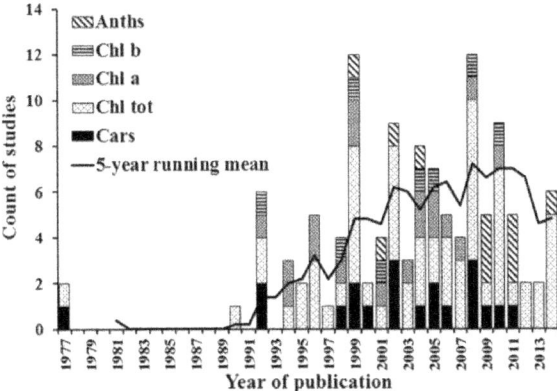

Figure 2: Histogram of numbers of selected studies published over time, showing the total in each year and the number focusing on each pigment type.

The solid line is a 5-year running mean of the total number of studies.

Despite this overall development in the field, there were substantial differences in research on different pigments. The first studies of total chlorophyll and carotenoids were published in 1977, followed by chlorophyll *a* and chlorophyll *b* in 1992 and anthocyanins in 1999. The growth rate of publications on chlorophyll *a*, chlorophyll *b*, carotenoids and anthocyanins has been significantly lower than that for total chlorophyll. These differential rates of growth are perhaps indicative of the increased difficulty in quantifying the concentrations of individual photosynthetic and protective pigments remotely.

The Relationships between Pigment Concentrations and Remotely Sensed Variables

The mean effect size for different pigments at the scales of leaf, canopy and landscape were calculated (Fig 3). At the leaf scale, the mean effect sizes were fairly consistent between different pigment types, varying from 0.87 to 0.93, while the difference in mean effect sizes between pigment types was statistically significant at the canopy and landscape scales. The mean effect size presented the highest value 0.93 (95% confidence interval, 0.92–0.95) for anthocyanins quantification at the leaf scale, far higher than the result of 0.35 (95% confidence interval, 0.18–0.51) at the landscape scale. The mean effect size for total chlorophyll quantification was 0.88 (95% confidence interval, 0.87–0.89) at the leaf scale, 0.73 (95% confidence interval, 0.69–0.77) at the canopy scale and 0.79 (95% confidence interval, 0.76–0.82) at the landscape scale. The mean effect size for carotenoids was the lowest of the various pigments at 0.87 (95% confidence interval, 0.84–0.90) at the leaf scale, still higher than the result 0.80 (95% confidence interval, 0.71–0.90) at the canopy scale and 0.85 (95% confidence interval, 0.76–0.94) at the landscape scale. The results show that these mean effect sizes varied across pigment types and scales. In general, the relationships are stronger at the leaf scale than those at the canopy and landscape scales.

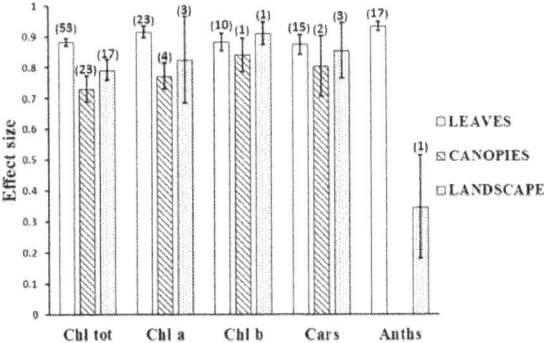

Figure 3: The mean effect size for pigment types at the scales of leaf, canopy and landscape.

(The numbers of reported relationships found in the literature are shown in brackets, error bars represent 95% confidence intervals).

Fig 3 shows that the highest number of relationships published was for pigment quantifications at the leaf scale. Pigment quantification at the canopy scale was less frequently reported in the literature and only a few studies were conducted at the landscape scale. This can be attributed to the limited availability and high costs of suitable airborne and spaceborne hyperspectral instruments [20]. For each scale, the highest number of relationships published was for total chlorophyll quantification, followed by chlorophyll *a*, carotenoids, chlorophyll *b* and anthocyanins. These findings are consistent with previous studies [10,11].

Wavelength Selection for Pigment Quantification Using Remotely Sensed Data

Optimal Wavelengths for Chlorophyll Quantification

There is a large quantity of studies on the relationships between chlorophyll concentration and remotely sensed data. The distributions of wavelengths used at the three scales are shown inFig 4. It should be noted that all of the wavelengths for pigment quantification were concentrated in the 350–950 nm region, except for total chlorophyll quantification at the canopy scale, which spread over 400–2400 nm. For comparison, wavelengths in the histograms and quantile plots were limited within the 350–950 nm region.

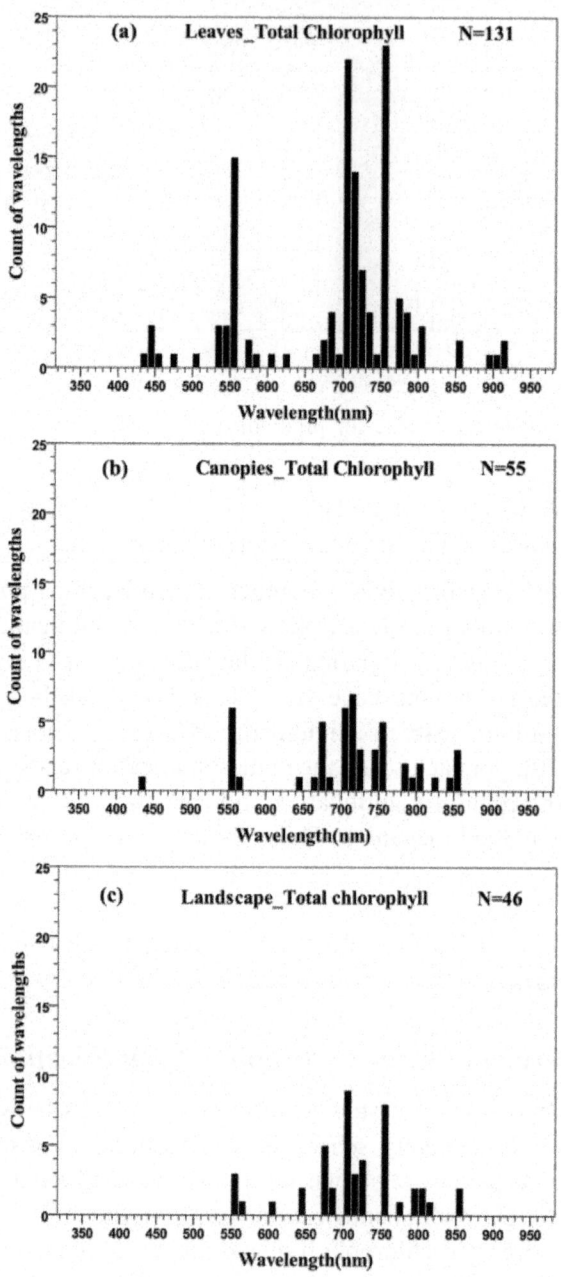

Figure 4: Histogram of wavelengths for total chlorophyll quantification using remotely sensed data at leaf (a), canopy (b) and landscape (c) scales using an interval width of 10 nm.

In general, the distribution of wavelengths displayed a double-peak feature, concentrated in the green (550–560 nm) and red edge (680–750 nm) regions rather than the main absorption wavelengths of chlorophyll (blue or red) (Fig 5). At the canopy scale, five wavelengths in the NIR to SWIR regions (1000–2400 nm) were also used for total chlorophyll quantification (not shown in Fig 4B). This is due to the major influence of canopy structure in canopy reflectance and because leaf chlorophyll concentration was relatively stable in the particular studies [76,77].

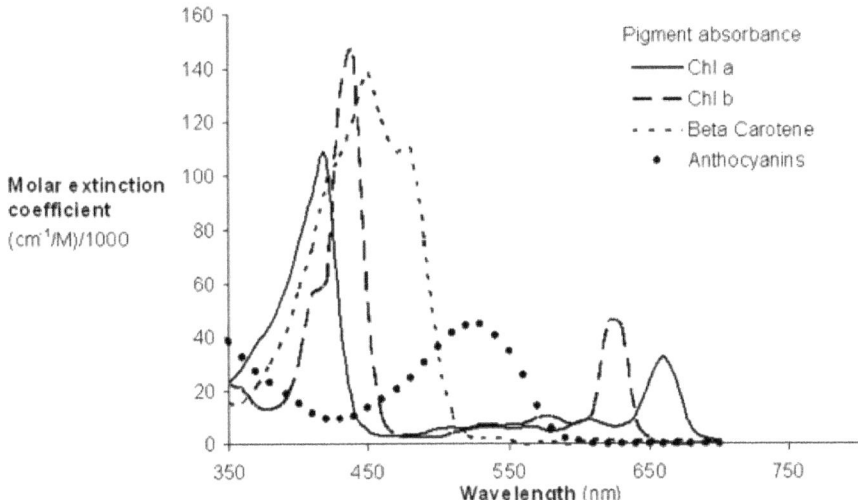

Figure 5: Absorption spectra of the major plant pigments (reproduced from Blackburn, 2007).

The distribution of wavelengths proposed for chlorophyll *a* quantification at the leaf scale was similar to that of total chlorophyll, concentrated in the green and red edge ranges (Fig 6A). At the canopy and landscape scales, the number of wavelengths is limited and is difficult to identify the central tendency of wavelength distribution (Fig 6B and Fig 6C).

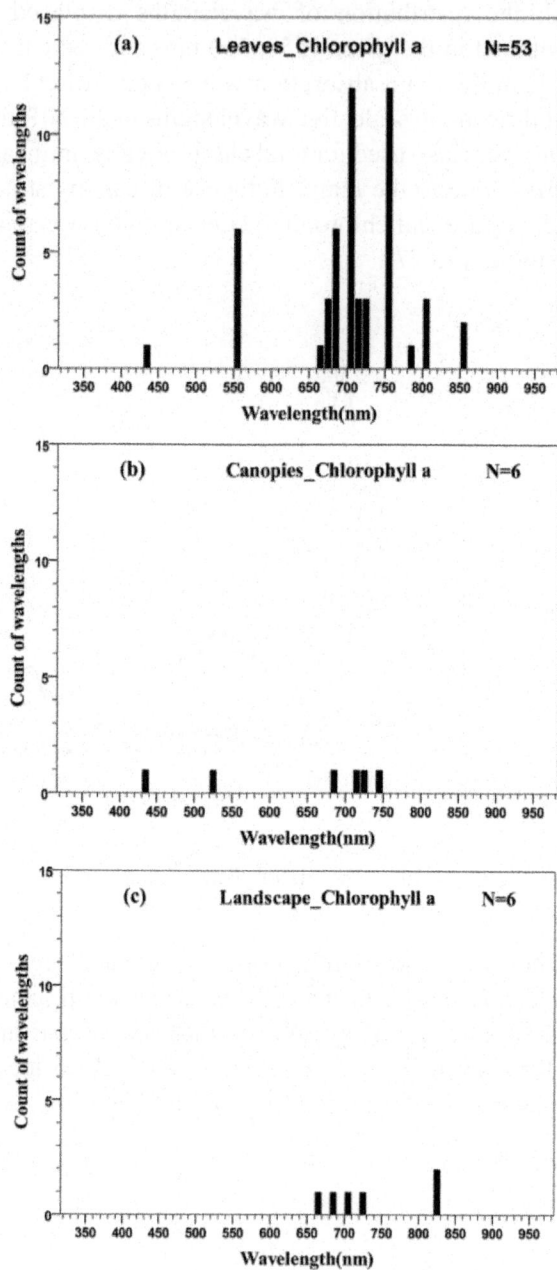

Figure 6: Histogram of wavelengths for chlorophyll *a* quantification using remotely sensed data at leaf (a), canopy (b) and landscape (c) scales by an interval width of 10 nm.

The distribution of wavelengths used for chlorophyll *b* quantification at the leaf scale were concentrated in the main absorption wavelength of chlorophyll *b* (red, 630–660 nm), the red edge (670–710 nm) and the NIR (800–810 nm) regions (Fig 7A). Only two wavelengths were selected at the landscape scale and could not be used for statistical inference (Fig 7B). The distributions of wavelengths used for quantification of different pigments at different scales can be compared in the quantile plots (Fig 8). There were similar wavelength distributions for total chlorophyll quantification at the scales of leaf, canopy and landscape (Fig 8A). For chlorophyll *a*there were similar wavelength distributions at the leaf and canopy scales, but the landscape scale differed (Fig 8B), while a comparison across scales for chlorophyll *b* was difficult due to a lack of data at scales other than the leaf (Fig 8C).

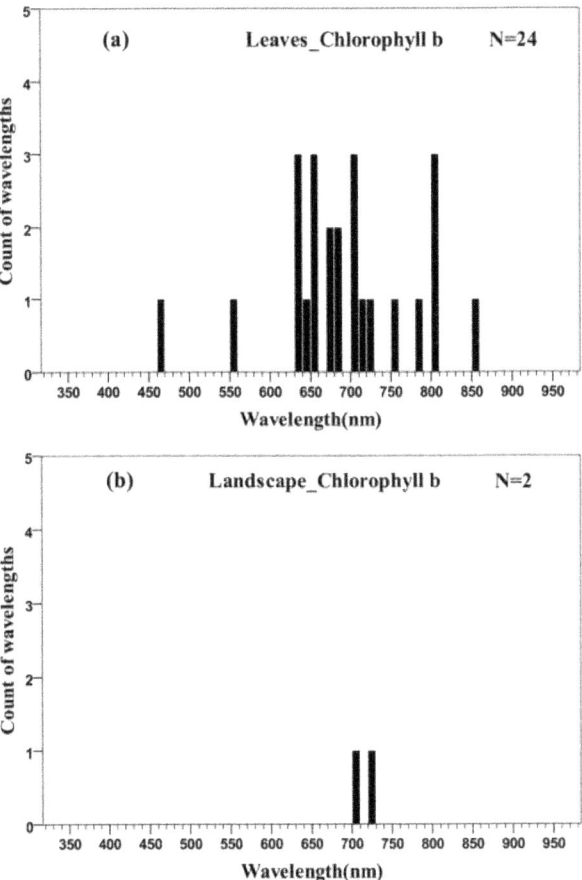

Figure 7: Histogram of wavelengths for chlorophyll *b* quantification using remotely sensed data at leaf (a) and landscape (b) scales using an interval width of 10 nm.

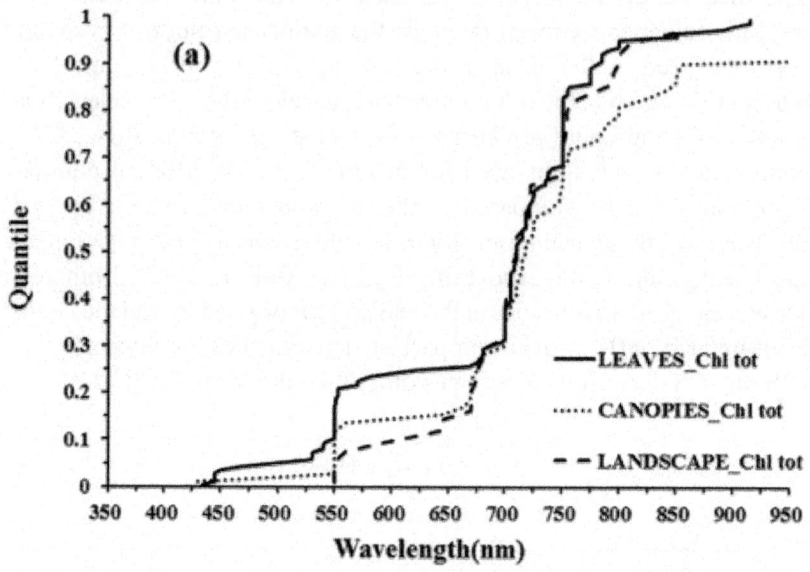

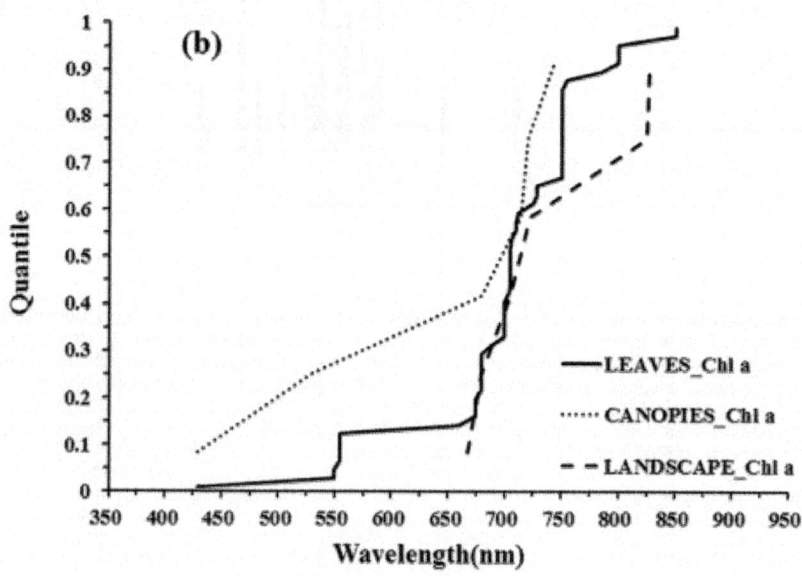

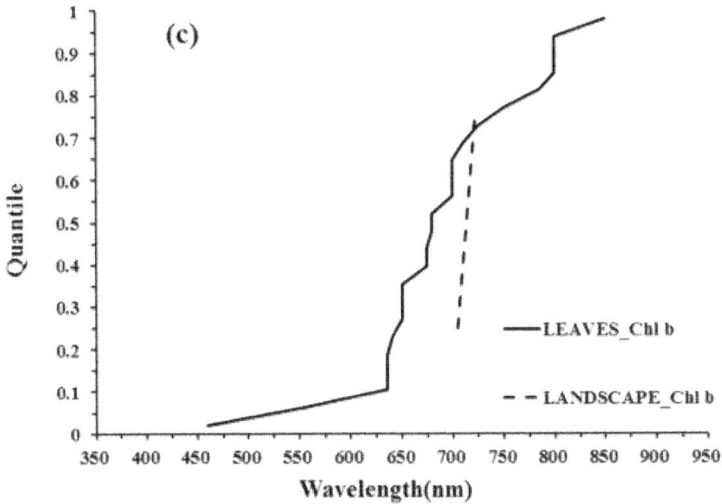

Figure 8: Quantile plots of the wavelengths used for the quantification of Chl tot (a), Chl a (b), and Chl b (c) at different scales.

At the leaf scale, the wavelength distributions for total chlorophyll and chlorophyll *a*quantification were relatively similar while there were notable differences for chlorophyll *b* (Fig 9A). In the region 425–625 nm, the wavelengths used for chlorophyll *a* quantification were concentrated in the region of the green peak in leaf reflectance (550nm), but the central tendency of wavelength distribution for chlorophyll *b* quantification was not obvious. In the red region, the wavelength distribution for chlorophyll *a* quantification was shifted to longer wavelengths than that of chlorophyll *b* (Fig 9A). The significant overlap in the absorption features of chlorophyll *a* and chlorophyll *b* (Fig 5) and the low concentrations of chlorophyll *b*with respect to chlorophyll *a* in most leaves can present difficulties in defining optimal wavelengths for chlorophyll *b* quantification. The absorption spectra of chlorophyll *a* and chlorophyll *b* both display a double-peak feature; the absorption maxima of chlorophyll *a* are at 430 and 662 nm, and chlorophyll *b* has peaks located at 453 and 642 nm (Fig 5). In the presence of carotenoids, it is difficult to separately assess chlorophyll *a* and chlorophyll *b* from reflectance data in the blue region. However, in the red region, the wavelength position of maximum absorption by chlorophyll *a* is longer than that of chlorophyll *b*, which can be exploited for chlorophyll *a* and chlorophyll *b* discrimination (as seen in Fig 9A). The capacity to use this approach to discriminate chlorophyll *a* and chlorophyll *b* is difficult to assess at the canopy and landscape scales due to the small number of studies on chlrophyll b (Fig 9B and 9C).

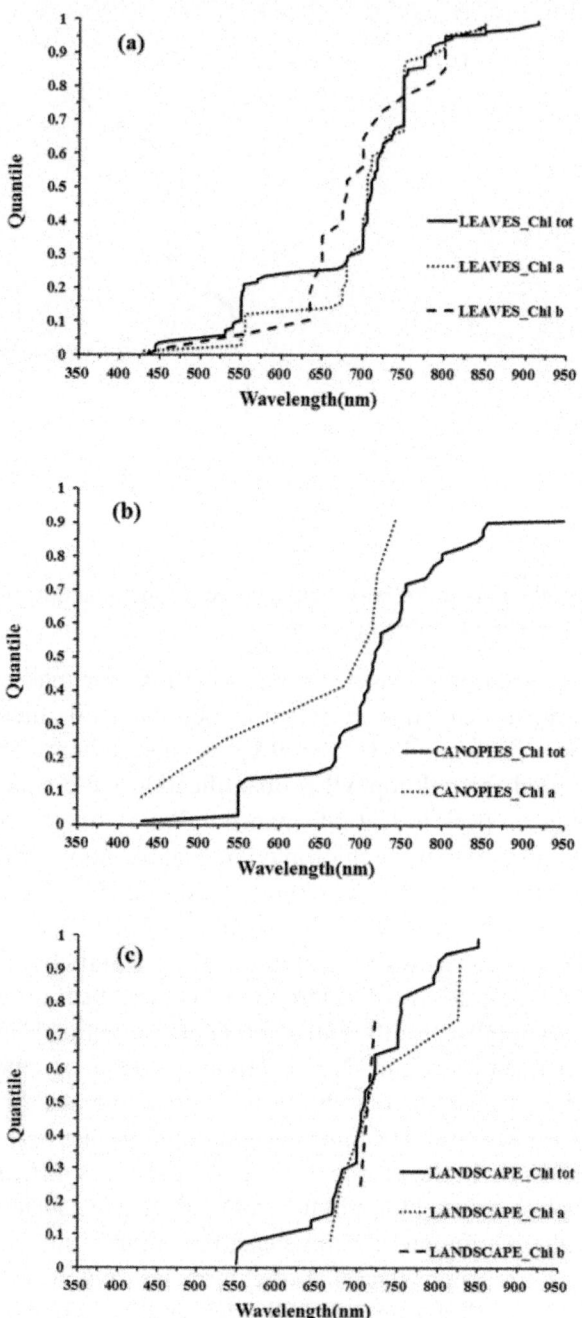

Figure 9: Quantile plot of the wavelengths used at leaf (a), canopy (b) and landscape (c) scales for the quantification of Chl tot, Chl a, and Chl b.

Optimal Wavelengths for Carotenoids Quantification.

At the leaf scale, the central tendency of wavelength distribution was not obvious but was mainly concentrated in the 500–580 nm region (Fig 10A). There were similar wavelength distributions for carotenoids quantification at the leaf and canopy scales (Fig 11) but at the landscape scale, the number of wavelengths was too small for statistical inference (Fig 10C).

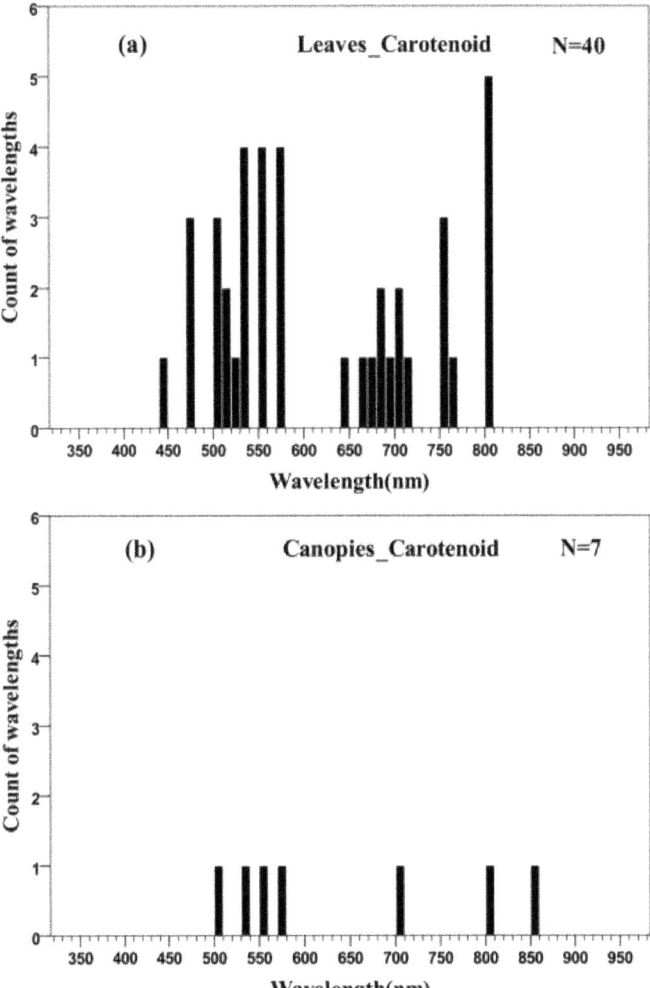

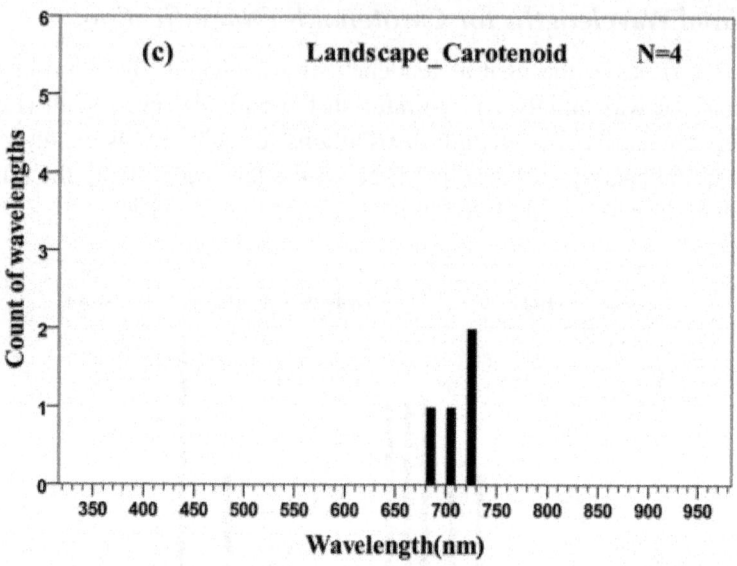

Figure 10: Histogram of wavelengths for carotenoids quantification using remotely sensed data at leaf (a), canopy (b) and landscape (c) scales using an interval width to 10 nm.

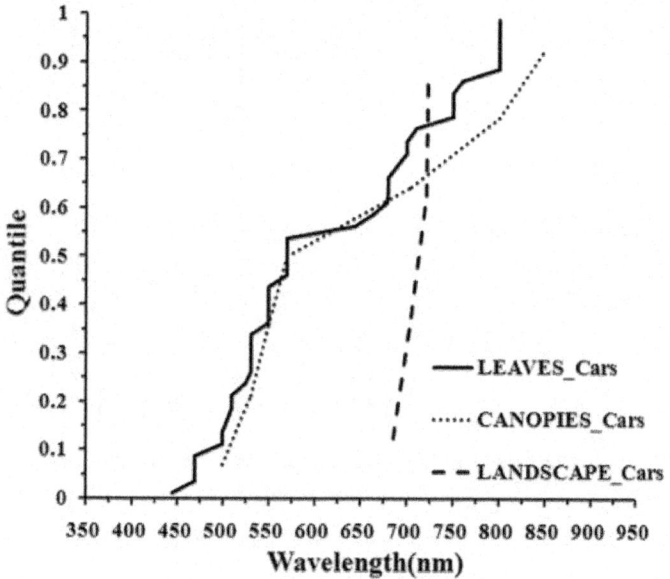

Figure 11: Quantile plot of the optimal wavelength for the quantification of Cars at different scales.

Optimal Wavelengths for Anthocyanins Quantification.

Quantification of anthocyanins from reflectance data has been given less attention by the passive optical hyperspectral remote sensing community than chlorophyll and carotenoids. Most studies have concentrated on the quantification of anthocyanins at the leaf scale, with some work at the landscape scale but nothing at canopy level. At the leaf scale, the distribution of wavelengths used for quantifying anthocyanins was concentrated in the main absorption wavelength of anthocyanins (green, 550–560 nm), the red edge (700–710 nm) and the NIR (780–790 nm) ranges (Fig 12A). Similarly, the two wavelengths used to estimate anthocyanin concentration at the landscape scale were distributed in the green and red edge regions, respectively (Fig 12B and Fig 13).

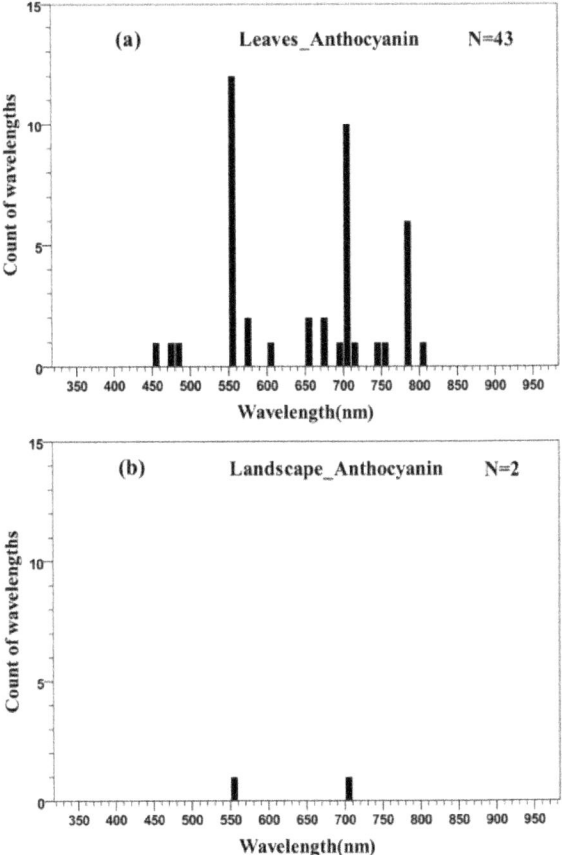

Figure 12: Histogram of wavelengths for anthocyanins quantification using remotely sensed data at leaf (a) and landscape (b) scales using an interval width to 10 nm.

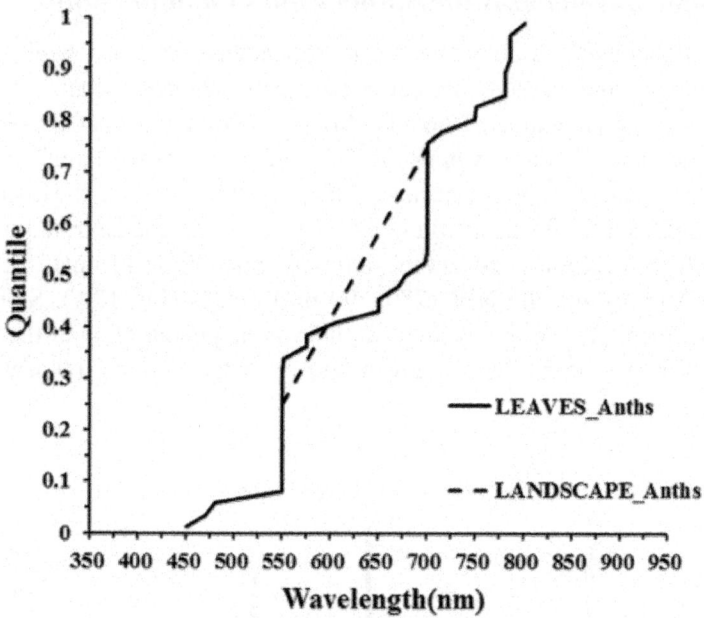

Figure 13: Quantile plot of the optimal wavelengths for the quantification of Anths at different scales.

DISCUSSION

This meta-analysis of 85 studies has demonstrated that remotely sensed variables are good estimators of plant pigment concentration. Most of the studies were conducted at the leaf scale, while pigment quantification at the canopy and landscape scales was less frequently reported. For each scale, most of the studies were conducted for total chlorophyll quantification, followed by chlorophyll *a*, carotenoids, chlorophyll *b* and anthocyanins. These findings are consistent with previous studies [10,11].

The strength of these relationships varied across pigments types and scales. In general, the relationships are stronger at the leaf scale than those at the canopy and landscape scales. At the leaf scale, the mean effect sizes were fairly consistent across different pigment types and were all greater than 0.87, while the difference in mean effect sizes between pigment types was statistically significant at the canopy and landscape scales. This result has been widely assumed, yet a quantitative evaluation has been lacking. At the leaf scale, the methodological basis for pigment quantification has been fully explored, which provides an important basis for developing estimation models at the canopy and landscape scales. The primary goal of most leaf scale passive optical hyperspectral remote sensing studies has been to develop analytical approaches for pigment quantification that can be applied to data from airborne and spaceborne sensors [11].

At the canopy and landscape scales, the experimental results are influenced by a number of factors, which obscures the relationships between spectral reflectance and concentrations of individual pigments. The reflectance spectrum of a whole canopy is subject to canopy biophysical attributes (e.g., orientation of leaves (leaf angle distribution; *LAD*), variations in number of leaf layers (*LAI*) and foliage clumping), presence of non-leaf elements (e.g., soil reflectance and the proportions of shadowed and sunlit background), anisotropic scattering of photons to interact with multiple surfaces such as leaves, woody material and soils, viewing geometry (e.g., sun and view zenith and azimuth angles) and illumination conditions (e.g., the ratio between direct and diffuse sunlight and atmospheric condition). It is the interaction of these factors, including their potential covariance or unique behavior that drive variation in canopy and landscape reflectance characteristics in three-dimensional space [10,22].

It should be noted that part of the variability in effect sizes at the canopy scale may be entirely artifactual. These artifacts are common in experimental studies: studies vary in terms of the quality of measurement; researchers make computational errors; people make typographical errors in copying numbers from handwritten tables to computer; and sampling errors. With the advent of airborne and spaceborne imaging spectrometers, there have been opportunities to measure plant pigment concentrations at the landscape scale. The reflectance spectrum from airborne and spaceborne sensors is subject to even more controlling factors, notably, soil/litter surface reflectance, and vegetation structure. The range of controlling factors should be taken into account in subsequent analyses.

Table 2 shows that the total sample size at the leaf scale is much more than that of canopy and landscape scales. The law of large numbers correctly states that large samples are reasonable representations of the population and parameter estimation is close to the real values when the sample size is large enough. Many researchers seem to believe that the same law applies to small samples and severely underestimate the amount of variability in findings that is caused by sampling errors. As a result, they erroneously expect statistics based on small samples to be close to the real values [13]. At the canopy and landscape scales, the number of studies and total sample size is limited, which influences the robustness and accuracy of effect sizes.

Despite the significant difference in effect sizes between different scales, it was found that the wavelength distribution for total chlorophyll quantification at the scales of leaf, canopy and landscape was similar, being concentrated in the green (550–560 nm) and red edge (680–750 nm) regions rather than the main absorption wavelength of chlorophyll (blue or red). The consistency in optimal wavelengths across scales can be attributed to several factors: (1) despite the many factors influencing reflectance at the canopy and landscape scales, it is the selective absorbance properties of pigments that determines the selection of wavelengths for pigment quantification, and (2) several estimation models derived at the leaf scale were directly applied to canopy and landscape scales. This suggests that the leaf-level study has provided an important basis for developing estimation models at the canopy and landscape scales.

At the leaf scale, the distribution of wavelengths used for chlorophyll a quantification was similar to that of total chlorophyll; the distribution of wavelengths for chlorophyll b quantification was concentrated in the main absorption wavelength of chlorophyll b (red, 630–660 nm), the red edge (670–710 nm) and the NIR (800–810 nm) regions; the central tendency of wavelength distribution for carotenoids quantification was not obvious, but was mainly concentrated in the 500–580 nm region; for the estimation of anthocyanins, the distribution of wavelengths was concentrated in the main absorption wavelength of anthocyanins (green, 550–560 nm), the red edge (700–710 nm) and the NIR (780–790 nm) ranges. In the present meta-analysis, the lack of studies reporting the quantification of carotenoids and anthocyanins at the canopy and landscape scales has hindered cross-scale comparisons (Fig 10; Fig 12). Consequently, it is not entirely clear if the optimal wavelengths for carotenoids and anthocyanins quantification at the leaf scale are necessarily the optimal wavelengths at the canopy and landscape scales, where multiple scattering and other confounding effects may alter the spectral response of individual pigments, much in the way that pigment absorption peaks can vary depending upon their chemical and scattering medium. Therefore, more work may be needed to determine the optimal algorithms for airborne or spaceborne platforms.

It should be noted that the lack of statistical information in the studies (e.g., sample size and coefficient of determination) has hindered a more comprehensive cross-study comparison in the present research. When selecting the final set of studies, 50 studies were excluded due to the lack of statistical information. Insufficient statistical information can not only limit the research population covered by meta-analysis but also render the findings of the original study somewhat suspect. Thus, it is suggested that when conducting primary research, such information should include, but not be limited to, the sample size, the pertinent test statistic (e.g., r, t, or F), the unit of pigment concentration/ content, the range of pigment concentrations/content, and estimation precision for pigment quantification (e.g. root mean squared error, RMSE).

This study has established the possibility of integrating the results of studies on the passive optical hyperspectral remote sensing of plant pigment concentrations across a range of vegetation types and scales using a meta-analysis approach. Despite the robust models for pigment prediction at the leaf scale, the continuing challenge is to properly account for the multiple factors introduced by scene components such as sunlit and shaded parts of tree crowns and gaps influencing the retrieved signal at the canopy and landscape scales. Recent work have illustrated that, in addition to other influencing factors such as illumination geometry and atmospheric conditions, canopy architecture had an important control on the applicability of models for pigment prediction. Scanning LIDAR systems have only recently become widely available which enable the estimation of the range between the sensor and a target by recording the time during which the emitted laser pulse is reflected off an object and returns to the sensor [21]. LIDAR systems have the ability to directly measure spatial variations in canopy height and other aspects of the vertical structure of canopies. Given the high degree of structural complexity at the canopy and landscape scales, it would appear that the integration of vertical canopy structural information provided by active LIDAR remote sensing with hyperspectral reflectance may has both a structural and physiological interpretation and improve the estimation of pigment concentrations over passive optical hyperspectral imagery alone [102].

ACKNOWLEDGMENTS

We acknowledge the contribution of Bao She, Weijiao Huang, Dilong Gan, Sujuan Wang and Zhewen Zhao for the literature database searches and associated support. The authors thank anonymous reviewers who provided very valuable comments also.

AUTHOR CONTRIBUTIONS

Conceived and designed the experiments: JH. Performed the experiments: Chen Wei YZ XW Chuanwen Wei JW. Analyzed the data: Chen Wei. Contributed reagents/materials/analysis tools: Chen Wei. Wrote the paper: Chen Wei GAB JH.

REFERENCES

1.   Curran PJ, Dungan JL, Gholz HL. Exploring the relationship between reflectance red edge and chlorophyll content in slash pine. Tree Physiol 1990; 7: 33–48. pmid:14972904 doi: 10.1093/treephys/7.1-2-3-4.33

2. Filella I, Serrano L, Serra J, Penuelas J. Evaluating wheat nitrogen status with canopy reflectance indices and discriminant analysis. Crop Sci 1995; 35: 1400–1405. doi: 10.2135/cropsci1995.0011183x003500050023x

3. Hendry GAF, Houghton JD, Brown SB. The degradation of chlorophyll: a biological enigma. New Phytol 1987; 107: 255–302. doi: 10.1111/j.1469-8137.1987.tb00181.x

4. Merzlyak MN, Gitelson A. Why and what for the leaves are yellow in autumn? On the interpretation of optical spectra of senescing leaves (Acerplatanoides L). J. Plant Physiol 1995; 145: 315–320. doi: 10.1016/s0176-1617(11)81896-1

5. Demmig—Adams B, Adams WW. The role of xanthophyll cycle carotenoids in the protection of photosynthesis. Trends Plant Sci 1996; 1: 21–26. doi: 10.1016/s1360-1385(96)80019-7

6. Peñuelas J, Filella I. Visible and near-infrared reflectance techniques for diagnosing plant physiological status. Trends Plant Sci 1998; 3: 151–156. doi: 10.1016/s1360-1385(98)01213-8

7. Merzlyak MN, Gitelson AA, Chivkunova OB, Rakitin VY. Non-destructive optical detection of pigment changes during leaf senescence and fruit ripening. Physiol Plantarum 1999; 106: 135–141. doi: 10.1034/j.1399-3054.1999.106119.x

8. Chalker-Scott L. Environmental significance of anthocyanins in plant stress responses. Photochem Photobiol 1999; 70: 1–9. doi: 10.1111/j.1751-1097.1999.tb01944.x

9. Carter GA, Knapp AK. Leaf optical properties in higher plants: linking spectral characteristics to stress and chlorophyll concentration. Am J Bot 2001; 88: 677–684. pmid:11302854 doi: 10.2307/2657068

10. Blackburn GA. Hyperspectral remote sensing of plant pigments. J. Exp Bot 2007; 58: 855–867. pmid:16990372 doi: 10.1093/jxb/erl123

11. Ustin SL, Gitelson AA, Jacquemoud S, Schaepman M, Asner GP, Gamon JA, et al. Retrieval of foliar information about plant pigment systems from high resolution spectroscopy. Remote Sens Environ 2009; 113: S67–S77. doi: 10.1016/j.rse.2008.10.019

12. Hunter JE, Schmidt FL. Methods of meta-analysis: correcting error and bias in research findings. Los Angeles, USA: SAGE Publications; 2004. p. 33–34.

13. Borenstein M, Hedges LV, Higgins JPT, Rothstein HR. Introduction to meta-analysis. West Sussex, United Kingdom: John Wiley & Sons Ltd; 2009. p. 12–13.

14. Curtis PS, Queenborough SA. Raising the standards for ecological meta–analyses. New Phytol 2012; 195: 279–281. doi: 10.1111/j.1469-8137.2012.04207.x. pmid:22702404

15. Glass GV. Primary secondary and meta-analysis of research. Educ Res 1976; 5: 3–8. doi: 10.3102/0013189x005010003

16. Verschuyl J, Riffell S, Miller D, Wigley TB. Biodiversity response to intensive biomass production from forest thinning in North American forests-A meta-analysis. Forest Ecol Manag 2011; 261: 221–232. doi: 10.1016/j.foreco.2010.10.010

17. Robinson EA Ryan GD Newman JA A meta-analytical review of the effects of elevated CO_2 on plant-arthropod interactions highlights the importance of interacting environmental and biological variables. New Phytol 2012; 194: 321–336. doi: 10.1111/j.1469-8137.2012.04074.x. pmid:22380757

18. Liao CZ Peng RH Luo YQ Zhou X H Wu X W Fang C M Chen J K Li B Altered ecosystem carbon and nitrogen cycles by plant invasion: a meta-analysis. New Phytol 2008; 177: 706–714. pmid:18042198 doi: 10.1111/j.1469-8137.2007.02290.x

19. Poorter H, Niinemets Ü, Poorter L, Wright IJ, Villar R. Causes and consequences of variation in leaf mass per area (LMA): a meta-analysis. New Phytol 2009; 182: 565–588. pmid:19434804 doi: 10.1111/j.1469-8137.2009.02830.x

20. Garbulsky MF, Peñuelas J, Gamon J, Inoue Y, Filella I. The photochemical reflectance index (PRI) and the remote sensing of leaf canopy and ecosystem radiation use efficiencies: a review and meta-analysis. Remote Sens Environ 2011; 115: 281–297. doi: 10.1016/j.rse.2010.08.023

21. Zolkos SG, Goetz SJ, Dubayah R. A meta-analysis of terrestrial aboveground biomass estimation using lidar remote sensing. Remote Sens Environ 2013; 128: 289–298. doi: 10.1016/j.rse.2012.10.017

22. Asner GP. Biophysical and biochemical sources of variability in canopy reflectance. Remote Sens Environ 1998; 64: 234–253. doi: 10.1016/s0034-4257(98)00014-5

23. Curran PJ, Dungan JL, Macler BA, Plummer SE, Peterson DL. Reflectance spectroscopy of fresh whole leaves for the estimation of chemical concentration. Remote Sens Environ 1992; 39: 153–166. doi: 10.1016/0034-4257(92)90133-5

24. Curran PJ, Windham WR, Gholz HL. Exploring the relationship between reflectance red edge and chlorophyll concentration in slash pine

leaves. Tree Physiol 1995; 15: 203–206. pmid:14965977 doi: 10.1093/treephys/15.3.203

25. Yoder BJ, Pettigrew-Crosby RE. Predicting nitrogen and chlorophyll content and concentrations from reflectance spectra (400–2500 nm) at leaf and canopy scales. Remote Sens Environ 1995; 53: 199–211. doi: 10.1016/0034-4257(95)00135-n

26. Gitelson AA, Merzlyak MN, Grits Y. Novel algorithms for remote sensing of chlorophyll content in higher plant leaves. In International Geoscience and Remote Sensing Symposium (IGARSS); Lincoln, NE, USA; May 1996. p. 2355–2357.

27. Gitelson AA, Kaufman YJ, Merzlyak MN. Use of a green channel in remote sensing of global vegetation from EOS-MODIS. Remote Sens Environ 1996; 58: 289–298. doi: 10.1016/s0034-4257(96)00072-7

28. Gitelson AA, Merzlyak MN. Remote estimation of chlorophyll content in higher plant leaves. Int J Remote Sens 1997; 18: 2691–2697. doi: 10.1080/014311697217558

29. Gitelson AA, Merzlyak MN. Remote sensing of chlorophyll concentration in higher plant leaves. Adv Space Res 1998; 22: 689–692. doi: 10.1016/s0273-1177(97)01133-2

30. Gitelson AA, Buschmann C, Lichtenthaler HK. The chlorophyll fluorescence ratio F735/F700 as an accurate measure of the chlorophyll content in plants. Remote Sens Environ 1999; 69: 296–302. doi: 10.1016/s0034-4257(99)00023-1

31. Adams ML, Philpot WD, Norvell WA. Yellowness index: an application of spectral second derivatives to estimate chlorosis of leaves in stressed vegetation. Int J Remote Sens 1999; 20: 3663–3675. doi: 10.1080/014311699211264

32. Sims DA, Gamon JA. Relationships between leaf pigment content and spectral reflectance across a wide range of species leaf structures and developmental stages. Remote Sens Environ 2002; 81: 337–354. doi: 10.1016/s0034-4257(02)00010-x

33. Richardson AD, Duigan SP, Berlyn GP. An evaluation of noninvasive methods to estimate foliar chlorophyll content. New Phytol 2002; 153: 185–194. doi: 10.1046/j.0028-646x.2001.00289.x

34. Gitelson AA, Gritz Y, Merzlyak MN. Relationships between leaf chlorophyll content and spectral reflectance and algorithms for non–destructive chlorophyll assessment in higher plant leaves. J Plant Physiol 2003; 160: 271–282. pmid:12749084 doi: 10.1078/0176-1617-00887

35. Zhao DL, Reddy KR, Kakani VG, Read JJ, Koti S. Selection of optimum reflectance ratios for estimating leaf nitrogen and chlorophyll concentrations of field-grown cotton. Agron J 2005; 97: 89–98. doi: 10.2134/agronj2005.0089

36. Kochubey SM, Kazantsev TA. Changes in the first derivatives of leaf reflectance spectra of various plants induced by variations of chlorophyll content. J. Plant Physiol 2007; 164: 1648–1655. pmid:17292510 doi: 10.1016/j.jplph.2006.11.007

37. Wang Q, Li PH. Hyperspectral indices for estimating leaf biochemical properties in temperate deciduous forests: comparison of simulated and measured reflectance data sets. Ecol Indic 2012; 14: 56–65. doi: 10.1016/j.ecolind.2011.08.021

38. Simic A, Chen JM, Leblanc SG, Dyk A, Croft H, Tian Han. Testing the top-down model inversion method of estimating leaf reflectance used to retrieve vegetation biochemical content within empirical approaches. IEEE J Sel Top Appl Earth Observ Remote Sens 2014; 7: 92–104. doi: 10.1109/jstars.2013.2271583

39. Gitelson A, Merzlyak MN. Quantitative estimation of chlorophyll-a using reflectance spectra: experiments with autumn chestnut and maple leaves. J. Photoch Photobio B 1994; 22: 247–252. doi: 10.1016/1011-1344(93)06963-4

40. Gitelson AA, Merzlyak MN. Spectral reflectance changes associated with autumn senescence of Aesculus hippocastanum L and Acer platanoides L Leaves Spectral features and relation to chlorophyll estimation. J. Plant Physiol 1994; 143: 286–292. doi: 10.1016/s0176-1617(11)81633-0

41. Gitelson AA, Merzlyak MN. Signature analysis of leaf reflectance spectra: algorithm development for remote sensing of chlorophyll. J. Plant Physiol 1996; 148: 494–500. doi: 10.1016/s0176-1617(96)80284-7

42. Thomas JR, Gausman HW. Leaf reflectance vs Leaf chlorophyll and carotenoid concentrations for eight crops. Agron J 1977; 69: 799–802. doi: 10.2134/agronj1977.00021962006900050017x

43. Gamon JA, Peñuelas J, Field CB. A narrow-waveband spectral index that tracks diurnal changes in photosynthetic efficiency. Remote Sens Environ 1992; 41: 35–44. doi: 10.1016/0034-4257(92)90059-s

44. Levizou E, Manetas Y. Photosynthetic pigment contents in twigs of 24 woody species assessed by in vivo reflectance spectroscopy indicate low chlorophyll levels but high carotenoid/chlorophyll ratios. Environ Exp Bot 2007; 59: 293–298. doi: 10.1016/j.envexpbot.2006.03.002

45. Gamon JA, Surfus JS. Assessing leaf pigment content and activity with a reflectometer. New Phytol 1999; 143: 105–117. doi: 10.1046/j.1469-8137.1999.00424.x

46. Merzlyak MN, Solovchenko AE, Gitelson AA. Reflectance spectral features and non–destructive estimation of chlorophyll carotenoid and anthocyanin content in apple fruit. Postharvest Biol Tec 2003; 27: 197–211. doi: 10.1016/s0925-5214(02)00066-2

47. Gitelson AA, Merzlyak MN. Non-destructive assessment of chlorophyll carotenoid and anthocyanin content in higher plant leaves: principles and algorithms. In Remote Sensing for Agriculture and the Environment. Stamatiadis S, LynchJ JM, Schepers JS, Eds. Ella, Greece: OECD; 2004. p. 78–94.

48. Garriga M, Retamales JB, Romero-Bravo S, Caligari PD, Lobos GA. Chlorophyll anthocyanin and gas exchange changes assessed by spectroradiometry in Fragaria chiloensis under salt stress. J. Integr Plant Biol 2014; 56: 505–15. doi: 10.1111/jipb.12193. pmid:24618024

49. Chappelle EW, Kim MS, McMurtrey JE III. Ratio analysis of reflectance spectra (RARS): an algorithm for the remote estimation of the concentrations of chlorophyll a chlorophyll b and carotenoids in soybean leaves. Remote Sens Environ 1992; 39: 239–247. doi: 10.1016/0034-4257(92)90089-3

50. Blackburn GA. Spectral indices for estimating photosynthetic pigment concentrations: a test using senescent tree leaves. Int J Remote Sens 1998; 19 657–675. doi: 10.1080/014311698215919

51. Chen L, Huang JF, Wang FM. Retrieval of pigment contents in rice leaves and panicles using hyperspectral data by artificial neuron network models. In International Geoscience and Remote Sensing Symposium (IGARSS); Seoul, Korea; July 2005. p. 1416–1419.

52. Blackburn GA. Relationships between spectral reflectance and pigment concentrations in stacks of deciduous broadleaves. Remote Sens Environ 1999; 70: 224–237. doi: 10.1016/s0034-4257(99)00048-6

53. Maccioni A, Agati G, Mazzinghi P. New vegetation indices for remote measurement of chlorophylls based on leaf directional reflectance spectra. J. Photoch Photobio B 2001; 61: 52–61. doi: 10.1016/s1011-1344(01)00145-2

54. Imanishi J, Nakayama A, Suzuki Y, Imanishi A, Ueda N, Morimoto Y, et al. Nondestructive determination of leaf chlorophyll content in two flowering cherries using reflectance and absorptance spectra. Landsc Ecol Eng 2010; 6: 219–234. doi: 10.1007/s11355-009-0101-8

55. Lichtenthaler HK, Gitelson A, Lang M. Non-destructive determination of chlorophyll content of leaves of a green and an aurea mutant of tobacco by reflectance measurements. J. Plant Physiol 1996; 148: 483–493. doi: 10.1016/s0176-1617(96)80283-5

56. Datt B. Visible/near infrared reflectance and chlorophyll content in eucalyptus leaves. Int J Remote Sens 1999; 20: 2741–2759. doi: 10.1080/014311699211778

57. Gitelson AA, Zur Y, Chivkunova OB, Merzlyak MN. Assessing carotenoid content in plant leaves with reflectance spectroscopy. Photochem Photobiol 2002; 75: 272–281. pmid:11950093 doi: 10.1562/0031-8655(2002)0750272accipl2.0.co2

58. Filella I, Porcar-Castell A, Munne-Bosch S, Back J, Garbulsky MF, Penuelas J. PRI assessment of long-term changes in carotenoids/ chlorophyll ratio and short–term changes in de-epoxidation state of the xanthophyll cycle. Int J Remote Sens 2009; 30: 4443–4455. doi: 10.1080/01431160802575661

59. Garrity SR, Eitel JUH, Vierling LA. Disentangling the relationships between plant pigments and the photochemical reflectance index reveals a new approach for remote estimation of carotenoid content. Remote Sens Environ 2011; 115: 628–635. doi: 10.1016/j.rse.2010.10.007

60. Gitelson AA, Merzlyak MN, Chivkunova OB. Optical properties and nondestructive estimation of anthocyanin content in plant leaves. Photochem Photobiol 2001; 74: 38–45. pmid:11460535 doi: 10.1562/0031-8655(2001)074<0038:opaneo>2.0.co;2

61. Steele MR, Gitelson AA, Rundquist DC, Merzlyak MN. Nondestructive estimation of anthocyanin content in grapevine leaves. Am J Enol Viticult 2009; 60: 87–92.

62. Gitelson AA, Chivkunova OB, Merzlyak MN. Nondestructive estimation of anthocyanins and chlorophylls in anthocyanic leaves. Am J Bot 2009; 96: 1861–1868. doi: 10.3732/ajb.0800395. pmid:21622307

63. Qin JL, Rundquist D, Gitelson A, Tan Z, Steele M. A non-linear model of nondestructive estimation of anthocyanin content in grapevine leaves with Visible/Red-infrared hyperspectral. In International Conference on Computer and Computing Technologies in Agriculture; Beijing, China; October 2011. p. 47–62.

64. Pappas CS, Takidelli C, Tsantili E, Tarantilis PA, Polissiou MG. Quantitative determination of anthocyanins in three sweet cherry varieties using diffuse reflectance infrared fourier transform spectroscopy. J. Food Compos Anal 2011; 24: 17–21. doi: 10.1016/j.jfca.2010.07.001

65. Vina A, Gitelson AA. Sensitivity to foliar anthocyanin content of vegetation indices using green reflectance. IEEE Geosci Remote Sens Lett 2011; 8: 464–468. doi: 10.1109/lgrs.2010.2086430

66. Ciganda V, Gitelson A, Schepers J. Non-destructive determination of maize leaf and canopy chlorophyll content. J. Plant Physiol 2009; 166: 157–167. doi: 10.1016/j.jplph.2008.03.004. pmid:18541334

67. Schlemmer M, Gitelson A, Schepers J, Ferguson R, Peng Y, Shanahan J, et al. Remote estimation of nitrogen and chlorophyll contents in maize at leaf and canopy levels. Int J Appl Earth Obs 2013; 25: 47–54. doi: 10.1016/j.jag.2013.04.003

68. Croft H, Chen JM, Zhang Y. The applicability of empirical vegetation indices for determining leaf chlorophyll content over different leaf and canopy structures. Ecol Complex 2014; 17: 119–130. doi: 10.1016/j.ecocom.2013.11.005

69. Ju CH, Tian YC, Yao X, Cao WX, Zhu Y, Hannaway D. Estimating leaf chlorophyll content using red edge parameters. Pedosphere 2010; 20: 633–644. doi: 10.1016/s1002-0160(10)60053-7

70. ZarcoTejada PJ. Hyperspectral remote sensing of closed forest canopies: estimation of chlorophyll fluorescence and pigment content. PhD thesis, York University, Toronto, Canada 2000.

71. Filella I, Penuelas J. The red edge position and shape as indicators of plant chlorophyll content biomass and hydric status. Int J Remote Sens 1994; 15: 1459–1470. doi: 10.1080/01431169408954177

72. Gitelson AA, Vina A, Ciganda V, Rundquist DC, Arkebauer TJ. Remote estimation of canopy chlorophyll content in crops. Geophys Res Lett 2005; 32: 1–4. doi: 10.1029/2005gl022688

73. Yang XH, Huang JF, Wang FM, Wang XZ, Yi QX, Wang Y. Science letters: a modified chlorophyll absorption continuum index for chlorophyll estimation. J. Zhejiang Univ 2006; 7: 2002–2006. doi: 10.1631/jzus.2006.a2002

74. Zhao DH, Huang LM, Li JL, Qi JG. A comparative analysis of broadband and narrowband derived vegetation indices in predicting LAI and CCD of a cotton canopy. Isprs J Photogramm 2007; 62: 25–33. doi: 10.1016/j.isprsjprs.2007.01.003

75. Wu CY, Niu Z, Tang Q, Huang WJ. Estimating chlorophyll content from hyperspectral vegetation indices: modeling and validation. Agr Forest Meteorol 2008; 148: 1230–1241. doi: 10.1016/j.agrformet.2008.03.005

76. Darvishzadeh R, Skidmore A, Schlerf M, Atzberger C, Corsi F, Cho M. LAI and chlorophyll estimation for a heterogeneous grassland using

hyperspectral measurements. ISPRS J Photogramm 2008; 63: 409–426. doi: 10.1016/j.isprsjprs.2008.01.001

77. Darvishzadeh R, Skidmore A, Schlerf M, Atzberger C. Inversion of a radiative transfer model for estimating vegetation LAI and chlorophyll in a heterogeneous grassland. Remote Sens Environ 2008; 112: 2592–2604. doi: 10.1016/j.rse.2007.12.003

78. Haboudane D, Tremblay N, Miller JR, Vigneault P. Remote estimation of crop chlorophyll content using spectral indices derived from hyperspectral data. IEEE T Geosci Remote 2008; 46: 423–437. doi: 10.1109/tgrs.2007.904836

79. Liu ML, Liu XN, Li M, Fang MH, Chi WX. Neural-network model for estimating leaf chlorophyll concentration in rice under stress from heavy metals using four spectral indices. Biosystems Eng 2010; 106: 223–233. doi: 10.1016/j.biosystemseng.2009.12.008

80. Xu X, Gu X, Song X, Li C, Huang W. Assessing rice chlorophyll content with vegetation indices from hyperspectral data. In International Conference on Computer and Computing Technologies in Agriculture; Beijing, China; October 2011. p. 296–303.

81. Clevers JGPW, Kooistra L. Using Hyperspectral Remote Sensing Data for Retrieving Canopy Chlorophyll and Nitrogen Content. IEEE J Sel Top Appl Earth Observ Remote Sens 2012; 5: 574–583. doi: 10.1109/jstars.2011.2176468

82. Clevers JGPW, Gitelson AA. Remote estimation of crop and grass chlorophyll and nitrogen content using red-edge bands on Sentinel-2 and -3. Int J Appl Earth Obs 2013; 23: 344–351. doi: 10.1016/j.jag.2012.10.008

83. Vincini M, Amaducci S, Frazzi E. Empirical Estimation of Leaf Chlorophyll Density in Winter Wheat Canopies Using Sentinel – 2 Spectral Resolution. IEEE T Geosci Remote 2014; 52: 3220–3235. doi: 10.1109/tgrs.2013.2271813

84. Cheng Q, Huang JF, Wang XZ, Wang RC. In situ hyperspectral data analysis for pigment content estimation of rice leaves. J Zhejiang Univ 2003; 4: 727–733. doi: 10.1631/jzus.2003.0727

85. Li J, Jiang JB, Chen YH, Wang YY, Su W, Huang WJ. Using hyperspectral indices to estimate foliar chlorophyll a concentrations of winter wheat under yellow rust stress. New Zeal J Agr Res 2007; 50: 1031–1036. doi: 10.1080/00288230709510382

86. Zhao X, Liu SH, Wang JD, Tian ZK. A method for estimating chlorophyll content of wheat from reflectance spectra. In International Geoscience

and Remote Sensing Symposium (IGARSS); Anchorage, AK, USA; September 2004. p. 4504–4507.

87. Bannari A, Khurshid KS, Staenz K, Schwarz J. Wheat crop chlorophyll content estimation from ground–based reflectance using chlorophyll indices. In International Geoscience and Remote Sensing Symposium (IGARSS); Denver, CO, USA; July 2006. p. 112–115.

88. Yang F, Li JL, Gan XY, Qian YR, Wu XL, Yang Q. Assessing nutritional status ofFestuca arundinacea by monitoring photosynthetic pigments from hyperspectral data. Comput Electron Agr 2010; 70: 52–59. doi: 10.1016/j.compag.2009.08.010

89. Peguero-Pina JJ, Morales F, Flexas J, Gil-Pelegrin E, Moya I. Photochemistry remotely sensed physiological reflectance index and de-epoxidation state of the xanthophyll cycle in Quercus coccifera under intense drought. Oecologia 2008 156: 1–11. doi: 10.1007/s00442-007-0957-y. pmid:18224338

90. Hall FG, Hilker T, Coops NC, Lyapustin A, Huemmrich KF, Middleton E, et al. Multi–angle remote sensing of forest light use efficiency by observing PRI variation with canopy shadow fraction. Remote Sens. Environ 2008; 112: 3201–3211. doi: 10.1016/j.rse.2008.03.015

91. Haboudane D, Miller JR, Tremblay N, Zarco-Tejada PJ, Dextraze L. Integrated narrow–band vegetation indices for prediction of crop chlorophyll content for application to precision agriculture. Remote Sens Environ 2002; 81: 416–426. doi: 10.1016/s0034-4257(02)00018-4

92. Coops NC, Stone C, Culvenor DS, Chisholm LA, Merton RN. Chlorophyll content in eucalypt vegetation at the leaf and canopy scales as derived from high resolution spectral data. Tree Physiol 2003; 23: 23–31. pmid:12511301 doi: 10.1093/treephys/23.1.23

93. Zarco-Tejada PJ, Miller JR, Harron J, Hu BX, Noland TL, Goel N, et al. Needle chlorophyll content estimation through model inversion using hyperspectral data from boreal conifer forest canopies. Remote Sens Environ 2004; 89: 189–199. doi: 10.1016/j.rse.2002.06.002

94. Dash J, Curran PJ. The MERIS terrestrial chlorophyll index. Int J Remote Sens 2004; 25: 5403–5413. doi: 10.1080/0143116042000274015

95. .Haboudane D, Tremblay N, Vigneault P, Miller JR. Indices-based approach for crop chlorophyll content retrieval from hyperspectral data. In International Geoscience and Remote Sensing Symposium (IGARSS); Barcelona, Spain; July 2007. p.3297–3300.

96. Rao NR, Garg PK, Ghosh SK, Dadhwal VK. Estimation of leaf total chlorophyll and nitrogen concentrations using hyperspectral satellite imagery. J Agr Sci 2008; 146: 65–75. doi: 10.1017/s0021859607007514

97. Delegido J, Fernandez G, Gandia S, Moreno J. Retrieval of chlorophyll content and LAI of crops using hyperspectral techniques: application to PROBA/CHRIS data. Int J Remote Sens 2008; 29: 7107–7127. doi: 10.1080/01431160802238401

98. Wu CY, Han XZ, Niu Z, Dong JJ. An evaluation of EO-1 hyperspectral hyperion data for chlorophyll content and leaf area index estimation. Int J Remote Sens 2010; 31: 1079–1086. doi: 10.1080/01431160903252335

99. Delegido J, Alonso L, González G, Moreno J. Estimating chlorophyll content of crops from hyperspectral data using a normalized area over reflectance curve (NAOC). Int J Appl Earth Obs 2010; 12: 165–174. doi: 10.1016/j.jag.2010.02.003

100. Delegido J, Van Wittenberghe S, Verrelst J, Ortiz V, Veroustraete F, Valcke R, et al. Chlorophyll content mapping of urban vegetation in the city of Valencia based on the hyperspectral NAOC index. Ecol Indic 2014; 40: 34–42. doi: 10.1016/j.ecolind.2014.01.002

101. Oppelt N, Mauser W. Hyperspectral monitoring of physiological parameters of wheat during a vegetation period using AVIS data. Int J Remote Sens 2004; 25: 145–159. doi: 10.1080/0143116031000115300

102. Blackburn GA. Remote sensing of forest pigments using airborne imaging spectrometer and LIDAR imagery. Remote Sens Environ 2002; 82: 311–321. doi: 10.1016/s0034-4257(02)00049-4

103. Guan YN, Guo S, Liu JG, Zhang X. Algorithms for the estimation of the concentrations of chlorophyll a and carotenoids in rice leaves from airborne hyperspectral data. In Computational Science-ICCS 2005; Atlanta, GA, USA; May 2005. p. 908–915.

104. Thomas V, Treitz P, McCaughey JH, Noland T, Rich L. Canopy chlorophyll concentration estimation using hyperspectral and lidar data for a boreal mixedwood forest in northern Ontario Canada. Int J Remote Sens 2008; 29: 1029–1052. doi: 10.1080/01431160701281023

105. Pena MA, Altmann SH. Use of satellite-derived hyperspectral indices to identify stress symptoms in an Austrocedrus chilensis forest infested by the aphid Cinara cupressi. Int J Pest Manage 2009; 55: 197–206. doi: 10.1080/09670870902725809

106. Jacquemoud S, Verdebout J, Schmuck G, Andreoli G. Hosgood B. Investigation of leaf biochemistry by statistics. Remote Sens Environ 1995; 54: 180–188. doi: 10.1016/0034-4257(95)00170-0

107. Cornell JA, Berger RD. Factors that influence the value of the coefficient of determination in simple linear and nonlinear regression models. Phytopathology 1987; 77: 63–70. doi: 10.1094/phyto-77-63

108. Gurevitch J, Curtis PS, Jones MH. Meta-analysis in ecology. Adv Ecol Res 2001; 32: 199–247. doi: 10.1016/s0065-2504(01)32013-5

109. Hedges LV. Estimation of effect size from a series of independent experiments. Psychol Bull 1982; 92: 490–499. doi: 10.1037//0033-2909.92.2.490

110. Higgins J, Thompson SG. Quantifying heterogeneity in a meta-analysis. Stat Med 2002; 21: 1539–1558. pmid:12111919 doi: 10.1002/sim.1186

111. Hedges LV, Olkin I. Statistical methods for meta-analysis. Orlando, FL, USA: Academic Press; 1985. p. 31–34.

112. Lipsey MW, Wilson DB. Practical meta-analysis. London, UK: SAGE Publications; 2000. p. 112–116.

113. Mosteller F, Colditz GA. Understanding research synthesis (meta-analysis). Annu Rev Publ Health 1996; 17 1–23. doi: 10.1146/annurev.pu.17.050196.000245

114. Han JW, Kamber M. Data mining:concepts and techniques second ed. San Francisco, CA, USA: Morgan Kaufmann; 2006. p. 53–54.

Chapter 7

STRUCTURAL COLOUR AND IRIDESCENCE IN PLANTS: THE POORLY STUDIED RELATIONS OF PIGMENT COLOUR

Beverley J. Glover[1], and Heather M. Whitney[2]

[1] Department of Plant Sciences, University of Cambridge, Downing Street, Cambridge CB2 3EA, UK

[2] School of Biological Sciences, University of Bristol, Woodland Road, Bristol BS8 1UG, UK

ABSTRACT

Background

Colour is a consequence of the optical properties of an object and the visual system of the animal perceiving it. Colour is produced through chemical and structural means, but structural colour has been relatively poorly studied in plants.

Scope

This Botanical Briefing describes the mechanisms by which structures can produce colour. In plants, as in animals, the most common mechanisms are multilayers and diffraction gratings. The functions of structural colour are then discussed. In animals, these colours act primarily as signals between members of the same species, although they can also play roles in camouflaging animals from their predators. In plants, multilayers are found predominantly in shade-plant leaves, suggesting a role either in photoprotection or in optimizing capture of photosynthetically active light. Diffraction gratings may be a surprisingly common feature of petals, and recent work has shown that they can be used by bees as cues to identify rewarding flowers.

Conclusions

Structural colour may be surprisingly frequent in the plant kingdom, playing important roles alongside pigment colour. Much remains to be discovered about its distribution, development and function.

INTRODUCTION: WHAT IS COLOUR?

The bright colours of flowers attract pollinating insects by making the floral tissue stand out against a background of vegetation. Analyses of insect visual acuity have shown that vegetation is visually very similar to bark, soil and stone from an insect's point of view, because all these materials weakly reflect light across the whole range of the insect visual spectrum (Kevan *et al.*, 1996). Flowers are different – they appear as bright colours because they selectively reflect certain wavelengths of light, which are perceptible to pollinating animals, and, usually, to humans as well.

Colour is a property of both the coloured object and the perception of the animal observing it (Fig. 1). Light arriving at an object can be transmitted through it, absorbed by it or reflected back from it. If an object reflects or transmits all wavelengths of light equally, then it is perceived as white (Fig. 1, top). If an object strongly absorbs all wavelengths of light, then it is perceived as black (Fig. 1, centre). However, if it absorbs all light except one set of wavelengths, such as the red, which it instead reflects or transmits, then it can be said to have a colour. What that colour is depends on the visual system of an animal observing the object. If it has photoreceptors that are strongly activated by red light, as vertebrates do, then the object will appear red (Fig. 1, bottom left). If it has no photoreceptors that respond to red light, the object will appear black – to that animal the object is indistinguishable from an object that absorbs all wavelengths of light. Because photoreceptors are triggered by a curve of wavelengths the situation can be more complex. So, for insects that do not have red-light receptors but whose green-light receptors respond to a curve of wavelengths with the tail of the curve in the red part of the spectrum, the object in question would appear dull green (Fig. 1, bottom right; Chittka and Raine, 2006).

Plants, like animals, achieve colour in two main ways. First, they use chemical- or pigment-based colour. Pigments are compounds which absorb subsets of the visible spectrum, transmitting and reflecting back only what they do not absorb and causing the tissue to be perceived as the reflected colours. Chlorophyll absorbs light in both the red and the blue parts of the spectrum, reflecting only green light, and causes leaves to appear green to humans. Similarly, a flower that humans perceive as red contains pigments which absorb yellow, green and blue light, leaving red light as the only wavelength visible to us which is reflected. Plant pigments have been thoroughly studied from a biochemical perspective, and their synthesis and regulation have also been characterized by molecular genetics.

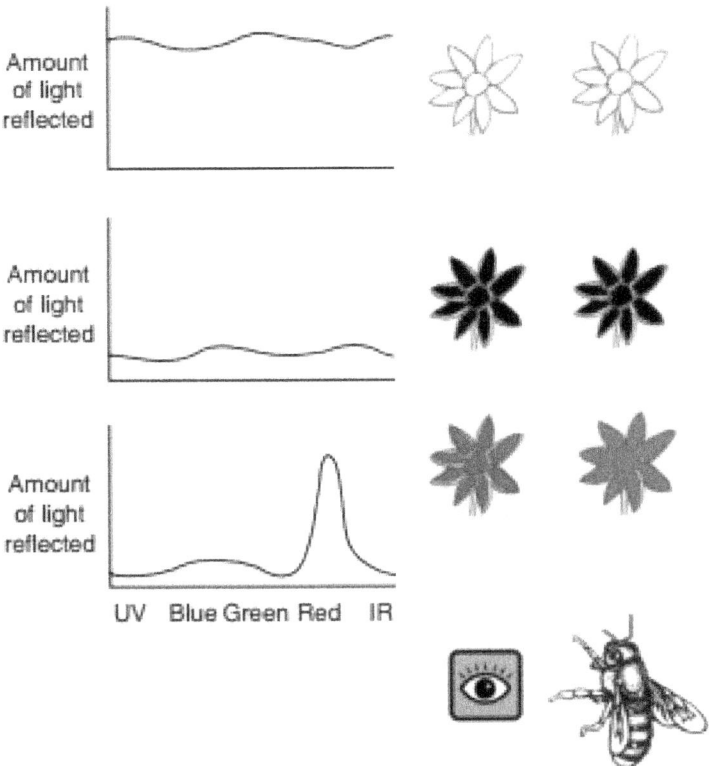

Figure. 1: Colour is a property of the light reflected by an object and the visual system of the animal observing it. If a flower reflects all wavelengths of light, it is perceived as white (top). If it absorbs all wavelengths then it appears black (centre). However, if it absorbs all wavelengths apart from one region of the spectrum, it has a colour. The flower shown in the bottom panel reflects red light. To the vertebrate eye, which has red-light receptors, the flower appears red. However, to the bee eye, which has no red-light receptors but whose green-light receptors are weakly stimulated by red light, the flower appears a dull green.

However, both plants and animals have also been shown to produce structural colours. A structural colour occurs when different wavelengths of light are selectively reflected from a substance, with the remaining wavelengths transmitted or absorbed. The famous blue butterflies of the genus *Morpho* have wing scales which selectively reflect a narrow bandwidth of blue light, allowing other wavelengths to be transmitted through the wing (Fig. 2A). The wings accordingly look intensely blue to humans, even though they contain no blue pigments (Vukusic *et al.*, 1999). Structural colour has been well characterized in animals, but very little studied in plants.

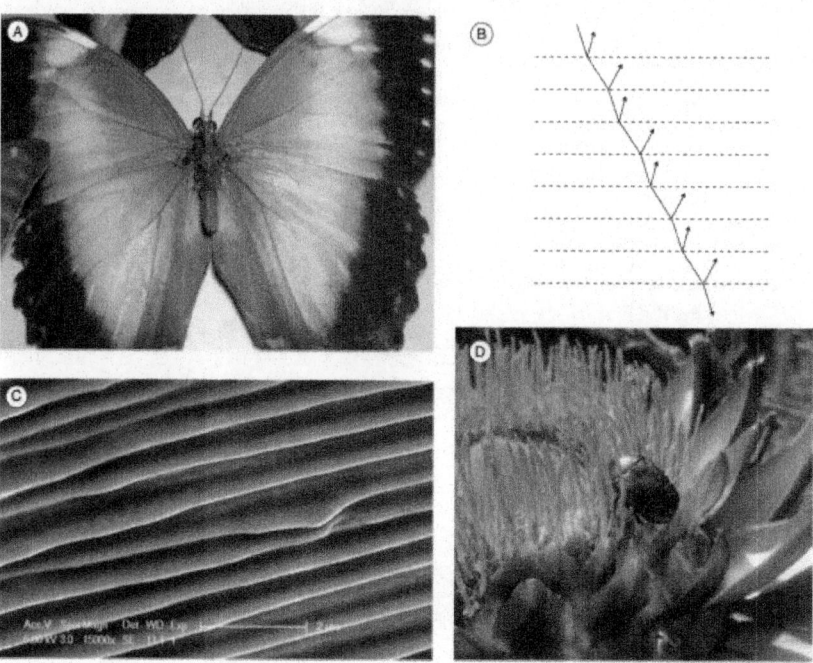

Figure. 2: Structural colour and iridescence. (A) The intense blue colour of the *Morpho*butterfly is due to reflection of light by multilayers. (B) Multilayers generate iridescence by reflecting different wavelengths of light at different angles at each boundary between layers. (C) Diffraction gratings consist of ordered parallel grooves at particular frequencies, like the cuticular striations on this tulip petal. (D) An iridescent beetle (rose chafer, *Cetonia aurata*) visits an artichoke flower.

IRIDESCENCE IS A UNIQUE PROPERTY OF STRUCTURAL COLOUR

Chemical and structural colours have several different properties. They differ first in the intensity of colour that they produce. Pigments are generally not very good at absorbing all but a very few wavelengths of light. Instead, they absorb most light of a number of wavelengths, but allow quite a broad range of wavelengths to be reflected or transmitted. This results in colours which can appear dull or muted, as they consist of a mixture of different colours of light. In contrast, structural colours can appear very intense, as reflective structures can be very precise in the bandwidths that they reflect.

Chemical and structural colours also differ in the patterns that they can produce. Chemical colours are diffuse, and look the same from all angles. To produce patterns of colour, different pigments must be localized to different cells or areas of a tissue. Commonly occurring pigment patterns in plants include different coloured venation on petals, and spots of dark pigment acting as targets at the bases of petals, near the nectaries. Structural colours have the potential to generate shifting patterns of colour as the viewer moves, rather than across different regions of the tissue. Reflective structures can reflect one particular peak wavelength of light at one angle, and another peak wavelength at a second angle. Thus, as an animal moves its position relative to the structure it will see the object change from the first colour to the second colour. The phenomenon of appearing different colours when viewed from different angles is called iridescence, and it is a unique attribute of structural colour. Iridescence can cover a few or many different colours, and can be in regions of the spectrum visible to a variety of animals, including in the ultraviolet (UV).

Structural Colour and Iridescence – Mechanisms Used By Animals

The mechanisms capable of producing structural colour in animals were described by both Hooke and Newton in the 17th and early 18th centuries, and a large body of literature has subsequently been produced, much of which is covered in several recent reviews (Parker, 2000; Vukusic and Sambles, 2003; Doucet and Meadows, 2009). A very brief overview shows that structural colour can be produced by either incoherent or coherent light scattering.

Incoherent light scattering takes place when individual light-scattering structures are randomly separated from one another by an average distance that is large when compared with the wavelength of the light. The light-scattering structures differentially scatter different visible wavelengths, but in such a way that the phase relationship of the scattered wavelengths is random. Although most structural colour in animals is produced by coherent light scattering, the blue colouration in many amphibians is attributed to incoherent scattering (Bagnara *et al.*, 2007), as is the blue colour of the sky.

The majority of structural colour, and all iridescence, in animals is produced by coherent light scattering, which occurs when the distribution of light-scattering elements, and the resulting phase relationship of reflected light waves, is precisely ordered. An ordered distribution of light scatterers can result in either constructive or destructive interference. If the phase difference between two waves is a multiple of exactly one full wavelength then the two waves constructively interfere with each other and there is a strong reflection of light at that particular wavelength. By contrast, if the phase of the reflected waves differs by half a wavelength, or an odd multiple of half wavelengths, then destructive interference occurs such that reflection of this wavelength is weak or absent.

The simplest type of coherent light scattering is that of thin-film interference, which gives the colour to soap bubbles and oil-slicked puddles. Thin-film interference occurs when two transparent layers of materials with different optical densities meet. The optical density of a material determines the extent to which light waves are slowed down as they pass through it. Light is also reflected at each side of the boundary between the two materials – both before and after passing through each individual layer. Optical density, the thickness of the material layer, and the angle and wavelength of the light all help to determine if the light reflecting from the bottom of a layer is in phase or out of phase with the light reflected from the top of the layer, which will in turn determine whether constructive or destructive interference occurs for each wavelength. Constructive interference for one wavelength and destructive interference for others results in the reflected light being of one colour. Multilayer reflectors that produce structural colour consist of ordered layers of these pairs of thin films layered in series, producing even stronger constructive interference for specific wavelengths and resulting in very pure, intense colours (Fig. 2B). The classic example of multilayered structural colour in animals is shown by the blue *Morpho rhetenor* butterfly, in which the multilayered structure on its wing scales produces a vivid blue colour of such intensity that it is said to have a visibility of up to half a mile (Vukusic *et al.*, 1999; Vukusic and Sambles, 2003).

A diffraction grating consists of a reflective surface over which runs a series of ordered and precisely spaced parallel grooves (Fig. 2C). Some of the light that hits the surface is reflected as normal, but light that hits the grooves is diffracted – split into its component wavelengths – and each wavelength is reflected at a different angle. Light with longer wavelengths has a higher diffraction angle than light with shorter wavelengths, so the light separates into its component parts, producing the rainbow effect that can be easily seen over the surface of a CD. Several beetle and spider species have been found to produce iridescence through this mechanism (Parker and Hegedus, 2003; Seago *et al.*, 2009).

Iridescence can also result from the presence of photonic crystals, which are ordered three-dimensional structures. The classic example of a photonic crystal is opal, which consists of tiny spheres of silica packed together. The diffraction of light through opal is determined by the size and regularity of the spheres, which in turn determines the colours shown. Three-dimensional structures generating iridescence have been found in a wide range of animals, including comb-jellies, several butterfly species, the feathers of a number of bird species and in the annelid *Aphrodita* sp. (Parker *et al.*, 2001; Vukusic and Sambles, 2003; reviewed in Welch and Vigneron, 2007). The spines of *Aphrodita* species show a multicoloured iridescence that is caused by a structure of holes ordered in hexagonal crystal structure within the spines (Parker *et al.*, 2001). Biological photonic crystals can vary greatly in both form and method of function.

Mechanisms of Plant Structural Colour and Iridescence

Structural colour and iridescence have arisen multiple times in the animal kingdom, so it is hardly surprising that they are also found in plants. All the general mechanisms used by animals to produce structural colour are also used by plants. Like animals, plants produce structural colour by both coherent and incoherent scattering. Incoherent 'Rayleigh' scattering (by particles smaller than the wavelength of light reflected) has been found in a number of plant species. The wax deposits on blue spruce (*Picea pungens*) and chalk dudleya (*Dudleya brittonii*) scatter shorter wavelengths of light preferentially, resulting in a blue colouration to the leaves (Vogelmann, 1993).

Iridescence has been shown to be produced by both multilayers (Fig. 2B) and diffraction gratings (Fig. 2C) in plants. The first example of multilayered iridescence in plants was found in the lycophyte *Selaginella*. Two species of *Selaginella*, *S. willdenowii* and *S. uncinata*, produce a vivid blue–green iridescence on their leaves when growing in shade. In the first detailed study into the mechanisms of plant iridescence, Hébant and Lee (1984) found that *Selaginella* leaves had two layers in the outer cell wall of their epidermal cells. These layers, visible under transmission electron microscopy, were each approx. 80 nm thick, the predicted thickness to cause multilayer interference that would result in the observed iridescence. These two layers were not found in ordinary green *Selaginella*leaves grown under higher light conditions and lacking iridescence (Hébant and Lee, 1984). Other plants with iridescent leaves are also found in low light environments, and all produce a similar blue–green iridescence. Although the multilayers in *Selaginella* appear to be relatively simple, with only a few layers producing the iridescence, other plant species produce more elaborate structures. The outer epidermal cell walls of the iridescent ferns *Danaea nodosa*, *Diplazium tomentosum* and*Lindsaea lucida* have many repeated dense layers alternating with arcs of cellulose microfibrils. The layers are of the correct thickness to cause iridescence through interference in the young iridescent leaves, but these layers are missing in the older leaves, which show no iridescence. The angle of the cellulose microfibrils changes gradually through the alternating layers up to a total 180 ° rotation (Graham *et al.*, 1993; Gould and Lee, 1996; Lee, 2007). The resulting helicoidal structure is remarkably similar to the helical stack of chitin microfibrils found in some iridescent beetle species and may therefore be an example of convergent evolution (Lee, 2007; Seago *et al.*, 2009). Leaf iridescence can also be caused by multilayers within the protoplast, not just within the cell wall. In the fern*Trichomanes elegans* and the angiosperms *Phyllagathis rotundifolia* and*Begonia pavonina*, specialized plastids called 'iridoplasts' are found in the iridescent leaves. These iridoplasts are much flatter than chloroplasts, and the thylakoid stacks within them are in such close contact that they form layers that cause the interference of light, resulting in the iridescent blue colouration (Graham *et al.*, 1993; Gould and Lee, 1996; Lee, 2007).

Multilayers generating iridescence are also found in the fruits of*Elaeocarpus angustifolius* and *Delarbrea michiana*, in this case arising from a structure called an 'iridosome'. This is secreted to the region outside the cell membrane of fruit epidermal cells, and consists of layers of cellulose that are of the predicted thickness to cause interference colouration (Lee, 1991; Lee *et al.*, 2000).

Diffraction gratings were identified in plants more recently, with the first report of their presence on the petals of species including *Tulipa* sp.,*Hibiscus trionum* (Fig. 3A) and *Mentzelia lindleyi* (Fig. 3E) published in 2009. In these species the petal epidermal cells are elongated and flat and the overlying cuticle produces a series of long, ordered ridges with a periodicity that acts as a diffraction grating and splits the light reflecting from the surface into component wavelengths (Fig. 3B, C; Whitney *et al.*, 2009a). The iridescence produced is often predominantly in the UV wavelengths, which, although invisible to the human eye, are easily visible to many animal pollinators including bees and birds. The cuticular striations creating floral iridescence can also occur in patterns overlying those caused by pigment colour (Whitney *et al.*, 2009a, b).

Figure. 3: Plant iridescence. (A) The inner part of the *Hibiscus trionum* petal has an oily iridescence overlying red pigmentation. (B) Scanning electron microscopy of this

region shows that the cells overlying the red pigment are covered with a diffraction grating made from cuticular striations, although the cells over the white region are smooth. (C) When petal diffraction gratings are replicated in transparent optical epoxy, light reflected from the epoxy is not white but shows a range of colours. (D) The iridescent labellum of *Ophrys speculum* is thought to mimic the wings of female pollinators. (E) *Mentzelia lindleyi* is iridescent as a result of diffraction gratings, but the iridescence is only detectable in the bee-visible UV region of the spectrum.

Flowers are also the site of the one example of a three-dimensional photonic structure that has been found in plants. The elongated hairs that cover the attractive bracts surrounding edelweiss flowers (*Leontopodium nivale* subsp. *alpinum*) have an internal structure that acts as a photonic crystal (Vigneron *et al.*, 2005). The hairs are hollow tubes with a series of parallel striations around the external surface. Through diffraction effects, the hairs absorb the majority of the UV light, effectively acting as an efficient sun-block. A variety of other epidermal cell morphologies are also known to influence light capture and reflection in petals (Kevan and Backhaus, 1998).

FUNCTIONS OF ANIMAL IRIDESCENCE

Iridescence appears to have as varied a range of functions as it does methods of production in the animal kingdom. The recent review byDoucet and Meadows (2009) gives a clear overview of the functions of animal iridescence. The most frequent role of animal iridescence appears to be in visual communication. Iridescence can relay information about the animal's species (Silberglied and Taylor, 1978), about its age if iridescence changes or deteriorates over time (Kemp, 2006; Bitton and Dawson, 2008), about sex, as in many species only one sex has iridescence (Rutowski, 1977), and nutritional status, as individuals with poor nutrition may lack the resources to produce very vivid colouration (Kemp and Rutowski, 2007). Iridescence has also been found to play an important role in mate choice in birds, butterflies and fish (Kodric-Brown and Johnson, 2002;Sweeney *et al.*, 2003; Kemp, 2007), while the depth of the blue structural colour on the testicles indicates the degree of dominance within the troop of a male vervet monkey (Prum and Torres, 2004).

As well as providing information for other animals, structural colour has also been implicated in helping animals avoid detection by their predators, either by mimicry or by camouflage. Colourful reef fish are well camouflaged against the equally colourful corals, while tiger beetles blend a range of structural colours together to produce a matt camouflage (Schultz, 1986; Schultz and Bernard, 1989; Seago *et al.*, 2009).

FUNCTIONS OF PLANT IRIDESCENCE

As with animals, structural colour in plants is important in both display and defence. However, in plants the targets of the display are not other plants but pollinating insects, and the defence may be against potentially damaging levels of light as well as animal predators.

The primary function of flower and fruit iridescence is likely to be the attraction of animals, particularly those species whose visual systems are attuned to iridescence for animal–animal communication. The fruits of*Elaeocarpus* and *Delarbrea michiana* (Lee, 1991; Lee *et al.*, 2000) have an iridescence that is thought to enhance animal attraction. Iridescence has also been shown to attract pollinating insects. It has been believed for some time that iridescence is used by pseudocopulatory flowers (such as species of *Ophrys*, Fig. 3D) to mimic female insects visually, but we were able to show that iridescence can act as an ordinary, learnable cue, in the same way that flower colour or shape might (Whitney *et al.*, 2009a). Foraging bumblebees were trained that iridescent targets (generated by an artificial diffraction grating) contained a reward, whether they had a basic pigment colour of purple, blue or yellow, and that non-iridescent targets in the same pigment colours did not. The bees learned the iridescent cue, and were able to use it when presented with red targets to identify correctly the rewarding ones. The diffraction gratings generating floral iridescence often occur in patterns overlying those caused by pigment colour (Whitney *et al.*, 2009b), suggesting that they might enhance pigment-based learnable cues.

The ability of structural colour to reflect strongly in specific wavelengths is thought to provide photoprotection to leaves. The Rayleigh scattering shown by *Picea pungens* and *Dudleya brittonii* is thought to result in enhanced reflection of shorter wavelengths, and thus to give protection against UV damage (Vogelmann, 1993). Protection against UV is also thought to be the primary function of the photonic crystal hairs overlying the surface of the edelweiss bracts, which protect the reproductive tissues against the potentially mutagenic UV levels found at the altitudes where this plant grows (Vigneron *et al.*, 2005). Photoprotection may also be the function of the blue multilayer iridescence produced by understorey plants such as *Danaea nodosa*, *Diplazium tomentosum*, *Lindsaea lucida* and*Begonia pavonina*. These plants are all adapted to low light conditions, and so might be at risk of photodamage if they encountered sunflecks or other high-intensity light. The iridescent blue leaves of *Begonia pavonina*recovered significantly more rapidly from light exposure than green non-iridescent leaves, although no difference was found between the iridescent and non-iridescent leaves of *Diplazium tomentosum* (Lee, 2007).

In contrast, it has been hypothesized that the iridescence of *Selaginella* species might aid the capture of photosynthetically active wavelengths in low light conditions because the leaf iridescence may act as a natural anti-reflective coating. Such coatings (on glasses and cameras) use thin film structures, analogous to those found in the iridescent *Selaginella* leaf, to produce constructive interference for certain wavelengths, increasing transmission of those wavelengths, but a side-effect is that the wavelengths not transmitted are strongly reflected because of destructive interference. In the same way, the iridescence in *Selaginella* could enhance blue-light reflection while enriching red-light absorption (Hébant and Lee, 1984).

Outlook

Our understanding of plant structural colour and iridescence lags some way behind the work in animals, perhaps because plant pigment biochemistry has been studied so successfully or perhaps because animal structural colours are so striking. It is not surprising that similar mechanisms to generate structural colour have evolved in both plants and animals, but it will be important in the years to come to establish the molecular mechanisms underlying the development of these structures, which are likely to be very different in organisms with such basic differences in body architecture. The identification of structurally coloured plant species that are amenable to a genetic or transgenic dissection of candidate genes will be necessary to allow such work to progress rapidly. Preliminary studies suggest that some members of the Compositae, a number of petaloid monocots and certain species of Solanaceae might represent good targets for molecular and developmental analysis. It is also apparent that plant structural colour has evolved to mediate plant responses to both biotic and abiotic factors. A primary role is for communication with animals, and structures are therefore likely to target colours visible to pollinating or predatory species. One immediate challenge is to investigate how many species show structural colour (or iridescence) restricted to the UV region of the spectrum, and therefore invisible to the human eye. Investigation of the UV reflectance of flowers pollinated by insects that are themselves iridescent might be fruitful, as the visual acuity of such animals is already entrained to shifting colours, rather than to static ones. Such a study will also provide an understanding of the evolutionary lability of structural colour, and of the extent to which it appears to have co-evolved in response to interactions with particular groups of insect. Given that we do not currently have a good understanding of which plants produce structural colour, how they produce it and what they produce it for, one of the most exciting aspects of plant structural colour is the amount that still remains to be learned.

ACKNOWLEDGEMENTS

We thank Murphy Thomas, Sean Rands, Matthew Dorling, Matt Box and Rosie Bridge for help with figure production, and the Insect Collection in the Department of Zoology, University of Cambridge, for access to iridescent animals. The ideas explored in this article were developed through many helpful discussions with Ulli Steiner, Mathias Kolle and Lars Chittka. H.M.W. is in receipt of a Lloyd's of London Tercentenary foundation fellowship.

REFERENCES

1. Bitton PP, Dawson RD. 2008. Age-related differences in plumage characteristics of male tree swallows Tachycineta bicolor: hue and brightness signal different aspects of individual quality. Journal of Avian Biology 39: 446 –452. Chittka L, Raine N. 2006. Recognition of flowers by pollinators. Current

2. Opinion in Plant Biology 9: 428–435.

3. Doucet SM, Meadows MG. 2009. Iridescence: a functional perspective. Journal of the Royal Society Interface 6: S115–S132.

4. Gould KS, Lee DW. 1996. Physical and ultrastructural basis of blue leaf iridescence in four Malaysian understory plants. American Journal of Botany 83: 45– 50.

5. Graham RM, Lee DW, Norstog K. 1993. Physical and ultrastructural basis of blue iridescence in two neotropical ferns. American Journal of Botany 80: 198– 203.

6. He'bant C, Lee DW. 1984. Ultrastructural and developmental control of iridescence in Selaginella leaves. American Journal of Botany 71: 216– 219.

7. Kemp DJ. 2006. Heightened phenotypic variation and age-based fading of ultraviolet butterfly wing coloration. Evolutionary Ecology Research 8: 515– 527.

8. Kemp DJ. 2007. Female butterflies prefer males bearing bright iridescent ornamentation. Proceedings of the Royal Society B 274: 1043–1047.

9. Kemp DJ, Rutowski RL. 2007. Condition dependence, quantitative genetics, and the potential signal content of iridescent ultraviolet butterfly coloration. Evolution 61: 168– 183.

10. Kevan PG, Backhaus WGK. 1998. Colour vision: ecology and evolution in making the best of the photic environment. In: Backhaus WGK, Kliegl R, Werner JS, eds. Colour vision – perspectives from different disciplines. Berlin: De Gruyter, 163 –183.

11. Kevan PG, Giurfa M, Chittka L. 1996. Why are there so many and so few
12. white flowers? Trends in Plant Sciences 1: 280–284.

13. Kodric-Brown A, Johnson SC. 2002. Ultraviolet reflectance patterns of male guppies enhance their attractiveness to females. Animal Behaviour 63: 391– 396.

14. Lee DW. 1991. Ultrastructural basis and function of iridescent blue colour of fruits in Elaeocarpus. Nature 349: 260–262.

15. Lee DW. 2007. Nature's palette, the science of plant colour. Chicago: The University of Chicago Press.

16. Lee DW, Taylor GT, Irvine AK. 2000. Structural fruit coloration in Delarbrea michieana (Araliaceae). International Journal of Plant Science 161: 297– 300.

17. Parker AR. 2000. 515 million years of structural colour. Journal of Optics A: Pure and Applied Optics 2: R15–R28.

18. Parker AR, Hegedus Z. 2003. Diffractive optics in spiders. Journal of Optics A: Pure and Applied Optics 5: S111–S116.

19. Parker AR, McPhedran RC, McKenzie DR, Botten LC, Nicorovici NA. 2001. Photonic engineering: Aphrodite's iridescence. Nature 409: 36.

20. Prum RO, Torres R. 2004. Structural colour of mammalian skin: convergent evolution of coherently scattering dermal collagen arrays. Journal of Experimental Biology 207: 2157– 2172.

21. Rutowski RL. 1977. The use of visual cues in sexual and species discrimination by males of the small sulphur butterfly Eurema lisa (Lepidoptera, Pieridae). Journal of Comparative Physiology 115: 61–74.

22. Schultz TD. 1986. Role of structural colors in predator avoidance by tiger beetles of the genus Cicindela (Coleoptera: Cicindelidae). Bulletin of the Entomological Society of America 32: 142–146.

23. Schultz TD, Bernard GD. 1989. Pointillistic mixing of interference colours in cryptic tiger beetles. Nature 337: 72–73.

24. Seago AE, Brady P, Vigneron J-P, Schultz TD. 2009. Gold bugs and beyond: a review of iridescence and structural colour mechanisms in beetles (Coleoptera). Journal of the Royal Society Interface 6: S165–S184.

25. Silberglied RE, Taylor OR. 1978. Ultraviolet reflection and its behavioral role in courtship of sulfur butterflies Colias eurytheme and Colias philodice (Lepidoptera, Pieridae). Behavioural Ecology and Sociobiology 3: 203– 243.

26. Sweeney A, Jiggins C, Johnsen S. 2003. Insect communication: polarized light as a butterfly mating signal. Nature 423: 31– 32.

27. Vigneron JP, Rassart M, Ve´rtesy Z, et al. 2005. Optical structure and function of the white filamentary hair covering the edelweiss bracts. Physics Review E 71: 011906.

28. Vogelmann TC. 1993. Plant tissue optics. Annual Review of Plant Physiology and Plant Molecular Biology 44: 489– 499.

29. Vukusic P, Sambles JR. 2003. Photonic structures in biology. Nature 424: 852– 855.

30. Vukusic P, Sambles JR, Lawrence CR, Wootton RJ. 1999. Quantified interference and diffraction in single Morpho butterfly scales. Proceedings of the Royal Society B 266: 1403– 1411.

31. Welch VL, Vigneron J-P. 2007. Beyond butterflies – the diversity of biological photonic crystals. Optical and Quantum Electronics 39: 295– 303.

32. Whitney HM, Kolle M, Andrew P, Chittka L, Steiner U, Glover BJ. 2009a. Floral iridescence, produced by diffractive optics, acts as a cue for animal pollinators. Science 323: 130–133.

33. Whitney HM, Kolle M, Alvarez-Fernandez R, Steiner U, Glover BJ. 2009b. Contributions of iridescence to floral patterning. Communicative and Integrative Biology 2: 230–232.

Chapter 8

BIODIVERSITY IN A TOMATO GERMPLASM FOR FREE AMINO ACID AND PIGMENT CONTENT OF RIPENING FRUITS

Guillermo Raúl Pratta[1,2], Gustavo Rubén Rodríguez[1,2], Roxana Zorzoli[2,3], Liliana Amelia Picardi[2,3,] Estela Marta Valle[1,4]

[1] National Council for Scientific and Technical Research, Buenos Aires, Argentina

[2] Chair of Genetics, Agronomic Sciences Faculty, National University of Rosario, Zavalla, Argentine

[3] Council for Research of the National University of Rosario, Zavalla, Argentina

[4] Institute of Molecular and Cell Biology of Rosario, CONICET/Biochemical and Pharmaceutical Sciences Faculty, Suipacha, Rosario, Argentine

ABSTRACT

Free amino acid and pigment composition in fruits at two ripening stages from a selected tomato germplasm was studied. The aims were contributing to knowledge on variability of ripening metabolism and identifying more consistently the genetic background of the plant material under analysis. Significant differences ($p < 0.05$) were found among ripening stages and among genotypes within ripening stage for all amino acids and pigments except by asparagine, alanine and chlorophyll b contents. The highest relative amino acid content corresponded to glutamate, glutamine, and GABA though some genotypes had relatively high asparagine content. Glutamate, glutamine and GABA performed oppositely: the former increased along ripening while the latter two decreased in their relative content. A Principal Components (PC) analysis was applied, determining that metabolites having the greatest contribution to general variability were threonine, serine, glutamate, glutamine, glycine, isoleucine, leucine, tyrosine, phenylalanine, lycopene and betacarotene, which showed the highest association with PC1. Alanine and chlorophylls a and b were highly associated to PC2. These two first PC explained the 62% of the total variation, and genotypes were distributed according to the ripening stage in their coordinates. Accordingly, a Hierarchical Clustering resulted in a dendrogram having a relatively high cophenetic correlation (0.70), in which two well defined groups were obtained according to ripening stage. These results verified the existence of variability in the metabolism of ripening fruit for amino acids and pigments, and allowed to identify unequivocally a set of selected tomato germplasm according to the fruit metabolic profiles in these two ripening stages.

INTRODUCTION

From an evolutionary viewpoint, tomato (Solanum lycopersicum) ripening could be considered as a transition from a green stage that prevents fruit consumption (hence protecting the developing seeds) to a ripe stage in which its attributes are optimum to attract predators, which consume fruit and help to disperse mature seeds [1]. This transition includes morphological, biochemical and physiological changes that lead to acquisition of appropriate color, texture, flavor, among other traits determining fruit quality. Some of these changes are variations in free amino acids and pigment composition [2,3]. Glutamate percent content noticeably increased between mature green and red ripe stages, simultaneously to a reduction in glutamine and GABA levels in tomato varieties [4]. From a productive viewpoint, the conservation of red ripe attributes during a longer time prolongs the opportunity of fruit commercialization, especially for the fresh market [5]. Shelf life is a measure for the period of tomato quality adequate maintenance, and has been reported as negatively correlated to glutamate relative molar content of ripe mature fruits [6]. Long shelf life tomatoes have been currently obtained by introgressing spontaneous ripening mutant genes such as nor, rin, alc and Nr, or by genetic transformation [7]. Both strategies have disadvantages, because spontaneous mutations present pleiotropic effects that diminish fruit quality and transgenic food are not well accepted by public opinion even presently [5]. Exotic germplasm of Solanum section Lycopersicon comprises S. lycopersicum var. cerasiforme, the closest relative S. pimpinellifolium, and other 10 wild species. They are invaluable plant genetic resources contributing abiotic and biotic resistance or tolerance genes but also for increasing fruit quality and prolonging shelf life [8,9]. Seventeen recombinant inbred lines from an interspecific cross S. lycopersicum cv. Caimanta x S. pimpinellifolium LA722 were obtained after six selfing cycles with antagonistic-divergent selection for fruit weight and shelf life [10] and characterized by quantitative fruit traits, total pericarp polypeptide profiles and AFLP markers [11,12]. A wide variation was found for all analyzed phenotypic and molecular attributes. The general goal of this research was to study free amino acid and pigment composition in this selected tomato germplasm, with the aims of contributing to know- ledge on variability of ripening metabolism, identifying more consistently the RILs genetic background, and verifying associations between glutamate content and fruit shelf life

MATERIALS AND METHODS

Plant Material

Ten seeds of RILs 4, 10, 12 and 15 together with the experimental testers, parents Caimanta and LA722 (Figure 1), were sown in seedling trays on August under a glasshouse. Then plants were grown at the experimental field station "José F. Villarino" located at latitude 33°S and longitude 61°W, from October to March under greenhouse conditions. Previous to the transplantation, the soil (a typical argiudol) was fertilized with poultry grit. The crop was watered twice a week, levels of irrigation that were sufficient to avoid water stress during the plants growing period. Mean values of fruit shelf life (in days) were: Caimanta = 13.00, LA722 = 18.60, RIL4 = 16.21, RIL 10 = 21.17, RIL12 = 17.41 and RIL15 = 12.79 (averaged from [10] and [12]).

Determination of Amino Acids and Pigment

Composition Samples of tomato fruits (four per line) were harvested at the mature green stage (MG, when fruit is green but stops growing) and red ripe stage (RR, when fruit is completely red but still firm). Pericarp tissue of harvested fruits was obtained by removing the locule tissue and seeds and then they were stored at −80°C until analysis [2]. Pericarp tissue (0.5 - 1 g fresh weight) was extracted with 5 ml chloro form/methanol (1.5/3.5 v/v) and the amino acids relative composition was determined in the methanolic phase by derivatization with ninhydrin or o-phthaldialdehyde using an amino acids analyzer following [4]. Pigments were determined according to [3].

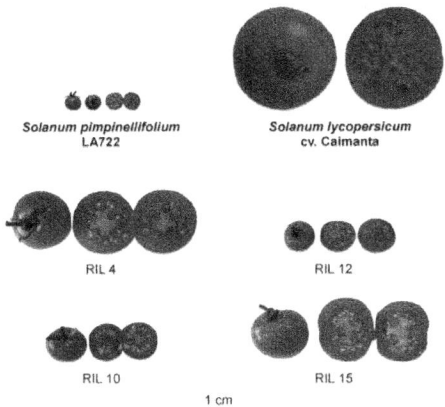

Figure : Fruits of four tomato Recombinant Inbred Lines (RIL) and their parents Solanum lycopersicum Caimanta and S. pimpinellifolium LA722, the experimental testers.

Statistical Analysis

Distribution's normality of total free amino acids and pigment contents (in $\mu mol \cdot mg^{-1}$ fresh weight and $mg \cdot 100 \ g^{-1}$ fresh weight, respectively) and each amino acid percent composition was analyzed following [13]. Comparisons were made by a hierarchical ANOVA for variables having normal distribution, in which the principal source of variation was ripening stage and the nested source of variation was genotype within ripening stage. Kruskall-Wallis test was applied for variables displaying not normal distribution. Mean values were calculated from three independent experiments in all cases. The Pearson correlation coefficients (r) among all variables (including the averaged shelf life values) were calculated [14]. Multivariate Analysis of Principal Components and Hierarchical Clustering with Ward method and Euclidean distances were used to identify metabolites (amino acids and pigments) mostly contributing to general variability and to assess the importance of each source of variation (genotype and ripening stage) in categorizing the obtained set of results [15] to get a data mining approach on metabolic changes occurring during tomato ripening.

RESULTS

Mean values of each amino acid relative molar content, and total amino acid and pigment contents of the pericarp fruit at MG and RR stages of the four RILs and the parents are in Table 1. All variables, except by total free amino acid content and the threonine and valine relative molar contents, were normally distributed (W > 0.95; ns). Significant differences (p < 0.05) were found among ripening stages and among genotypes within ripening stage in all cases except by the relative contents of asparagine and alanine and chlorophyll b content. For these variables, significant differences were found among genotypes within ripening stage but not among ripening stages. In all genotypes, the total amino acid content at RR stage was higher than at MG stage except by the tester Caimanta, in which no difference was detected among stages. In most cases, the highest relative amino acid content corresponded to glutamate, glutamine, and GABA though some genotypes (LA722 and RIL4) had relatively high relative asparagine content. Additionally, glutamate, glutamine and GABA performed oppositely given that the former increased from MG to RR while the latter two decreased in their relative molar content. As expected, chlorophyll a content diminished from MG to RR, chlorophyll b also decreased or remained constant, according to the genotype source of variation, while lycopene and beta-carotene increased among ripening stages, the magnitude of such changes widely depending on the genotype. In general, the wild tester LA722 showed the greatest values for all pigments. Just a few (34 of 190, i.e., nearly 18%)

correlation coefficients among metabolites were significant (data not shown). Some of the remarkable correlation coefficients were those between glutamate and glutamine ($r = -0.79$, $p < 0.01$), threonine, glycine, isoleucine, leucine, tyrosine and phenylalanine ($r > 0.75$, $p < 0.001$ in all cases), serine, alanine and valine ($r > 0.72$, $p < 0.01$ in all cases), isoleucine and lycopene ($r = -0.72$, $p < 0.01$), and lycopene and beta-carotene ($r = 0.87$, $p < 0.001$). On the other hand, GABA, asparagine, and chlorophyll a and b had no significant correlation with any metabolite, though chlorophyll b at MG was positively associated with fruit shelf life ($r = 0.90$, $p < 0.01$). Other metabolites associated with this latter trait were leucine at RR ($r = -0.92$, $p < 0.01$), alanine at RR ($r = -0.85$, $p < 0.05$) and lycopene at RR ($r = 0.88$, $p < 0.05$) but the previously reported correlation with glutamate at RR [6] was not confirmed in this experiment ($r = 0.21$, ns). The two first Principal Components (PC1 and PC2) explained the 62% of the total variation in metabolite composition of MG and RR tomato fruits. If PC3 was also considered, this percentage increases to 72% but only PC1 and PC2 will be further analyzed. Metabolite contributions to PC1 and PC2 (i.e., the respective eigenvalues and the association among each original variable and both PCs) are shown in Table 2. The metabolites having the more remarkable correlation coefficients (threonine, serine, glutamate, glutamine, glycine, isoleucine, leucine, tyrosine, phenylalanine, lycopene and beta-carotene) also had the greatest contribution to general variability and the highest association with PC1. Alanine and chlorophylls a and b were highly associated to PC2. Figure 2 shows the distribution of genotypes by ripening stage (and also of each metabolite) in the PC1 and PC2 coordinates. A clear separation by ripening stage was obtained in the axis of PC1, which explained the greatest proportion of total variability (46%, Table 2). It should be considered as an extent of variability according to ripening metabolism, indicating that threonine, serine, valine, glutamine, glycine, isoleucine, leucine, tyrosine, and phenylalanine relative molar content were higher at MG while glutamate relative content and lycopene and beta-carotene contents were higher at RR, though there were some genotype exceptions (RIL10 MG and RIL15 RR, Figure 2). PC2, which explained a lesser proportion of general variability (16%), better accounted for genotype biodiversity, and separations in this axis are more noticeable at the MG stage (Figure 2). Hence, genotypic differences are less important at RR. Hierarchical Clustering confirmed PC results. The dendrogram had a relatively high cophenetic correlation (0.70) and two well defined groups were obtained according to ripening stage, one including all genotypes at MG and RIL15 at RR, and the other, the remaining genotypes at RR (Figure 3). In the first group, separation among genotypes occurred at a higher distances that in the second group, RIL10 being a discrepant genotype at MG. RIL15 had the minor differences between ripening stages, while the most discrepant genotype at RR was the tester Caimanta, the cultivated parent.

Table 1: Relative molar composition of free amino acids, and total amino acid (AA) and pigment contents in the pericarp tissue of tomato fruits at two ripening stages in different selected genotypes (four Recombinant Inbred Lines—RIL—and their parents, Caimanta of Solanum lycopersicum and LA722 of S. pimpinellifolium)

Genotypes	Caimanta		LA722		RIL4		RIL10		RIL12		RIL15	
Metabolite/Ripening Stage	MG	RR	MG	RR	MG	RR	MG	RR	MG	RR	MG	RR
mg·100 g⁻¹ fresh weight												
Chlorophyll a	1.36	0.32	6.73	1.77	1.47	0.40	3.36	0.85	2.17	2.56	1.57	0.30
Chlorophyll b	0.67	1.39	2.27	1.99	0.85	1.05	4.37	0.55	1.68	3.56	0.79	0.43
Lycopene	0.75	15.15	0.96	22.68	0.64	12.94	1.27	24.47	1.71	19.90	0.17	7.93
Beta-carotene	0.24	3.98	0.52	6.55	0.49	3.70	0.72	3.22	0.43	2.56	0.46	1.45
Relative molar content of free amino acids (%)												
Aspartate	0.38	9.18	1.21	1.68	0.59	2.34	0.99	2.13	2.15	3.56	1.83	1.13
Threonine	7.58	1.25	3.76	0.97	6.78	1.72	1.59	0.21	1.92	0.24	2.48	2.74
Serine	8.02	1.74	8.69	2.81	5.34	3.28	2.05	0.81	14.09	1.08	5.59	1.83
Asparagine	10.40	3.53	17.68	34.31	6.63	19.72	9.44	4.31	2.33	1.17	13.48	6.04
Glutamate	13.35	58.26	16.09	44.74	19.88	36.28	4.86	61.74	26.37	78.83	6.72	47.11
Glutamine	25.58	13.66	24.98	3.57	29.78	23.16	17.80	3.47	14.93	6.10	44.24	20.08
Glycine	2.02	0.79	1.47	0.96	1.94	0.51	0.52	0.19	1.35	0.38	0.98	1.61
Alanine	2.48	2.75	5.03	2.44	1.74	2.46	1.97	1.47	6.20	2.61	1.56	2.76
Valine	4.16	1.24	3.39	2.34	3.30	0.86	2.54	1.14	7.44	0.98	1.92	1.31
Isoleucine	2.62	0.38	1.10	0.20	2.31	0.77	0.60	0.12	0.93	0.31	1.38	0.58
Leucine	1.54	0.53	0.46	0.33	1.57	0.51	0.43	0.11	0.75	0.22	1.05	0.63
Tyrosine	1.88	0.41	1.05	0.23	1.20	0.10	0.45	0.12	0.60	0.29	0.52	0.13
Phenylalanine	3.88	1.05	1.58	0.96	3.92	0.68	0.82	0.48	1.25	0.57	1.80	2.14
GABA	12.38	3.83	8.61	1.81	10.84	4.04	53.95	20.95	15.56	1.80	13.32	6.53
Others	3.73	1.40	4.90	2.65	4.18	3.56	1.99	2.75	4.13	1.86	3.13	5.38
Total AAs (μmol·mg⁻¹ fresh weight)	25.03	24.66	8.36	20.59	15.41	40.53	17.47	33.97	1.52	17.71	3.51	5.46

MG: fruits at mature green ripening stage, RR: fruits at red ripe ripening stage, Caimanta and LA722 were the experimental testers.

Table 2: Eigenvalue composition of Principal Components 1 and 2 (PC1 and PC2, respectively) and associations among metabolites and each Principal Component (a1 and a2, respectively)

Metabolite	PC1	a1	PC2	a2
Aspartate (%)	−0.18	−0.54	−0.05	−0.09
Threonine (%)	0.29	0.89	−0.20	−0.35
Serine (%)	0.24	0.72	0.24	0.43
Asparagine (%)	−0.02	−0.07	−0.04	−0.06
Glutamate (%)	−0.25	−0.77	−0.11	−0.20
Glutamine (%)	0.24	0.72	−0.10	−0.17
Glycine (%)	0.29	0.88	−0.10	−0.18
Alanine (%)	0.09	0.27	0.39	0.69
Valine (%)	0.21	0.65	0.27	0.49
Isoleucine (%)	0.30	0.91	−0.18	−0.32
Leucine (%)	0.29	0.86	−0.24	−0.42
Tyrosine (%)	0.27	0.83	−0.07	−0.12
Phenylalanine (%)	0.28	0.85	v0.25	−0.45
GABA (%)	0.03	0.09	0.24	0.42
Others (%)	0.21	0.64	0.01	0.02
Total AA ($\mu mol \cdot mg^{-1}$)	−0.16	−0.48	−0.28	−0.50
Chlorophyll a ($mg \cdot 100\ g^{-1}$)	0.09	0.26	0.41	0.72
Chlorophyll b ($mg \cdot 100\ g^{-1}$)	−0.09	−0.28	0.38	0.68
Lycopene ($mg \cdot 100\ g^{-1}$)	−0.29	−0.89	−0.13	−0.24
Beta-carotene ($mg \cdot 100\ g^{-1}$)	−0.26	−0.79	−0.15	−0.26

DISCUSSION

As previously reported, a wide range of variability was found in this experiment for tomato pericarp free amino acid composition, and total amino acids and pigments content [6,7,8]. Greatest variability was detected among mature green and red ripe stages than among genotypes, indicating that these attributes are primarily ripeningregulated [16] and constitute one of the biological changes taking place during such a transition [17]. In fact, as demonstrated by Hierarchical Clustering, ripening stages could be identified by each amino acid relative molar composition, and total free amino acids and pigments content. For the latter attributes, association is visually noticeable, but for amino acids composition, different physiological functions were proposed. Glutamate was proposed as a precursor of chlorophyll [18], hence the increase in its percent composition towards the end of ripening could be due to the cessation of chlorophyll biosynthesis. On the other hand, another role was claimed for glutamate [17], given that even a mutant defective in chlorophyll degradation also shown the increase at ripe stage [3]. As an "umami taste" is provoked by glutamate, the higher glutamate relative molar content in all tomato germplasm analyzed ([19,20] in addition to the already mentioned authors) could have an attracting to mammalian predators function [21]. Interestingly, genotype biodiversity was higher in fruits at mature green stage than at red ripe stage. As wild germplasm was analyzed here, this finding suggesting an evolutive conservation across different genetic backgrounds of optimum attributes to attract fruit predators so assuring seed dispersion [7]. Correlation between glutamate composition at RR and fruit shelf life proposed by [6] was not confirmed in this experiment, probably due to differences in gene frequencies among populations studied in both reports. Spontaneous mutant homozygote genotypes having marked pleiotropic effects on tomato many ripening associated traits were studied in that report, which could explain the significant correlation found in that study. However, significant associations of shelf life with pigments (chlorophyll b at mature green and lycopene at red ripe stages) and other amino acids (leucine and alanine at red ripe stages) were detected in this set of recombinant inbred lines and their parents. This interesting finding should be further explored, since the correlation coefficients were high and positive with pigments, and high and negative with amino acids. Accordingly, the wild parent LA722 was the genotype contributing the highest values of both pigments and fruit shelf life, and the lesser values of both amino acids at the corresponding ripening stages. Other important modification in amino acid metabolism during tomato ripening transition was the increase in asparagine levels in the wild tester LA722, also shown by RIL4. In previous reports [2-4,6], asparagine decreased from mature green to red ripe in different tomato genotypes (normal and mutant for ripening cultivars, the wild variety S. lycopersicum var. cerasiforme) as well as glutamine and GABA. These two latter were the most abundant amino acids together with the previously mentioned glutamate. The different performance of asparagine in LA722 and RIL4 was similar to the currently observed performance of glutamate in most tomato genotypes. Such an increase in glutamate was even observed in LA722 and RIL4, though the ratio glutamate at red ripe: glutamate at mature green stages was lower in both genotypes than in the remaining ones, hence alternative metabolic functions related to the ammonium assimilation/dissimilation cycle [16] during ripening transition could be taking place in the wild germplasm. The alternative function and parallel performance of glutamate/asparagine in these genotypes would also contribute to identify more consistently the RILs genetic background, since one of them were more similar to the wild parent while the others performed as the cultivated Caimanta, which evidences gene segregation and recombination during the selection process. Additionally, the lack of association among glutamate content and fruit shelf life reported by [6] could also be explained by this evident variability in ripening amino acids metabolism. Finally, the great contribution of threonine, serine, valine, glycine, isoleucine, tyrosine, and phenylalanine relative molar content to variability among ripening stages, detected by both single correlation and Principal Component analyses, deserves further studies to elucidate its physiological meaning, since there are no antecedents in their contribution to tomato ripening.

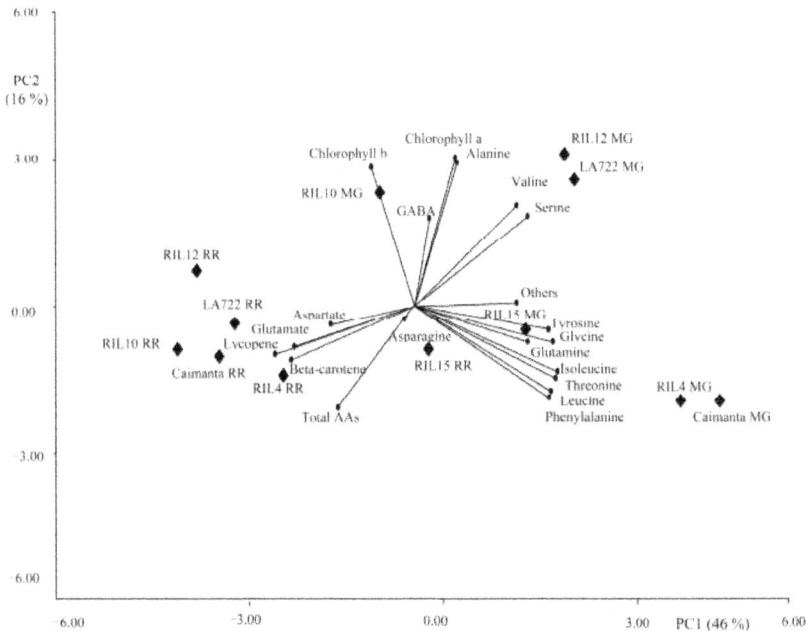

Figure 2: Distribution of metabolites (circles) and tomato genotypes (rhombus, four Recombinant Inbred Lines—RIL—and the parents Solanum lycopersicum cv. Caimanta and S. pimpinellifolium LA722, the experimental testers) at mature green (MG) and red ripe (RR) stages in the coordinates of Principal Components 1 and 2 (PC1 and PC2, respectively).

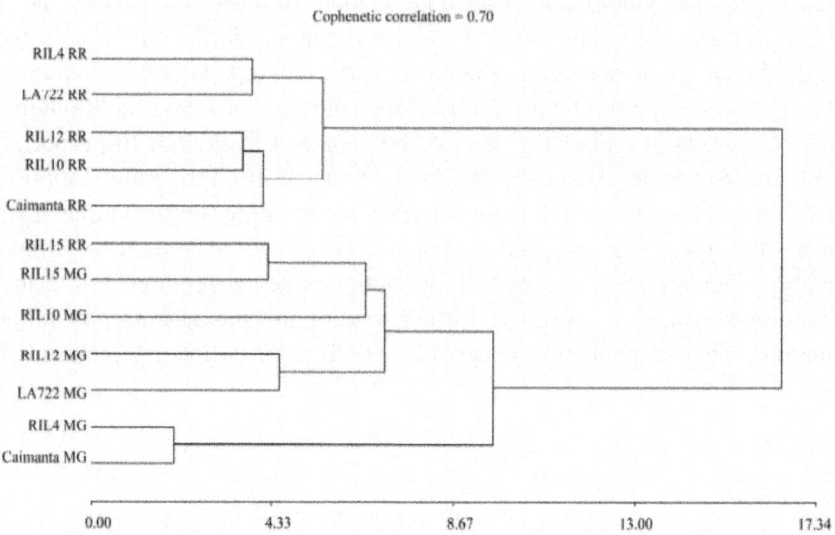

Figure 3: Dendrogram of the four tomato Recombinant Inbred Lines (RIL) and the parents Solanum lycopersicum cv. Caimanta and S. pimpinellifolium LA722, the experimental testers, at two ripening stages: mature green (MG) and red ripe (RR).

REFERENCES

1.   J. J. Giovannoni "Genetic Regulation of Fruit Development and Ripening," Plant Cell, Vol. 16, Supplement 1, September 2004, pp. 170-180. doi:10.1105/tpc.019158

2.   E. M. Valle, S. B. Boggio and H. W. Heldt, "Free Amino Acids Composition of Phloem Sap and Growing Fruit of Lycopersicon Esculentum," Plant Cell Physiology, Vol. 39, No. 4, June 1998, pp. 458-461.

3.   S. Bortolotti, S. B. Boggio, L. Delgado, E. G. Orellano and E. M. Valle, "Different Induction Patterns of Glutamate Metabolising Enzymes in Ripening Fruits of the Tomato Mutant Green Flesh," Physiologia Plantarum, Vol. 119, No. 3, May 2003, pp. 384-391. doi:10.1034/j.1399-3054.2003.00184.x

4.   S. B. Boggio, J. F. Palatnik, H. W. Heldt and E. M. Valle, "Changes in the Aminoacid Composition and Nitrogen Metabolizing Enzymes in Ripening Fruits of Lycopersicon Esculentum Mill," Plant Science, Vol. 159, No. 1, March 2000, pp. 125-133. doi:10.1016/S0168-9452(00)00342-3

5.   G. R. Rodríguez, G. R. Pratta, D. R. Liberatti, R. Zorzoli and L. A. Picardi, "Inheritance of Fruit Shelf Life and Other Tomato Fruit Quality

Traits Estimated from Crosses and Backcrosses of Selected Parents," Euphytica, Vol. 176, July 2010, pp. 137-147.

6.   G. Pratta, R. Zorzoli, S. B. Boggio, L. A. Picardi and E. M. Valle, "Glutamine and Glutamate Levels and Related Metabolizing Enzymes in Tomato Genotypes with Different Shelf-Life," Scientia Horticulturae, Vol. 100, No. 1-4, April 2004, pp. 341-347. doi:10.1016/j.scienta.2003.08.004

7.   J. A. Labate, S. Grandillo, T. Fulton, S. Muños, A. L. Caicedo, I. Peralta, Y. Ji, R. T. Chetelat, J. W. Scout, M. J. Gonzalo, D. Francis, W. Yang, E. van der Knaap, A. M. Baldo, B. Smith-White, L. A. Mueller, J. P. Prince, N. E. Blanchard, D. V. Storey, M. R. Stevens, M. D. Robbins, J. F. Wang, B. E. Liedl, M. A. O'Connell, J. R. Stommel, K. Auki, Y. Iijima, A. J. Slade, S. R. Hurst, D. Loeffler, M. N. Steine, D. Vafeados, C. McGuire, C. Freeman, A. Amen, J. Goodstal, D. Facciotti, J. van Eck and M. Causse, "Tomato," In: C. Kole, Ed., Genome Mapping and Molecular Breeding in Plants, Vol. 5: Vegetables, Springer-Verlag, Berlin-Heidelberg, June 2007.

8.   N. Schauer, D. Zamir and A. Fernie, "Metabolic Profiling of Leaves and Fruits of Wild Species Tomato: A Survey of the Solanum Lycopersicum Complex," Journal of Experimental Botany, Vol. 56, No. 410, April 2005, pp. 297-307 (Special Issue).

9.   R. Zorzoli, G. Pratta and L. A. Picardi, "Variabilidad Para la Vida Postcosecha y El Peso de Los Frutos en Tomate Para Familias F3 de un HÍBrido InterespecÍFico," Pesquisa Agropecuaria Brasileira, Vol. 35, November 2000, pp. 2423-2427. doi:10.1590/S0100-204X2000001200013

10. G. R. Rodriguez, G. R. Pratta, R. Zorzoli and L. A. Picardi, "Recombinant Lines Obtained from an Interspecific Cross among Lycopersicon Species Selected by Fruit Weight and Fruit Shelf Life," Journal of the American Society for Horticultural Science, Vol. 131, No. 5, October 2006, pp. 651-656.

11. M. Gallo, G. R. Rodríguez, R. Zorzoli, L. A. Picardi and G. R. Pratta, "Proteómica de la Madurez del Tomate," Revista de Investigaciones de la Facultad de Ciencias Agrarias de la Universidad Nacional de Cuyo, Vol. 42, December 2010, pp. 119-133.

12. G. R. Pratta, G. R. Rodríguez, R. Zorzoli, E. M. Valle and L. A. Picardi, "Phenotypic and Molecular Characterization of Selected Tomato Recombinant Inbred Lines Derived from a Cross Solanum Lycopersicum × S. Pimpinellifolium," Journal of Genetics, 2001, in Press.

13. S. S. Shapiro and M. B. Wilk, "An Analysis of Variance Test for Normality Complete Samples," Biometrika, Vol. 52, No. 3-4, June 1965, pp. 591-611.

14. M. Kearsey and H. Pooni, "The Genetical Analysis of Quantitative Traits," Chapman and Hall, London, May 1996.

15. Q. S. Du, Z. Q. Jiang, W. Z. He, D. P. Li and K. C. Chou, "Amino Acid Principal Component Analysis (AAPCA) and Its Applications in Protein Structural Class Prediction," Journal of Biomolecules Structucture and Dynamics, Vol. 23, No. 6, August 2006, pp. 635-640.

16. B. G. Forde and P. J. Lea, "Glutamate in Plants: Metabolism, Regulation, and Signaling," Journal of Experimental Botany, Vol. 58, No. 9, November 2007, pp. 2339-2358. doi:10.1093/jxb/erm121

17. A. Sorrequieta, G. Ferraro, S. B. Boggio and E. M. Valle, "Free Amino Acid Production during Tomato Fruit Ripening: A Focus on L-Glutamate," Amino Acids, Vol. 38, No. 5, September 2009, pp. 1523-1532. doi 10.1007/s00726-009-0373-1.

18. F. Carrari and A. Fernie, "Metabolic Regulation Underlying Tomato Fruit Development," Journal of Experimental Botany, Vol. 57, No. 9, February 2007, pp. 1883- 1897.

19. F. Mounet, M. Lemaire-Chamley, M. I. Maucourt, C. Cabasson, J. L. Giraudel, C. Deborde, R. Lessire, P. Gallusci, A. Bertrand, M. Gaudillere, C. Rothan, D. Rolin and A. Moinga, "Quantitative Metabolic Profiles of Tomato Flesh and Seeds during Fruit Development: Complementary Analysis with ANN and PCA," Metabolomics, Vol. 3, No. 3, April 2007, pp. 273-287. doi:10.1007/s11306-007-0059-1

20. Y. Semel, N. Schauer, U. Roessner, D. Zamir and A. R. Ferni, "Metabolite Analysis for the Comparison of Irrigated and Non-Irrigated Field Grown Tomato of Varying Genotype," Metabolomics, Vol. 3, No. 3, April 2007, pp. 289-295.doi:10.1007/s11306-007-0055-5

21. S. D. Roper and N. Chaudhar, "Processing Umami and Other Tastes in Mammalian Taste Buds," Annals of the New York Academy of Sciences, Vol. 1170, July 2009, pp. 60-65. doi:10.1111/j.1749-6632.2009.04107.x.

Chapter 9

CHARACTERIZATION OF PHOTOSYNTHETIC PIGMENT COMPOSITION, PHOTOSYSTEM II PHOTOCHEMISTRY AND THERMAL ENERGY DISSIPATION DURING LEAF SENESCENCE OF WHEAT PLANTS GROWN IN THE FIELD

Congming Lu[1,3], Qingtao Lu[1], Jianhua Zhang[2] and Tingyun Kuang[1]

[1] Photosynthesis Research Centre, Institute of Botany, Chinese Academy of Sciences, Beijing 100093, PR China

[2] Department of Biology, Hong Kong Baptist University, Kowloon, Hong Kong, PR China

ABSTRACT

Photosynthetic pigment composition and photosystem II (PSII) photochemistry were characterized during the flag leaf senescence of wheat plants grown in the field. During leaf senescence, neoxanthin and β-carotene decreased concomitantly with chlorophyll, whereas lutein and xanthophyll cycle pigments were less affected, leading to increases in lutein/chlorophyll and xanthophyll cycle pigments/chlorophyll ratios. The chlorophyll a/b ratio also increased. With the progression of senescence, the maximal efficiency of PSII photochemistry decreased only slightly in the early morning (low light conditions), but substantially at midday (high light conditions). Actual PSII efficiency, photochemical quenching and the efficiency of excitation capture by open PSII centres decreased significantly both early in the morning and at midday and such decreases were much greater at midday than in the early morning. At the same time, non-photochemical quenching, zeaxanthin and antheraxanthin contents at the expense of violaxanthin increased both early in the morning and at midday, with a greater increase at midday. The results in the present study suggest that a down-regulation of PSII occurred in senescent leaves and that the xanthophyll cycle plays a role in the protection of PSII from photoinhibitory damage in senescent leaves by dissipating excess excitation energy, particularly when exposed to high light.

INTRODUCTION

Leaf senescence is the sequential degradation process that leads to a massive mobilization and export of nitrogen and minerals and eventually to leaf death (Buchanan-Wolleston, 1997). The most remarkable event in leaf senescence is the disassembly of the photosynthetic apparatus within chloroplasts and thus the concomitant decrease in photosynthetic activity (Woolhouse, 1984; Grover and Mohanty, 1992). A key event leading to the decrease in photosynthetic activity during leaf senescence is associated with the loss of RuBP carboxylase/oxygenase (Crafts-Brandner *et al.*, 1990; Grover, 1993). The loss of major thylakoid proteins is usually delayed to the later senescence phases relative to stromal enzymes and is less directly correlated to the decrease in photosynthetic activity (Okada *et al.*, 1992; Mae *et al.*, 1993). Thus, the decrease in photosynthetic CO_2 fixation during leaf senescence normally occurs earlier than that in the maximal efficiency of photosystem II (PSII) photochemistry (Humbeck *et al.*, 1996). Previous studies in this laboratory have shown that PSII activity seemed to be affected only slightly whereas a substantial decrease in photosynthetic capacity occurred during leaf senescence (Lu and Zhang, 1998*a*, *b*).

A substantial decrease in photosynthetic capacity accompanied by only a slight decrease in PSII photochemistry in senescent leaves can potentially expose the senescent leaves to excess excitation energy, which, if not safely dissipated, may result in photodamage to PSII because of an overreduction of reaction centres (Demmig-Adams and Adams, 1992). Excess excitation energy can be harmlessly dissipated in the antennae complexes of PSII as heat through a process which involves the xanthophyll cycle and a low thylakoid pH. The xanthophyll cycle pigments zeaxanthin (Z) and antherxanthin (A) are formed from violaxanthin (V) under conditions of excess excitation energy and are both thought to be involved in the photoprotective dissipation process (Demmig-Adams and Adams, 1992; Gilmore, 1997).

The role of the xanthophyll cycle under conditions of cold-temperature stress (Verhoeven *et al.*, 1999), water stress (Munné-Bosch and Alegre, 2000), UV-B elevation (Levall and Bornman, 2000), nitrogen deficiency (Verhoeven *et al.*, 1997), and iron deficiency (Morales *et al.*, 2000), has been widely investigated. However, the changes in the xanthophyll cycle pigments and other photosynthetic pigments during leaf senescence have not been fully characterized. Moreover, it is not yet clear whether the xanthophyll cycle plays any role in the dissipation of excess light energy in senescent leaves with decreased photosynthetic capacity.

Senescence is usually studied in some model systems, in which leaf senescence is induced by either incubating detached leaves or whole plants in the dark. Although these systems provide many practical advantages and valuable information on the mechanisms of leaf senescence phenomena, effects observed in these model systems are complicated by the wounding of the tissues and/or of light/dark transition. Even experiments with intact plants grown under a diurnal light/dark cycle in a controlled growth chamber may not represent all aspects of senescence occurring in a natural habitat, where simultaneous changes in environmental factors affect the senescent process (Smart, 1994). Until now, only few reports on the senescence of field-grown plants have been published (Adams *et al.*, 1990; Humbeck *et al.*, 1996; Murchie*et al.*, 2000). Moreover, the characterization of PSII photochemistry, thermal energy dissipation and the xanthophyll cycle in either senescent leaves induced by above model systems or in naturally senescent leaves of plants grown under natural habitat has received little attention. It is of interest to perform a detailed characterization of photosynthetic pigment composition, in particular these pigments related to the xanthophyll cycle, and PSII photochemistry during leaf senescence occurring in natural habit.

In the present study, flag leaves were chosen to investigate the changes in the photosynthetic characteristics during leaf senescence of wheat plants grown in the field since the photosynthesis of flag leaves is most important for grain-filling. The objectives of this study were (1) to characterize fully photosynthetic pigment composition, (2) to examine if and how down-regulation of PSII happens in senescent leaves when exposed to excess light energy, and (3) to determine if the xanthophyll cycle plays a role in dissipating excess light energy during leaf senescence. To these ends, the changes in the xanthophyll cycle pigments and their de-epoxidation status have been investigated as well as other photosynthetic pigment composition and fluorescence quenching parameters during leaf senescence in response to low (early morning) and high light conditions (at midday).

MATERIALS AND METHODS

Plant Material

Winter wheat (*Triticum aestivum* L. cv. Beijing 3348) was grown in 1999–2000 in a field situated at Beijing. Seeds were sown on 2 October, 1999 and a clay soil was used. Nutrients and water were supplied sufficiently throughout and thus potential nutrients and drought stress were avoided.

Flag leaves were used for all analyses beginning on 12 May at the stage of flowering and ending on 10 June. During this period, the weather showed the typical Beijing spring weather with a mean daily air temperature between

19–23 °C and an average PPFD at the leaf level around $1400\,\mu mol\,m^{-2}s^{-1}$ at midday.

Modulated Chl fluorescence measurements were made in attached leaves in the field with a PAM-2000 portable fluorometer (Walz, Effeltrich, Germany) connected to a notebook computer with data acquisition software (DA-2000, Heinz, Walz). Essentially, the experimental protocol of Demmig-Adams et al. was followed (Demmig-Adams et al., 1996). The minimal fluorescence level (F_o) in dark-adapted state was measured by the measuring modulated light which was sufficiently low ($<0.1\,\mu mol\,m^{-2}s^{-1}$) not to induce any significant variable fluorescence. To determine the minimal fluorescence level during illumination (F'_o), a black cloth was rapidly placed around the leaf and the leaf-clip holder in the presence of far-red light ($7\,\mu mol\,m^{-2}s^{-1}$) in order to oxidize the PSII centres fully. Upon darkening of the leaf, fluorescence dropped to the F'_o level and immediately rose again within several seconds. The maximal fluorescence level in the dark-adapted state (F_m) and the maximal fluorescence level during natural illumination (F'_m) were measured by a 0.8 s saturating pulse at $8000\,\mu mol\,m^{-2}\,s^{-1}$. F_m was measured after 30 min of dark adaptation. F'_m was measured when morning and midday PPFDs were approximately 200 and $1400\,\mu mol\,m^{-2}s^{-1}$, respectively. The steady-state fluorescence level during exposure to natural illumination (F_s) was also measured when morning and midday PPFDs were approximately 200 and $1400\,\mu mol\,m^{-2}s^{-1}$, respectively. All measurements of F_o and F'_o were performed with the measuring beam set to a frequency of 600 Hz, whereas all measurements of F_m and F'_m were performed with the measuring beam automatically switching to 20 kHz during the saturating flash.

The actual PSII efficiency (Φ_{PSII}) and the efficiency of excitation capture by open PSII centres were calculated as $(F'_m - F_s)/F'_m$ and F'_v/F'_m, respectively (Genty et al., 1989). Photochemical quenching (q_p) was calculated as $(F'_m - F_s)/(F'_m - F'_o)$ (van Kooten and Snel, 1990). Non-photochemical quenching (NPQ) was calculated as $(F_m/F'_m)-1$ (Bilger and Björkman, 1990). It should be noted that the F_m values at predawn which were fully recovered after the 30 min dark adaptation were used for the calculation of NPQ.

Pigment Analyses

Leaf samples were taken and immediately frozen in liquid nitrogen. Leaf samples were extracted in ice-cold 100% acetone and the pigment extracts were filtered through a 0.45 μm membrane filter. Pigments were separated and quantified by HPLC essentially as described earlier (Thayer and Björkman, 1990).

RESULTS

Pigment Composition

Total chlorophyll (Chl) content and the ratio of Chl *a/b* did not change considerably until 20 d after flowering. From the 20th to the 28th day, the Chl content decreased remarkably from 524 to 110 μmol m⁻² and the ratio Chl *a/b* increased from 2.4 to 3.3. No difference was observed in the total Chl content and the ratio Chl *a/b* between samples taken early in the morning and at midday (Fig. 1).

Figure 2 shows the changes in carotenoid contents expressed on a total chlorophyll basis during the senescence of flag leaves. Neoxanthin and β-carotene contents were largely unchanged after flowering, both early in the morning and at midday (Fig. 2A, B). Lutein content remained unchanged until the 20th day after flowering and then increased significantly from 70 to 150 mmol mol⁻¹ Chl on the 28th day. There was no significant difference in the lutein content between the samples taken early in the morning and at midday (Fig. 2C). V, A and Z contents also remained unchanged until 20 d after flowering and then increased significantly starting from 20 d after flowering. However, the V content was lower and the A and Z contents were higher in samples taken at midday than in samples taken early in the morning (Fig. 2E, F, G).

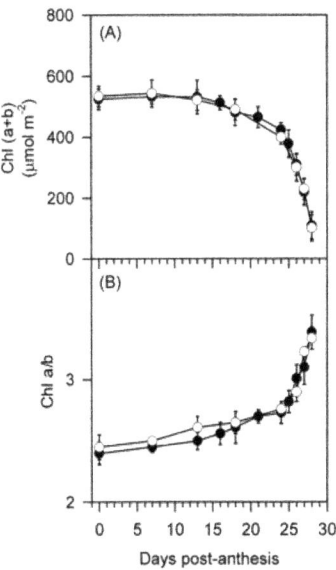

Figure. 1: (A) Total chlorophyll (*a+b*) content and (B) the ratio Chl*a/b* during leaf senescence of wheat grown in the field. Samples were harvested early in the morning

(•) and at midday (○). Morning and midday PPFDs were approximately 200 and 1400 µmol photons m⁻²s⁻¹. Data are means±SE of 3–5 independent measurements.

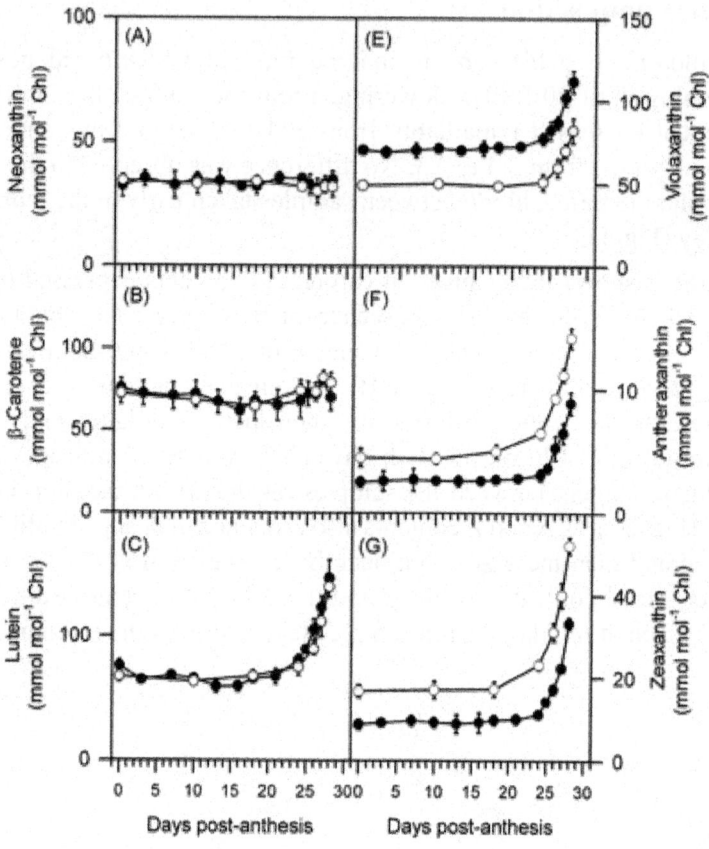

Figure. 2: (A) Neoxanthin, (B) β-carotene, (C) lutein, (E) violaxanthin, (F) antheraxanthin, and (G) zeaxanthin contents expressed on a total chlorophyll basis during leaf senescence of wheat grown in the field. Samples were harvested early in the morning (•) and at midday (○). Morning and midday PPFDs were approximately 200 and 1400 µmol photons m⁻² s⁻¹. Data are means±SE of 3 independent measurements.

Figure 3 shows the changes in several fluorescence parameters during the senescence of flag leaves. When measured early in the morning, the maximal efficiency of PSII photochemistry (F_v/F_m) was virtually kept at a high value of around 0.82 until 26 d after flowering and decreased slightly only on the 28th day with a value of 0.72. When measured at midday, F_v/F_m maintained a value of around 0.82 until 20 d after flowering; thereafter it decreased significantly with senescence progressing to a value of 0.45 on the 28th day (Fig. 3A). The actual PSII efficiency (Φ_{PSII}), the efficiency of excitation capture by open PSII centres (F'_v/F'_m), and the photochemical quenching coefficient (q_p) decreased significantly only from 20 d after flowering and their decreases were much greater when measured at midday than early in the morning (Fig. 3B, C, D). Non-photochemical quenching (NPQ) was largely unchanged until around 20 d after flowering both in the morning and at midday and increased significantly thereafter. The NPQ values were much higher at midday than early in the morning (Fig. 3E).

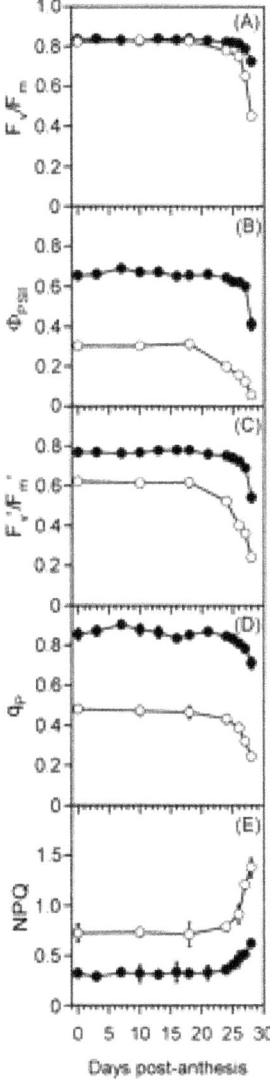

Figure. 3: (A) The maximal efficiency of PSII photochemistry (F_v/F_m), (B) the actual PSII efficiency (Φ_{PSII}), (C) the efficiency of excitation capture by open

PSII reaction centres (F'_v/F'_m), (D) the photochemical quenching coefficient (q_p), and (E) the non-photochemical quenching (NPQ) during leaf senescence of wheat grown in the field. Measurements of F_v/F_m were made in attached leaves in the early morning (•) and at midday (○) after a 30 min dark adaptation. For Φ_{PSII}, F'_v/F'_m, q_p, and NPQ, measurements were made on attached leaves in the early morning (•) and at midday (○) during natural illumination. Morning and midday PPFDs were approximately 200 and 1400 μmol photons m^{-2} s^{-1}. Data are means±SE of 5–6 independent measurements.

DISCUSSION

Characterization of Pigment Composition

In the present study, changes in the photosynthetic pigment composition were investigated during the senescence of flag leaves of wheat plants grown in the field. The changes in pigment contents demonstrate that senescence of flag leaves was not induced until around 20 d after flowering. The leaves in this period are defined here as 'the control leaves'. Senescence of flag leaves started 20 d after flowering. With senescence progressing, absolute photosynthetic pigment contents expressed on a leaf area basis decreased dramatically. However, they degraded to different extents. Neoxanthin and β-carotene decreased equally with chlorophyll, whereas lutein and the carotenoids within the xanthophyll cycle were much less affected. Thus, a relative enrichment in lutein and the xanthophyll cycle pigments was observed (Fig. 2). In addition, an increase in the Chl a/b ratio in senescent leaves suggests that Chl b was more affected than Chl a (Fig. 1). These results indicate that there was a change in the photosynthetic pigment stoichiometry during leaf senescence.

The carotenoids within the xanthophyll cycle underwent epoxidations and de-epoxidations to different extents in control and senescent leaves in response to different light conditions. During the night, most of the xanthophyll cycle pool was in the epoxidated form V in both control and senescent leaves with (Z+A)/(V+Z+A) ratio around 0.05 and 0.12, respectively. Either in the early morning or at midday, the de-epoxidated forms A and Z increased significantly with senescence progressing at the expense of V. Such an increase in the de-epoxidated forms A and Z was much greater at midday than in the morning (Fig. 2E, F, G).

Down-Regulation of PSII Efficiency

The maximal efficiency of PSII photochemistry (F_v/F_m) decreased only slightly when measured in the early morning, but decreased substantially in severely senescent leaves at midday. Such a midday decrease in F_v/F_m became larger with senescence progressing and could recover to the values that were similar to those in the morning (Fig. 3A). Similarly, compared to the control leaves, the actual PSII efficiency (Φ_{PSII}) decreased markedly in senescent leaves both early in the morning and at midday. Such a decrease was more evident at midday than early in the morning and could also recover to the values that were similar to those in the morning (Fig. 3B). These reversible changes in F_v/F_m and Φ_{PSII} can be ascribed as a down-regulation of PSII that may reflect the protective or regulatory mechanisms to avoid photodamage to the photosynthetic apparatus (Krause, 1988; Demmig-Adams, 1990; Demmig-Adams and Adams, 1992). Such a down-regulation of PSII in senescent leaves can be explained by the increase in the proportion of closed PSII centres (estimated from decreased q_p) and the decrease in the efficiency of excitation energy capture (F'_v/F'_m) (Fig. 3C, D).

A considerable decrease in F'_v/F'_m in senescent leaves both in the morning and at midday, but only a slight decrease in F_v/F_m measured in the morning, suggests that the decreased F'_v/F'_m in senescent leaves could be associated with an increase in energy dissipation in the PSII antennae (Demmig-Adams et al., 1995, 1996; Demmig-Adams and Adams, 1996). The data from this study show that under low light conditions (early in the morning), an increase in non-photochemical quenching (NPQ) with the progression of senescence was accompanied by an increase in A and Z contents, when compared to the control leaves (Figs 2F, G; 3E), suggesting that the xanthophyll cycle-related thermal dissipation was significantly enhanced already at low light in senescent leaves (Horton et al., 1996; Gilmore, 1997). Under high light conditions (at midday), senescent leaves had a greater increase in NPQ, which was again associated with a greater increase in A and Z contents (Figs 2F, G; 3E), indicating that a higher level of thermal dissipation involved in the xanthophyll cycle occurred in senescent leaves when exposed to high light. The results in the present study suggest that xanthophyll cycle plays a role in dissipating excess light energy during leaf senescence, particularly when exposed to high light.

In summary, the photosynthetic pigment composition and PSII photochemistry during flag leaf senescence of wheat plants grown in the field has been investigated. These results indicate that there was a change in the photosynthetic pigment stoichiometry during leaf senescence and that a down-regulation of PSII occurred in senescent leaves particularly when exposed to high light. The present results demonstrate that the xanthophyll cycle-related thermal dissipation in the PSII antennae was enhanced significantly in senescent leaves, which may protect the photosynthetic apparatus from photoinhibitory damage in senescent leaves when exposed to high light.

ACKNOWLEDGMENTS

This work was supported by the Program of 100 Distinguished Young Scientists of the Chinese Academy of Sciences, the State Key Basic Research and Development Plan of China (No. G1998010100), the Innovative Foundation of the Laboratory of Photosynthesis Research, Institute of Botany, Chinese Academy of Sciences (CL), as well as the Hong Kong Baptist University (FRG grant) (CL, JZ).

REFERENCES

1. Adams III WW, Winter K, Schreiber U, Schramel P. 1990. Photosynthesis and chlorophyll characteristics in relation to changes in pigment and element composition of leaves of Platanus occidentalis L. during autumnal leaf senescence. Plant Physiology 93, 1184–1190.

2. Bilger W, Bjo¨rkman O. 1990. Role of the xanthophyll cycle in photoprotection elucidated by measurements of light-induced absorbance changes, fluorescence and photosynthesis in Hedera canariensis. Photosynthesis Research 25, 173–185.

3. Buchanan-Wollaston V. 1997. The molecular biology of leaf senescence. Journal of Experimental Botany 48, 181–199.

4. Crafts-Brandner SJ, Salvucci ME, Egli DB. 1990. Changes in ribulose bisphosphate carboxylaseuoxygenase and ribulose 5-phosphate kinase abundance and photosynthetic capacity during leaf senescence. Photosynthesis Research 23, 223–230.

5. Demmig-Adams B. 1990. Carotenoids and photoprotection: a role for the xanthophyll zeaxanthin. Biochimica et Biophysica Acta 1020, 1–24.

6. Demmig-Adams B, Adams III WW. 1992. Photoprotection and other responses of plants to high light stress. Annual Review of Plant Physiology and Plant Molecular Biology 43, 599–626.

7. Demmig-Adams B, Adams III WW. 1996. Xanthophyll cycle and light stress in nature: uniform response to excess direct sunlight among higher plant species. Planta 198, 460–470.

8. Demmig-Adams B, Adams III WW, Barker DH, Logan BA, Bowling DR, Verhoeven AS. 1996. Using chlorophyll fluorescence to assess the fraction of absorbed light allocated to thermal dissipation of excess excitation. Physiologia Plantarum 98, 253–264.

9. Demmig-Adams B, Adams III WW, Barker DH, Logan BA, Verhoeven AS. 1995. Xanthophyll cycle-dependent energy dissipation and flexible PSII efficiency in plants acclimated to light stress. Australian Journal of Plant Physiology 22, 249–260.

10. Genty B, Briantais J-M, Baker NR. 1989. The relationship between the quantum yield of photosynthetic electron transport and quenching of chlorophyll fluorescence. Biochimica et Biophysica Acta 990, 87–92.

11. Gilmore AM. 1997. Mechanistic aspects of xanthophyll cycledependent photoprotection in higher plant chloroplast and leaves. Physiologia Plantarum 99, 197–209.

12. Grover A, Mohanty P. 1992. Leaf senescence-induced alterations in structure and function of higher plant chloroplasts. In: Abrol YP, Mohanty P, Govindjee, eds. Photosynthesis: photoreactions to plant productivity. The Netherlands: Kluwer Academic Publishers, 225–255.

13. Grover A. 1993. How do senescing leaves lose photosynthetic activity? Current Science 64, 226–234.

14. Horton P, Ruban AV, Walters RG. 1996. Regulation of light harvesting in green plants. Annual Review of Plant Physiology and Plant Molecular Biology 47, 655–684.

15. Humbeck K, Quast S, Krupinska K. 1996. Functional and molecular changes in the photosynthetic apparatus during senescence of flag leaves from field-grown barley plants. Plant, Cell and Environment 19, 337–344.

16. Krause GH. 1988. Photoinhibition of photosynthesis. An evaluation of damaging and protective mechanisms. Physiologia Plantarum 74, 566–574.

17. Levall M, Bornman JF. 2000. Differential response of a sensitive and tolerant sugar beet line to Cercospora beticola infection and UV-B radiation. Physiologia Plantarum 109, 21–27.

18. Lu C-M, Zhang J. 1998a. Modifications in photosystem II photochemistry in senescent leaves of maize plants. Journal of Experimental Botany 49, 1671–1679.

19. Lu C-M, Zhang J. 1998b. Changes in photosystem II function during senescence of wheat leaves. Physiologia Plantarum 104, 239–247.

20. Mae T, Thomas H, Gay AP, Makino A, Hidema J. 1993. Leaf development in Lolium temulentum: photosynthesis and photosynthetic proteins in leaves senescing under different irradiances. Plant and Cell Physiology 34, 391–399.

21. Morales F, Belkhodja R, Abadı́a A, Abadı́a J. 2000. Photosystem II efficiency and mechanisms of energy dissipation in irondeficiency, field-grown pear tress (Pyrus communis L.). Photosynthesis Research 63, 9–21.

22.   Munne'-Bosch S, Alegre L. 2000. The xanthophyll cycle is induced by light irrespective of water status in field-grown lavender (Lavandula stoechas). Physiologia Plantarum 108, 147–151.

23.   Murchie EH, Chen Y, Hubbart S, Pen S, Horton P. 2000. Interactions between senescence and leaf orientation determine in situ patterns of photosynthesis and photoinhibition in field-grown rice. Plant Physiology 119, 553–563.

24.   Okada K, Inoue Y, Satoh K, Katoh S. 1992. Effects of light on degradation of chlorophyll and proteins during senescence of detached rice leaves. Plant and Cell Physiology 33, 1183–1191.

25.   Smart CM. 1994. Gene expression during leaf senescence. The New Phytologist 126, 419–448.

26.   Thayer SS, Bjo"rkman O. 1990. Leaf xanthophyll content and composition in sun and shade determined by HPLC.Photosynthesis Research 23, 331–343.

27.   van Kooten O, Snel JFH. 1990. The use of chlorophyll fluorescence nomenclature in plant stress physiology.Photosynthesis Research 25, 147–150.

28.   Verhoeven AS, Adams III WW, Demmig-Adams B. 1997. Enhanced employment of the xanthophyll cycle and thermal energy dissipation in spinach exposed to high light and N stress. Plant Physiology 113, 817–824.

29.   Verhoeven AS, Adams III WW, Demmig-Adams B, Groce R, Bassi R. 1999. Xanthophyll cycle pigment localization and dynamics during exposure to low temperatures and light stress in Vinca major. Plant Physiology 120, 727–737.

30.   Woolhouse HW. 1984. The biochemistry and regulation of senescence in chloroplasts. Canadian Journal of Botany 62, 2934–2942.

Chapter 10

A SMALL INDEL MUTATION IN AN ANTHOCYANIN TRANSPORTER CAUSES VARIEGATED COLOURATION OF PEACH FLOWERS

Jun Cheng[1,2], Liao Liao[1], Hui Zhou[1,2], Chao Gu[1], Lu Wang[1] and Yuepeng Han[1]

[1]Key Laboratory of Plant Germplasm Enhancement and Specialty Agriculture, Wuhan Botanical Garden of the Chinese Academy of Sciences, Wuhan, 430074, P.R. China

[2]Graduate University of Chinese Academy of Sciences, 19A Yuquanlu, Beijing, 100049, P.R. China

ABSTRACT

The ornamental peach cultivar 'Hongbaihuatao (HBH)' can simultaneously bear pink, red, and variegated flowers on a single tree. Anthocyanin content in pink flowers is extremely low, being only 10% that of a red flower. Surprisingly, the expression of anthocyanin structural and potential regulatory genes in white flowers was not significantly lower than that in both pink and red flowers. However, proteomic analysis revealed a GST encoded by a gene—regulator involved in anthocyanin transport (*Riant*)—which is expressed in the red flower, but almost undetectable in the variegated flower. The *Riant* gene contains an insertion-deletion (indel) polymorphism in exon 3. In white flowers, the *Riant* gene is interrupted by a 2-bp insertion in the last exon, which causes a frameshift and a premature stop codon. In contrast, both pink and red flowers that arise from bud sports are heterozygous for the *Riant* locus, with one functional allele due to the 2-bp deletion or a novel 1-bp insertion. Southern blot analysis indicated that the *Riant* gene occurs in a single copy in the peach genome and it is not interrupted by a transposon. The function of the*Riant* gene was confirmed by its ectopic expression in the *Arabidopsis tt19*mutant, where it complements the anthocyanin phenotype, but not the proanthocyanidin pigmentation in seed coat. Collectively,these results indicate that a small indel mutation in the *Riant* gene, which is not the result of a transposon insertion or excision, causes variegated colouration of peach flowers.

INTRODUCTION

Colouration is an important agronomic trait that contributes to the ornamental value of plants, which is determined by three major classes of plant pigments: anthocyanin, chlorophyll, and carotenoid. In flowers, anthocyanin is the major pigment and confers red, violet, or blue colours (Tanaka *et al.*, 2008). The anthocyanin biosynthetic pathway has been well studied in a variety of plants (Winkel-Shirley, 2001; Grotewold, 2006). This pathway starts with a condensation of malonyl-CoA and 4-coumaroyl CoA, and is catalysed by multiple enzymes including chalcone synthase (CHS), chalcone isomerase (CHI), flavanone 3-hydroxylase (F3H), flavonoid 3′-hydroxylase (F3′H), flavonoid 3′5′-hydroxylase (F3′5′H), dihydroflavonol 4-reductase (DFR), and UDPG-flavonoid glucosyltransferase (UFGT), to generate anthocyanins. Subsequently, water-soluble anthocyanin is transported into the vacuole.

Two mechanisms have been proposed for anthocyanin transport from the site of synthesis to the vacuole: vesicle-mediated transport and transporter-mediated transport (Grotewold and Davis, 2008; Zhao and Dixon, 2010). The evidence for the former mechanism comes from microscopy observations that anthocyanin accumulates initially in vesicle-like structures alongside the tonoplast that merge with the central vacuole (Zhang *et al.*, 2006; Poustka *et al.*, 2007; Conn *et al.*, 2010). The other mechanism is supported by transporter proteins located in the tonoplast, including multidrug resistance-associated proteins (MRP) and multidrug and toxic compound extrusion (MATE) transporters, shown to mediate anthocyanin transport (Debeaujon *et al.*, 2001; Goodman *et al.*, 2004;Grotewold, 2004; Gomez *et al.*, 2009; Zhao and Dixon, 2009; Francisco *et al.*, 2013). In addition, GSTs also have an essential role in transport of anthocyanins from the ER to the vacuole (Larsen *et al.*, 2003; Conn *et al.*, 2008; Gomez *et al.*, 2011). Initially, GST was proposed to conjugate anthocyanins with glutathione to form stable water-soluble conjugates, which were then transported into vacuoles by ABC transmembrane transporters (Marrs *et al.*, 1995). However, no anthocyanin–glutathione conjugates have been found *in vivo*. Instead, evidence suggests that GST functions as an anthocyanin carrier that may escort anthocyanins from the ER to the tonoplast (Mueller *et al.*, 2000; Sun *et al.*, 2012).

Variegation (variable colouration with patches of different colours) is one of the quality parameters sought after by the plant ornamental industry. Variegation is a common phenomenon, with the mechanism of variegation being studied in numerous plant species. In maize, insertion and excision of transposable elements located in the promoter region of the bronze gene encoding GST causes unstable pigmentation in kernels (Schiefelbein*et al.*, 1988). Transposable elements are also responsible for the unstable pigmentation phenotype in ornamental plants such as petunia (Quattrocchio *et al.*, 1999; Spelt *et al.*, 2000), snapdragon (Noda *et al.*, 1994), carnation (Itoh *et al.*, 2002; Nishizaki *et al.*, 2011), and morning glory (Habu *et al.*, 1998; Inagaki *et al.*, 1994). Besides transposable elements, RNA interference (RNAi) and DNA methylation are involved in unstable pigmentation in plants. For example, silencing of the *CHS* gene results in flower colour variegation in petunia (Koseki *et al.*, 2005). High methylation levels in the promoter region of *MYB10* inhibit gene expression, which causes variable colour patterns in the peel of apple (Telias *et al.*, 2011).

Peach [*Prunus persica* L. (Batsch)] is one of the most popular fruit trees in the world. Peach belongs to the Rosaceae family and serves as a model species of woody perennial angiosperms due to its small genome size of about 230Mb/haploid (The International Peach Genome Initiative, 2013). Peach trees are primarily grown for their fruit, but some cultivars are selected for their ornamental value. In China, the cultivation of ornamental peach has a long history. The ancient book Luoyanghuamuji written by Shihou Zhou in 1082 records ornamental peach cultivars such as Ersetao (variegated flower) and Ziyetao (red-leaved); and the variegated flower was highly praised in an ancient poem entitled 'Er Se Tao' written by Yong Shao in Song Dynasty, approximately 1000 years ago. Modern varieties have been bred for flower variegation such as Hongbaihuatao (HBH), which produces red, pink, and variegated flowers. Several studies have been conducted to investigate the genetic basis for peach flower colour. Chen *et al.* (2014) found that anthocyanin pathway genes such as *CHS*, *CHI*, and *F3H* show higher level of expression in red flowers than in white flower in ornamental peach (*Prunus persica* f. *versicolor* [Sieb.] Voss). The expression of a *MYB*-like gene (*Peace*) controls the pigmentation of flowers in the flowering peach 'Genpei'. However, the mechanism underlying variegation in peach remains unclear.

In this study, cv. HBH was selected to investigate the molecular basis for peach flower variegation. A *GST* gene—regulator involved in anthocyanin transport (*Riant*)—was found to be associated with variegation. Levels of the Riant protein were high in red flower, but barely detectable in variegated flowers. This was due to small insertions and deletions (indels) in the last exon, resulting in a frameshift mutation. Anthocyanin accumulation in the flower of cv. HBH therefore appears to be regulated at the post-transcriptional level, unlike previous reports of transcriptional regulation in flowers of other peach varieties. This study demonstrates that the gene encoding GST is critical for anthocyanin accumulation in peach, and is helpful in understanding the mechanism underling the variegation in peach flowers.

MATERIALS AND METHODS

Plant Material

Peach (*P. persica*) cultivars used in this study, including HBH, 'Mantianhong', 'Hongcuizhi', and 'Sahongtao', are maintained at Wuhan Botanical Garden of the Chinese Academy of Sciences (Hubei Province, China). Cv. HBH has variegation in flower colouration, while the other three cultivars bear only red flowers. Young leaves and flower buds with a diameter ranging from 0.5 to 0.8cm were collected in spring. Petals and sepalswere removed from flower samples and individually put into separate aluminium bags. All samples were immediately frozen in liquid nitrogen, and then stored at −80 °C until use.

HPLC Analysis of Anthocyanin in Peach Flower

Anthocyanin was extracted according to a previously reported protocol (Cheng *et al.*, 2014). Approximately 0.5g of tissue was ground in liquid nitrogen and then added to 25ml of extraction solution (80:20 v/v methanol/water mixture containing 1.18mM HCl). The mixture was centrifuged at 10000 g for 10min. An aliquot of 10ml of supernatant was collected and evaporated under vacuum at 30 °C using a rotary evaporator. The residual was resuspended in acidified water (1.18mM HCl).

Anthocyanin was dissolved in 1ml methanol, filtered through a 0.22 μm Millipore membrane, and analysed using an HPLC-ESI-MS/MS system (ThermoFisher Scientific, Pittsburgh, PA, USA). The analytical column was a ZORBAX Extend C18, 4.6×250mm, with a particle size of 5 μm (Agilent Technologies, Waldbronn, Germany). The analytical column was sequentially eluted using mobile phase A (formic acid:water, 5:95, v/v) and mobile phase B (methanol) with a flow rate of 0.8ml/min. The linear gradient of phase B was as follows: 0–2min, 5%; 2–7min, 5–15%; 7–20min, 15–20%; 20–25min, 20–27%; 25–32min, 27%; 32–41min, 27–35%; 41.01–43min, 5%. UV-visible light detector wavelength was set at 520nm.

Analysis of Gene Expression using Quantitative Real-Time PCR

Total RNA was extracted using ZP401 kit (Zoman, Beijing, China) according to the manufacturer's instructions. Total RNA was treated with DNase I (Takara, Dalian, China) to remove any contamination of genomic DNA. Approximately 3 μg of total RNA was used for cDNA synthesis using PrimeScriptTM RT-PCR Kit (Takara, Dalian, China). An SYBR green-based real-time PCR assay was carried out in a total volume of 20 μl reaction mixture containing 10.0 μl of 2× SYBR Green I Master Mix (Takara, Dalian, China), 0.2 μM of each primer, and 100ng of template cDNA. A peach actin gene *PpGAPDH* (*ppa008812m*) was used as a constitutive control. Primer sequences of genes involved in anthocyanin biosynthesis and transport are listed, available at *JXB* online.

Amplification was conducted using StepOnePlus Real-Time PCR System (Applied Biosystems, Foster, CA, USA). The amplification programme consisted of an initial denaturing step at 95°C for 30 s, followed by 40 cycles of 95°C for 30 s, and 60°C for 34 s. The fluorescent product was detected at the second step of each cycle. Melt curve analysis was performed at the end of 40 cycles to ensure the proper amplification of target fragments. Fluorescence readings were consecutively collected during the melting process from 60 to 90 °C at the heating rate of 0.5°C/sec. All analyses were repeated three times using biological replicates.

Genomic DNA Blot Analysis

For cv. HBH, total DNA was separately extracted from red and variegated flowers at balloon stage, while total DNA of three cultivars used as controls (Mantianhong, Hongcuizhi, and Sahongtao), was extracted from young leaves. DNA extraction was performed using a cetyltrimethylammonium bromide (CTAB) method. Approximately 5 μg of genomic DNA was digested with *Hind*III, *Spe*I, and *Xba*I, separated on a 1.0% agarose gel, and transferred onto nylon membranes (Hybond-N, Amersham, UK) using the capillary transfer method. A pair of primers (5'-CTCAGTTCCTCTCCCGTCAG-3'/5'-CCAGCCAGATAGCTGCTCTT-3') was designed to synthesize DNA probes using cDNA from leaves of cv. HBH as a template. The probe consisted of the last 17bp in exon 1, a complete exon 2, and a partial segment of exon 3 of *Riant*. Hybridization was carried out using the DIG Easy Hyb kit (Roche Applied Science, Indianapolis, IN, USA) according to the manufacturer's instructions. Blots were exposed to a Lumi-Film X-ray film (Hyperfilm, Amersham) at room temperature for 25min.

Phylogenetic Analysis

The amino acid sequences of genes encoding GST from different plants were used for phylogenetic analysis. Sequence alignment was performed using CLUSTAL X, and the resulting data matrix was analysed using equally weighted neighbour joining (NJ). An NJ tree was generated using MEGA (version 5.0). Bootstrap values were calculated from 1000 replicate analyses.

2D Electrophoresis (2-DE)

The protein extraction protocol was based on the phenol method (Hurkman and Tanaka, 1986; Ahsan *et al.*, 2008). Briefly, 2g of flower petal was ground in liquid nitrogen and added to 10ml extraction buffer containing 0.5M Tris-HCl (pH 8.3), 0.1M KCl, 50mM EDTA, 2% (v/v) 2-mercaptoethanol, and 0.7M sucrose. An equal volume of Tris-HCl-saturated phenol (pH 8.0) was subsequently added and mixed by vigorous vortexing for 2min followed by centrifugation at 3500*g* for 15min. After centrifugation, the top phenol phase was collected and proteins were precipitated by adding four volumes of cold methanol containing 0.1M ammonium acetate at −20 °C for 2h. The precipitated proteins were recovered by centrifugation at 3500 *g* for 10min followed by three washes with cold methanol containing 0.1M ammonium acetate. The protein pellet was dried at room temperatureuntil needed for solubilization.

Protein concentration was quantified according to the Bradford method (Bradford, 1976). A total of 1.2mg of protein was loaded into the IPG strips (pH4–7, 24cm, Bio-Rad, USA) through rehydration for 12h at room temperature. Isoelectric focusing (IEF) electrophoresis was conducted using the following procedure: 200V for 1h (step and hold); 500V for 1.5h (step and hold); 1000V for 1h (gradient), 8000V for 2h (gradient), and 8000V for 6h (step and hold) for a total of 42000 volt-hoursusing a Protean IEF Cell (Bio-Rad). Then, the IPG strips were equilibrated for 15min in the equilibration buffer (50mM Tris-HCl pH 6.8, 6M urea, 10% v/v glycerol, 2.5% w/v SDS, and 5% 2-mercaptoethanol) containing 0.5% DTT, followed by 15min in the equilibration buffer containing 25mg/ml iodoacetamide. The second dimensional electrophoresis was run on 12% SDS-PAGE and conducted using the following procedure: 100V, 30min; 250V, 5h. The 2-DE gels were stained with Coomassie Brilliant Blue (CBB R-250).

Trypsin Digestion and MALDI-TOF-MS Analysis

The differentially expressed protein spots were excised from the gel manually and washed with double-distilled water twice. The gel slices were destained, dehydrated, and digested with trypsin. The digested protein peptides were analysed over a mass range of 800–4,000Da using an Autoflex speed™ MALDI-TOF-TOF mass spectrometer (Bruker Daltonics, Bremen, Germany). Subsequently, the obtained PMF data were searched against the NCBI nr database and Swiss-Port database using MASCOT software (Mascot Wizard 1.2.0, Matrix Science Ltd., www.matrixscience.com). The parameters were set as follows: carbamidomethylation of cysteine and oxidation of methionine; peptide charge state of +1 and peptide mass tolerance of 0.5Da; a maximum of one for missed cleavages and monoisotopic.

PAGE

PCR products were mixed with an equal volume of formamide loading buffer (98% formamide, 10mM EDTA pH 8.0, 0.025% Bromophenol Blue and Xylene Cyanol). The mixture was denatured at 94°C for 3min, and then immediately chilled on ice. An aliquot of 2 μl mixture was loaded on a 6% polyacrylamide gel, and electrophoresed for 1.5h at 1200V. Bands were visualized after silver staining, and recorded on a ScanMaker 3830 (Microtek, Shanghai, China).

Expression Vector Construction and Plant Transformation

A pair of primers, 5'-GAATTCATGGTTGTGAAAGTGTAT GGTCC-3'/5'-CTCGAGTGGGGGTATCTCATATCTAGTAGTC-3', was designed to amplify the whole coding region of the *Riant* gene using cDNA synthesized from flowers of cv. HBH as templates. The forward and reverse primers contained *Eco*RI and *Xho*I sites at the 5' end, respectively. The PCR product was digested with *Eco*RI and *Xho*I, and inserted into *Eco*RI/*Xho*I-digested pSAK277 (Hellens *et al.*, 2005). The *Arabidopsis tt19* mutant (CS60000) with the Columbia genetic background was obtained from the Arabidopsis Biological Resource Center (Ohio State University, OH, USA). *Arabidopsis*transformation was performed according to the floral dip method (Clough and Bent, 1998). For transgenic plant selection, T_0 seeds were sterilized and germinated on Murashige and Skoog (MS) medium containing 12 μg ml^{-1} kanamycin and 3% (w/v) Suc. Following 1 week of selection, kanamycin-resistant plants with red hypocotyls were transplanted to soil and placed in a growth chamber at 25°C and 50–80% relative humidity.

RESULTS

Colouration and Anthocyanin Composition in Red, Pink, and Variegated Petals of cv. HBH

Flowering peach cv. HBH produces red, pink, and variegated flowers on a single tree (Fig. 1A). The variegated flowers show a great variation in petal colouration (Fig. 1B), and can be classified into four types: white and red/pink spotted, white and pink somatic sectors, white and red somatic sectors, and pink and red somatic sectors. Most variegated flowers belong to type 1, while the other three types of variegated flowers are occasionally produced. Besides the petal tissue, variegation also appears in the colouration of sepal, stamen, and pistil tissues (Fig. 1C). Based on the flower colouration, the branches of cv. HBH can be grouped into types: red-, pink-, and white-flower branches. Red-flower branches produce exclusively red flowers with no variegated flowers.

Pink-flower branches bear predominantly pink flowers, and occasionally produce pink flowers with red somatic sectors. White-flower branches bear predominantly white flowers with red/pink spotted, and occasionally produce white flowers with pink or red somatic sectors, pink flowers, and red flowers. Since the white-flower branch is the principal branch in cv. HBH, the pink- and red-flower branches are deemed to arise from bud sports.

Figure 1: Flower colouration of ornamental cv.HBH.(A) Peach tree cv. HBH bears pigmented and variegated flowers at bloom stage. (B) Three kinds of coloured flowers

within a single tree.(C) The colouration in sepal, pistil, and stamen. The arrow indicates the stamen. (A colour version of this figure is available at *JXB* online.).

HPLC analysis was conducted to investigate differences in anthocyanin content among red petal, pink petal, and white petal with red/pink spotted (termed variegated petalhereinafter). The red petal consists of six peaks. Based the authors' previous study (Cheng *et al.* 2014), peaks 1–6 correspond to cyanidin 3-galactoside, cyanidin 3-glucoside, cyanidin 3-rutinoside, peonidin 3-glucoside, cyanidin 3-rhamnoside, and peonidin 3-rutinoside, respectively. The main component in peach flower is cyanidin 3-glucoside, and its content varies greatly among red, pink, and variegated petals. The content of cyanidin 3-glucoside in red and pink petals is 12.1 and 1.26 µg/100g fresh weight (FW), respectively. However, cyanidin 3-glucoside is almost undetectable in variegated petals. In addition, the flavonol content was also determined. Overall, there is no striking difference in flavonol content among red, pink, and variegated petals. The amount of flavonol was slightly higher in variegated petals (4.8 µg/100g FW) than in red petals (4.1 µg/100g FW) and pink petals (4.6 µg/100g FW). Taken together, these results show that anthocyanin is responsible for the red and pink pigmentation in petals.

Microscopic analysis of fresh hand-cut sections of flower petals showed that red petals had several layers of coloured cells, with anthocyanin accumulation in both epidermal and sub-epidermal layers (Fig. 2). Pink petals only accumulated anthocyanins in the upper and lower epidermal layers. However, white petals accumulated no anthocyanins in either epidermal or sub-epidermal layers.

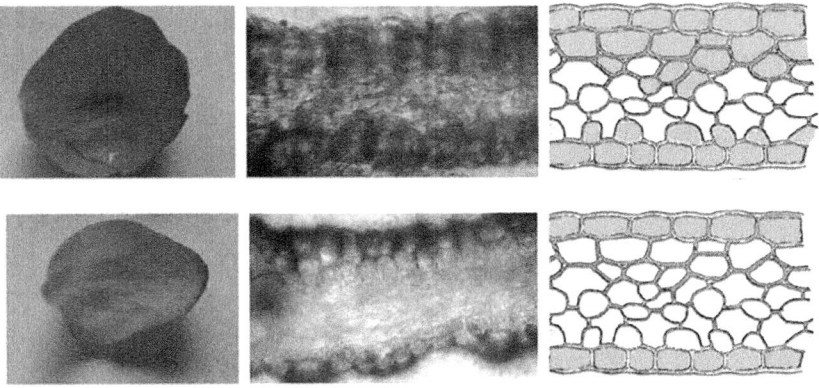

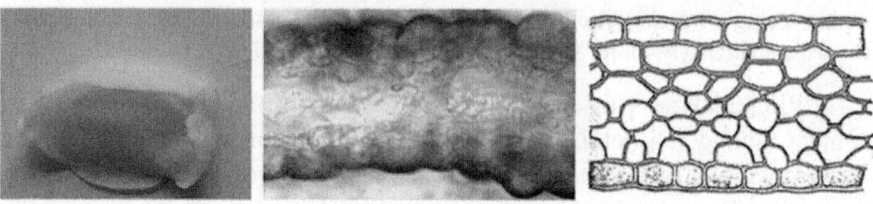

Figure 2: Examination of anthocyanin accumulation in petal cell layers. The middle column represents cross-sections of red, pink, and white petals (photo taken with microscope), while the right column is a diagrammatic representation of the cross-sections of red, pink, and white petals. (A colour version of this figure is available at *JXB*online.).

Expression Profiling of Genes Involved in Anthocyanin Biosynthesis and Transport

Red, pink, and variegated petals of flowers at balloon stage were collected to investigate the expression profile of anthocyanin biosynthesis genes using real-time PCR, including *PpCHS, PpCHI, PpF3 H, PpF3H, PpDFR,PpLDOX*, and *PpUFGT*. No striking difference in expression level was observed for any of these anthocyanin biosynthesis genes when comparing red, pink, and variegated petals (Fig. 3). Surprisingly, two genes—*PpCHS* and *PpCHI*—showed a significantly higher level of expression in variegated petals than in red and pink petals ($P<0.05$). Similarly, all the anthocyanin regulatory genes of *MYBs* also showed no significant difference in expression levels among pink, red, and white petals.

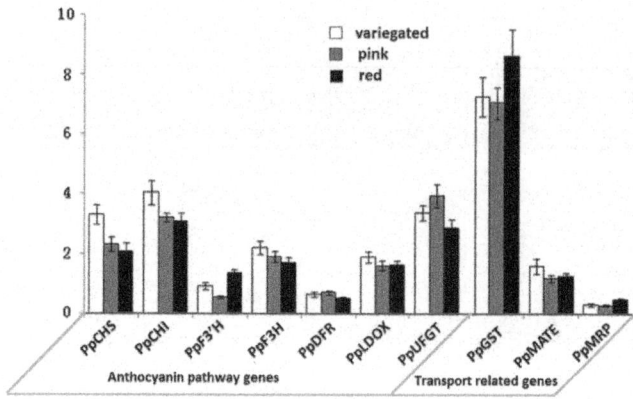

Figure 3: Expression level of genes involved in biosynthesis and transport of anthocyanin in petals of cv. HBH.

Subsequently, the expression level of three genes involved in anthocyanin transport, e.g. *PpMRP*, *PpMATE*, and *PpGST*, was also measured. All these anthocyanin transport-related genes, like the anthocyanin biosynthesis genes, showed no significant difference in expression level among red, pink, and variegated petals (Fig. 3). These results suggested that anthocyanin accumulation is not regulated at the transcriptional level in the flower of cv. HBH.

Proteomic Analysis Reveals a Candidate GST Correlated with Flower Colour Variegation in cv. HBH

To determinate whether anthocyanin accumulation in the flower of cv. HBH is controlled at the post-transcriptional level, 2-DE protein analysis was performed. Petals from red and variegated flowers at balloon stage were chosen for 2-DE analysis. As a result, a total of 84 protein spots displaying differential abundance (>1.6-fold change) were identified. Of these proteins, 40 and 44 were up- and down-regulated in red petals, respectively.

These differentially abundant proteins were digested with trypsin and analysed using MALDI TOF/TOF MS/MS. Of the 84 proteins surveyed, 53 were successfully identified. Among these identified proteins (Table 1), two, PpCHS-like and PpGST with spot numbers B41 and B54, respectively, are potentially related to anthocyanin pigmentation. The PpCHS-like protein was highly expressed in red petals but undetectable in variegated petals. Coding sequences of the *PpCHS*-like gene were cloned from both red and variegated petals, but no difference was identified. Phylogenetic analysis indicated that the *PpCHS*-like gene is closely related to genes encoding a biphenyl synthase in *Malus* . These results suggest that the *PpCHS*-like gene is unlikely involved in regulation of anthocyanin pigmentation. In contrast, the PpGST protein was found in both red and variegated petals, but its level was much higher in red petals than in variegated petals. Phylogenetic analysis showed that the *PpGST* gene is closely related to*VvGST* (Fig. 4). Since the *VvGST* gene is known for anthocyanin transport in grapevine (Gomez *et al.*, 2011), the *PpGST* gene, *Riant*, is a strong candidate responsible for variegation of flower colouration.

Table 1: *Proteins differentially expressed in red and white flower from* P. persica *cv.* HBH*

No.	Spot no.	Protein name	MW(Da)	p/	PSC(%)
1	A27	Geranylgeranyl pyrophosphate synthase family protein (*Populus trichocarpa*)	37305	5.06	23%
2	A42	Coatomer subunit epsilon-2-like (*Fragaria vesca* subsp. vesca)	32395	5.16	9%
3	A49	Triose phosphate isomerase cytosolic isoform-like protein (*Capsicum annuum*)	27433	6.00	37%
4	A68	PREDICTED: peroxiredoxin-2E, chloroplastic-like (*F. vesca*)	24230	8.96	20%
5	A48	S-locus lectin protein kinase family protein (*Theobroma cacao*)	87288	6.64	1%
6	A22	Calcium-binding EF hand family protein (*T. cacao*)	31298	4.8	18%
7	A62	Adenine nucleotide hydrolases-like superfamily protein (*T. cacao*)	17992	6.2	48%
8	A64	Hypothetical protein	17685	4.77	14%
9	A33	PREDICTED: 14-3-3-like protein-like (*F. vesca* subsp. vesca)	29687	4.77	29%
10	A55	Chalcone-flavanone isomerase family protein (*T. cacao*)	32276	7.77	35%
11	A26	Fructose-bisphosphate aldolase 4 (*Camellia oleifera*)	42688	8.15	16%
12	A38	EF hand family protein, expressed isoform 1 (*T. cacao*)	30276	6.44	13%
13	A28	Temperature-induced lipocalin (*P. persica*)	21450	5.60	38%
14	A44	ATP-dependent Clp protease proteolytic subunit 4 (*T. cacao*)	32028	5.97	19%
15	A13	Actin 7 (*Arabidopsis thaliana*)	41954	5.31	50%
16	A58	Cyclophilin peptidyl-prolyl cis-trans isomerase family (*T. cacao*)	15270	5.61	45%
17	A23	Annexin-like protein RJ4 (*P. trichocarpa*)	35923	6.19	49%
18	A24	RNA-binding protein Nova-1-like (*Vitis vinifera*)	30818	6.01	38%
19	B15	Full=UDP-sugar pyrophosphorylase	67671	5.71	21%
20	B29	26S proteasome non-ATPase regulatory subunit 4-like (*F. vesca*)	43042	4.48	19%
21	B35	Monodehydroascorbate reductase (*Malus domestica*)	47111	6.51	26%
22	B40	Isovaleryl-CoA dehydrogenase 1, mitochondrial-like (*F. vesca*)	43644	6.20	23%
23	B41	Chalcone synthase 1-like (*F. vesca* subsp. vesca)	43390	5.97	32%
24	**B54**	**GST-like protein (M. domestica)**	**24389**	**5.34**	**37%**
25	B28	TCP domain class transcription factor (*M. domestica*)	57477	5.72	40%
26	B23	RNA-binding KH domain-containing protein isoform 1 (*T. cacao*)	59072	6.12	35%
27	B09	Lipoxygenase (*M. domestica*)	90278	5.40	20%
28	B21	Starch synthase isoform I (*Manihot esculenta*)	71556	5.38	22%
29	B48	Papain family cysteine protease (*T. cacao*)	40868	5.86	17%
30	B17	Mediator of RNA polymerase II transcription subunit 37e-like (*F. vesca*)	69534	5.25	20%
31	B03	Patellin-3-like (*F. vesca* subsp. vesca)	65404	4.85	6%
32	B01	Heat shock protein 70 (Hsp 70) family protein isoform 1 (*T. cacao*)	100326	5.40	20%
33	B27	TCP domain class transcription factor (*M. domestica*)	57477	5.72	44%
34	B05	Patellin-3-like (*F. vesca* subsp. vesca)	65404	4.85	10%
35	B16	NADP-malic protein (*Prunus armeniaca*)	65358	5.73	24%
36	B33	3-Ketoacyl-acyl carrier protein synthase I (*T. cacao*)	52446	6.38	23%
37	B19	DC1 domain-containing protein (*T. cacao*)	65260	4.80	31%
38	B06	Cytosolic aconitase (*Pyrus pyrifolia*)	108637	6.98	10%
39	B58	PREDICTED: allene oxide cyclase 4, chloroplastic-like (*F. vesca*)	20527	5.41	19%
40	B55	Chaperonin 20 isoform 1 (*T. cacao*)	26399	7.79	11%
41	B44	Caffeic acid 3-O-methyltransferase 1 (*T. cacao*)	42020	5.40	21%
42	B25	TCP domain class transcription factor (*M. domestica*)	45351	5.26	26%
43	B18	UDP-sugar pyrophosphorylase (*T. cacao*)	67671	5.71	14%
44	B31	Hypothetical protein	39546	6.77	7%
45	B20	Nucleoporin nup211-like (*Glycine max*)	54774	5.99	16%
46	B24	Glucose-6-phosphate 1-dehydrogenase	59358	6.13	24%
47	B50	Prolyl 4-hydroxylase alpha subunit, putative (*Ricinus communis*)	33559	6.26	36%
48	B13	Oligopeptidase A-like (*F. vesca* subsp. vesca)	90582	6.42	12%
49	B49	Probable rhamnose biosynthetic enzyme 1-like (*Citrus sinensis*)	33636	6.17	21%
50	B67	Peroxiredoxin-2B-like (*F. vesca* subsp. vesca)	17480	5.70	57%
51	B60	Regulator of ribonuclease-like protein 2-like (*F. vesca* subsp. vesca)	18008	5.69	64%
52	B51	Protein PPLZ12, putative (*R. communis*)	31730	5.27	41%
53	B65	Ribonuclease UK114-like (*F. vesca* subsp. vesca)	19977	8.99	44%

* The protein associated with the variegated colouration of the peach flower is highlighted in bold.

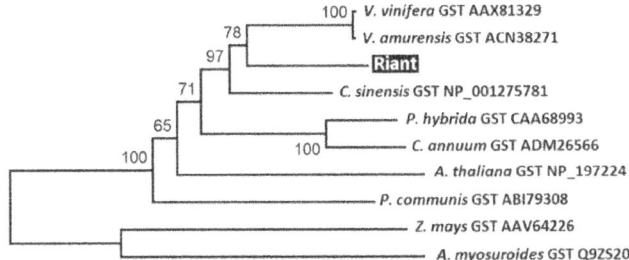

Figure 4: A phylogenetic tree derived from amino acid sequences of GST genes from plants. GenBank accession numbers are listed after the gene name. The *Riant* gene isolated in this study is highlighted. The numbers indicate bootstrap values calculated from 1000 replicate analyses.

Small Indels in the *Riant* Gene and their Association with Variegation in Petal Colouration

To confirm that the *Riant* gene is responsible for petal variegation, its coding sequences from red, pink, and variegated petals were cloned and sequenced. Comparison of the coding sequences revealed a 2-bp insertion in the third exon of *Riant* (Fig. 5A). The 2-bp insertion causes a frameshift and a premature stop codon. Subsequently, a pair of primers flanking the 2-bp insertion, 5'-CTCTGGTGGATCAGTGGCT-3' (GIF) and 5'-TATCCCTGGAAGATGGCTC-3' (GIR), was designed to amplify red, pink, and variegated petals from different clones of cv. HBH (Fig. 5B). Interestingly, all the red or pink petals contained two bands, suggesting they are heterozygous at the *Riant* locus. Three alleles, designated *Riant1*,*Riant2*, and *riant3*, were detected among the red and pink petals. Of the 13 pigmented petals tested, 12 have a *Riant1*/*riant3* genotype. One (the second sample in Fig. 5B) has a *Riant2*/*riant3* genotype. Sequencing of the PCR products revealed that both *Riant1* and *Riant2* have an intact ORF, while the *riant3* allele has a frameshift mutation due to a 2-bp insertion in the third exon (Fig. 5A). It is worth noting that *Riant2* contains a 3-bp insertion in the third exon, so should not induce a frame shift. All the variegated petals amplified only one band that corresponds to the *riant3*allele, suggesting they are homozygous at the *Riant* locus.

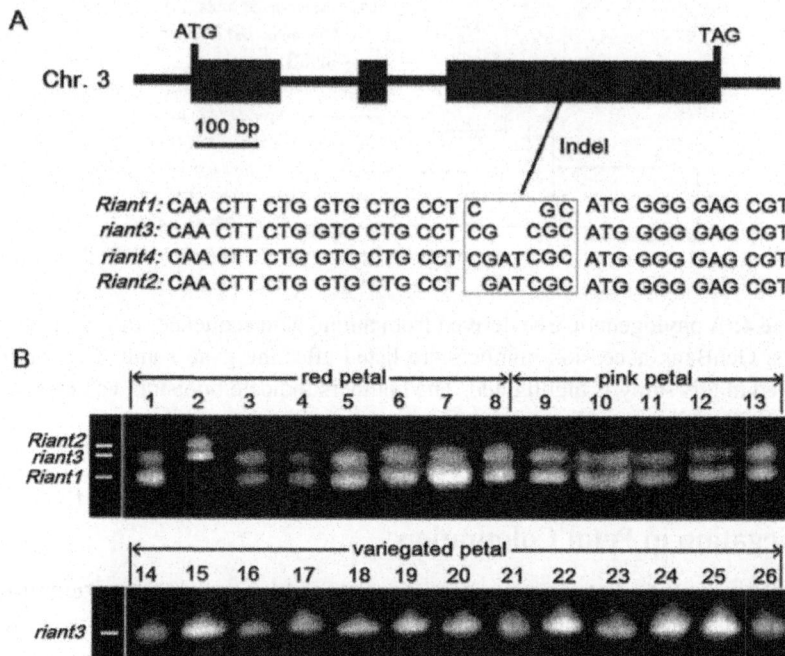

Figure 5: The *Riant* gene isolated from peach cv. HBH. (A) Genomic structure and genetic variation highlighted in a square box. (B) Genotyping of red, pink, and variegated flowers based on the indel in the last exon of the *Riant* gene; the detected alleles are indicated.

To determine whether there are other alleles of the *Riant* gene, 15 more flowers (five with red petals, five with pink petals, and five with variegated petals) were randomly collected from different clones of cv. HBH. The petals of these flowers were individually subjected to genomic DNA extraction, and the extracted DNA was subsequently amplified using the primers GIF/GIR as mentioned above. Cloning and sequencing of the PCR products revealed one more frameshift mutant allele from the variegated flower, designated *riant4*, which contains a 4-bp insertion (Fig. 5A). All four alleles—*Riant1* to *Riant4*—were deposited in GenBank under accession nos. KT312847 to KT312850, respectively.

Copy Number of the *Riant* Gene in the Peach Genome

In many species, variation in flower colouration is caused by transposon activity. To determine whether the small indels in the third exon of the *Riant* gene have also arisen from a transposon or other DNA fragment that was not amplified by the primers GIF and GIR, aDNA gel blot analysis was performed.

Three cultivars (Mantianhong, Hongcuizhi, and Sahongtao) that bear red flowers were used as controls. Genomic DNA was digested with*Spe*I, *Hind*III, and *Xba*I. The digested DNA was hybridized with a probe covering the indel site in the third exon of the *Riant* gene. Both *Hind*III and*Xba*I digestions yielded a single hybridizing DNA band in all the tested samples, while there were two hybridizing bands for *Spe*I digestion in all the tested samples (Fig. 6). This result indicates that only one copy of the*Riant* gene is present in the peach genome, and it is not interrupted by a transposon. In other words, the small indel mutation in the *Riant* gene is not due to insertion and excision of a transposable element.

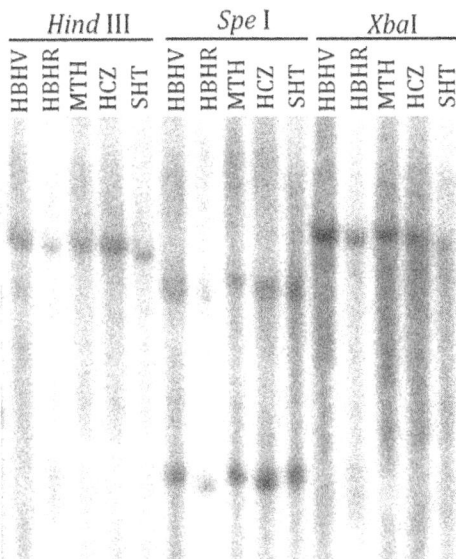

Figure 6: Southern blot analysis of peach genomic DNA. The*Riant*-specific probe consists of a partial sequence of the first exon, whole fragment of the second exon, and a partial sequence of the third exon that covers the indel site. HBHV and HBHR represent variegated and red flowers from cv. HBH, respectively. MTH, Mantianhong; HCZ, Hongcuizhi; SHT, Sahongtao.

Functional Analysis of the *Riant* Gene in the *Arabidopsis* Transparent Testa19 Mutant

The *Arabidopsis* transparent testa19 (*tt19*) mutant, lacking GST, was selected to investigate the functionality of the *Riant* gene. The coding sequences of both *Riant1* and *riant3* alleles were separately transferred into the *Arabidopsis tt19* mutant under the control of the cauliflower mosaic virus 35S promoter, and several transgenic lines were generated for each construct. Seeds of the *Arabidopsis tt19* mutant, T1 transgenic lines, and wild-type *Arabidopsis* were germinated and grown on MS medium. Germinating seedlings of wild-type plants and transgenic lines expressing *Riant1* had red hypocotyls, whereas hypocotyls of the *Arabidopsis tt19* mutant and transgenic lines expressing *riant3* were green (Fig. 7A). Moreover, seeds collected from kanamycin-resistant T1 plants and the *Arabidopsis tt19* mutant were pale brown in colour, while seeds of wild-type plants were dark brown in colour (Fig. 7B). This suggests that Riant, like petunia AN9, complements the anthocyanin accumulation in vegetative tissues, but not the brown pigmentation in the seed coat (Kitamura *et al.*, 2004). In addition, reverse transcription (RT)-PCR analysis showed that both *Riant1* and *riant3* were highly expressed in transgenic lines (Fig. 7C). Taken together, these results demonstrated that, the *Riant1* allele is involved in the transport of anthocyanins from cytosol to vacuole, but the *riant3* allele is nonfunctional.

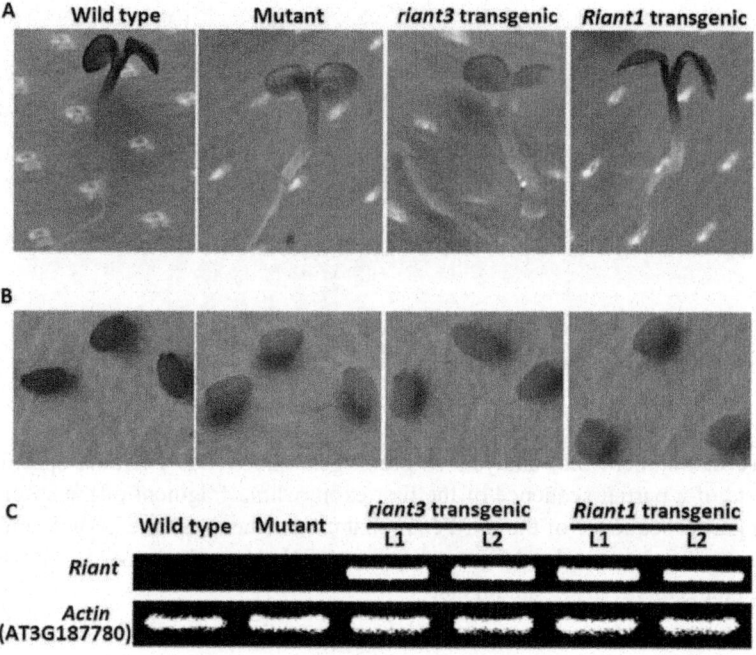

Figure 7: Complementation of the pigmentation of *Arabidopsis tt19* mutant seedlings of the ecotype Columbia with the *Riant* gene. (A) Phenotypes of wild-type, mutant, and transgenic *Arabidopsis* seedlings. (B) Phenotypes of wild-type, mutant, and transgenic *Arabidopsis* seeds. (C) Expression level of the *Riant* gene in wild-type, mutant, and transgenic *Arabidopsis* seedlings. Two transgenic lines each of *Riant1* and *riant3* were analysed, and these exhibited similar phenotypes, as shown. (A colour version of this figure is available at *JXB* online.).

DISCUSSION

The *Riant* Gene is Involved in Anthocyanin Transport and is Critical for Flower Colouration in Peach cv. HBH

In plants, flower pigmentation is mainly attributed to anthocyanin accumulation (Bogs *et al.*, 2005; Cutanda-Perez *et al.*, 2009). In this study, HPLC analysis revealed that cyanidin 3-glucoside is the main component of anthocyanin in flowers of cv. HBH, which is consistent with previous reports (Chaparro *et al.*, 1995a; Cheng *et al.*, 2014; Uematsu *et al.*, 2014). Both red and pink petals accumulate cyanidin 3-glucoside, and its content is approximately 10-fold higher in red petals than in pink petals. In contrast, cyanidin 3-glucoside is almost undetectable in white sectors of the variegated petal. This suggests that anthocyanin accumulation contributes to flower colouration in cv. HBH and its flower colour variegation is related to a change in anthocyanin accumulation.

In contrast to anthocyanin, flavonol shows the highest level of accumulation in the variegated petal, followed by pink and red petals. This is consistent with previous findings that blocking anthocyanin accumulation strengthens the metabolic flux towards flavonols (Gou *et al.*, 2011; Sun *et al.*, 2012). Real-time PCR analysis reveals that early biosynthetic genes such as *CHS*, *CHI*, and *F3H* show higher levels of expression in variegated petals than in red and pink petals, whereas, the expression levels of late biosynthetic genes such as *DFR*, *LDOX*, and *UFGT* are not significantly different among red, pink, and variegated petals. Thus, it seems that anthocyanin and flavonol biosynthesis is coordinately regulated by anthocyanin pathway genes (Owens *et al.*, 2008; Han *et al.*, 2010). However, all the potential *MYB* regulatory genes tested in this study showed no significant difference in expression level among red, pink, and white petals of cv. HBH.

The vacuole is the cellular compartment where anthocyanins accumulate. Increasing evidence shows that GST is indispensable for the transport of anthocyanins from the ER to the vacuole (Marrs *et al.*, 1995; Alfenito *et al.*, 1998; Zhao and Dixon, 2010; Gomez *et al.*, 2011; Gou *et al.*, 2011). This study also demonstrates that the *Riant* gene encoding GST is essential for anthocyanin pigmentation in peach cv. HBH. In variegated petals, two alleles of the *Riant* gene were identified, but both of them encode truncated nonfunctional proteins due to 2- or 4-bp insertions in the third exon that result in frameshift mutations. Interestingly, the mutation of the *Riant* gene does not alter the expression level of genes involved in anthocyanin biosynthesis. Similar results have been reported in the *Arabidopsis tt19* mutant (Sun *et al.*, 2012). In *Arabidopsis*, anthocyanin accumulates first in vesicles and then is transported to the vacuole via fusion with the tonoplast (Poustka *et al.*, 2007). Knockout of the *GST* gene results in weak accumulation of anthocyanins in vesicles, but not in the vacuole (Goodman *et al.*, 2004; Gomez *et al.*, 2011; Li *et al.*, 2011; Sun *et al.*, 2012). In peach,no anthocyanin was foundin white flowers, but some was present in variegated petals. Anthocyanins may be temporarily accumulated in vesicles, but subsequently degraded in variegated flowers of cv. HBH. In addition, the *PpMATE* gene shows a higher level of expression in variegated petals than in red and pink petals. This suggests a relationship between the *PpMATE* and *Riant* genes, to coordinately transport anthocyanins from the ER to the vacuole. MATE is involved in anthocyanin transport via the transporter-mediated mechanism, suggesting that both vesicle-mediated trafficking and MATE transporter-mediated mechanisms are involved in the sequestration of anthocyanins to vacuoles in peach.

Genetic mapping reveals that two loci, *B* and *Fc*, are responsible for flower colour in peach. The *B* locus has been mapped to an interval flanked by two markers, Pr1-12 and BPPCT028, on the bottom of chromosome 1 (Martínez-Gómez *et al.*, 2007), while the *Fc* locus is anchored to an interval flanked by two markers, OPJ01 and MA039a, on the upper region of chromosome 3 (Yamamoto *et al.*, 2005). The *Peace* gene that regulates petal pigmentation in peach 'Genpei' is located at the bottom of chromosome 1 (Uematsu *et al.*, 2014). The peach reference genome (The International Peach Genome Initiative, 2013) has been searched, and the*Riant* gene was found to be located at the top of chromosome 3. Thus, it is worthy of further study to clarify whether the *Peace* and *Riant* genes are actually candidates of the *B* and *Fc* loci, respectively.

In *Arabidopsis*, TT19 participates in both anthocyanin accumulation in vegetative tissues and proanthocyanidin (PA) accumulation in seed coats (Kitamura *et al.*, 2004). The accumulation of PA pigments is responsible for brown colouration in *Arabidopsis* seed coats. Petunia anthocyanin 9 (AN9) is an orthologue of *Arabidopsis* TT19. However, its ectopic expression in the *Arabidopsis* *tt19* mutant complements the anthocyanin accumulation in vegetative tissues, but not the brown pigmentation in the seed coat (Kitamura *et al.*, 2004). Like *AN9*, *Riant* also complements the anthocyanin phenotype, but not the PA defect of the *tt19* mutant.

Potential Mechanism Underlying the Mutation of the *Riant* Gene in Flowering Peach cv. HBH

Transposable elements are often responsible for the phenotype of colour variation in plants (Habu *et al.*, 1998; van Houwelingen *et al.*, 1998;Pooma *et al.*, 2002; Schwinn *et al.*, 2006; Nishizaki *et al.*, 2011; Lazarow *et al.*, 2012). In this study, variegation in peach was shown to be associated with small indels in the last exon of the *Riant* gene. Both red and pink petals are heterozygous at the *Riant* locus, with a functional allele and a frameshift mutant allele, whereas variegated petals contain two nonfunctional alleles (homozygous, Fig. 5). Moreover, DNA blot analysis shows that there is no polymorphism between red and variegated petals. These results strongly suggest that there are no transposable elements in the *Riant* locus in cv. HBH.

Flowering peach cv. HBH is quite similar in variegation to the previously reported peach cultivar Pillar, which bears dark pink, light pink, and white flowers on the same tree (Chaparro *et al.*, 1995*a*). The phenotype in cv. Pillar is assumed to be controlled by an active transposable element in the*W* locus (Chaparro *et al.*, 1995*b*). If the dark pink flowers carry a functional*W* allele that is reverted by excision of the transposable element, its self-pollinated progeny are expected to segregate for flower colouration. However, self-pollinated seeds of the dark pink flowers on Pillar trees yield only anthocyanin-deficient progeny (Chaparro *et al.*, 1995*b*). This suggests that the unstable phenotype in cv. Pillar cannot be ascribed to an active transposable element. Similarly, no transposable element is identified in the *Peace* gene responsible for the variegated phenotype in flowering peach 'Genpei' that produces pink and variegated flowers on the same tree (Uematsu *et al.*, 2014). Thus, it appears that flower colour variegation is not due to transposable elements. DNA methylation and RNAi can also cause colour variegation in plants (Koseki *et al.*, 2005;Telias *et al.*, 2011). These two mechanisms lead to down-regulation of the anthocyanin pathway. However, expression levels of genes were similar among red, pink, and variegated flowers. This suggests colour variegation in cv. HBH is controlled at the translational level, and is unlikely to be related to DNA methylation or RNAi.

A total of four alleles were identified at the *Riant* locus. Of these alleles,*Riant1* has the same coding sequence as the *GST* gene (*ppa011307m*) retrieved from the peach reference genome of cv. Lovell (The International Peach Genome Initiative 2013). All the tested flowers contain the *riant3*allele. Thus, the *riant3* allele probably represents an original allele in cv. HBH, whereas, *Riant1*, *Riant2*, and *riant4* are mutants of the *riant3* allele. These small indels appear responsible for the DNA variation at the *Riant*locus.

Strand slippage during DNA replication is a well-understood mechanism of the small indel mutagenesis (Garcia-Diaz and Kunkel 2006; Montgomery*et al.* 2013), and the G–C pairing surrounding the indels contributes to the rate of small indels (Tanay and Siggia, 2008). The *Riant1* allele has a 2-bp (GC) deletion compared with the *riant3* allele. Interestingly, the indel polymorphic locus of the *riant3* allele contains a (GC)2 sequence, which has the potential to induce strand slippage. Thus, strand slippage may be responsible for the DNA variation at the *Riant* locus. Besides strand slippage, DNA single- or double-stranded breaks are also required for the generation of small indels as specialized translesion synthesis polymerases are capable of bypassing DNA lesions without repairing them, which allows replication on damaged DNA substrates and, in some cases, to promote mutagenic DNA synthesis (Waters *et al.*, 2009; De and Babu, 2010; Roerink *et al.*, 2012). For example, *Sulfolobus solfataricus* DNA polymerase IV (Dpo4), a member of the Y family of DNA polymerases, can generate deletions and mismatches at an unusually high average rate and preferentially at cytosine flanked by 5′-template guanine (Goodman, 2002). In this study, small indels also occur at the site with GC nucleotide sequences. Therefore, it cannot be excluded that the DNA variation of the*Riant* gene may be caused by similar mechanisms.

Relationship between Somatic Chimerism and Flower Colour Variegation in cv. HBH

Chimerism is one of the factors that causes variegated colouration in plants (Stewart and Dermen, 1979; Mandal *et al.*, 2000; Walker *et al.*, 2006; Pelsy, 2010). Dicotyledonous plants usually have stratified apical meristems containing three layers of dividing cells, L1, L2, and L3, and each layer contributes to different tissues of the developing organs (Carles and Fletcher, 2003). The word chimera indicates that one or more layers consist of genetically distinct cells, and chimerascan be classified as periclinal, mericlinal, or sectorial chimeras. For floral meristems, layers L1 and L2 contribute to epidermis and internal tissues, respectively, while layer L3 forms the innermost tissues such as vascular tissues (Brand *et al.*, 2001; Filippis *et al.*, 2013). Thus, layers L1 and L2 play an important role in determining flower colouration (Stewart and Dermen, 1979; Carles and Fletcher, 2003).

Anthocyanin content in pink-coloured petals is extremely low—only approximately 10% of that in red-coloured petals. A similar result is also observed in grapevine. 'Cabernet Sauvignon' and its bud sport 'Malian' bears dark red and pink berries, respectively, with Malian berry containing only 10% of the anthocyanin content of the Cabernet Sauvignon berry (Boss *et al.*, 1996). In Cabernet Sauvignon, both the L1 and L2 layers carry one red and one white allele, giving rise to a coloured epidermis derived from the L1 layer and several sub-epidermal coloured layers in the skin derived from L2 (Walker *et al.*, 2006). In contrast, Malian is a periclinal chimera with the L1 layer carrying one red and one white allele of the berry colour locus while two white alleles for the L2 layer, resulting in only the epidermis containing anthocyanin. The pink flower of peach cv. HBH is a periclinal chimera with the L1 layer capable of accumulating anthocyanin while the L2 layer cannot. Since both pink and red flowers belong to bud sports and are heterozygous for the *Riant* gene responsible for anthocyanin accumulation, the L1 layer in the pink flower should be heterozygous with one functional allele such as *Riant1* or *Riant2*, whereas, both the L1 and L2 layers in the red flower are heterozygous. The variegated flowers, including pink and red somatic sectors (type 1), pink and white somatic sectors (type 2), and red and white somatic sectors (type 3), can be attributed to the existence of genetically distinct cells within the same layer. Thus, a model is proposedfor the variegated phenotype in flower colouration of peach cv. HBH (Fig. 8).

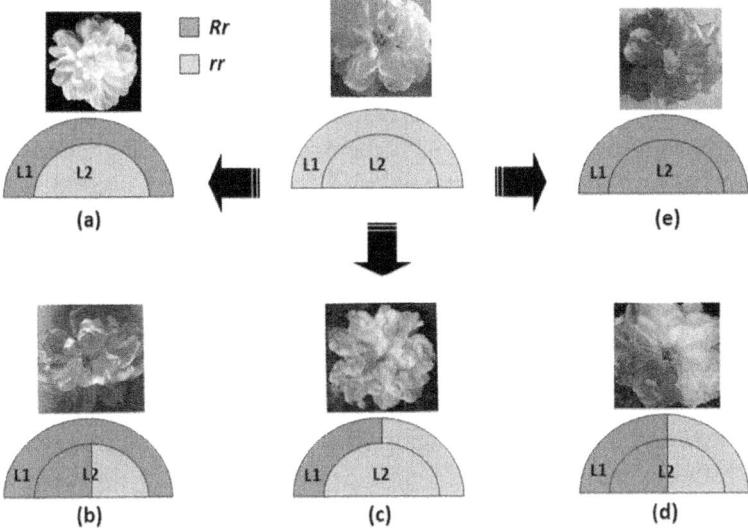

Fig. 8: A proposed model for the variegated phenotype in flower colouration of peach cv. HBH. L1 and L2 indicate different layers of floral meristems, and *R* and *r*represent functional and nonfunctional alleles of the*Riant* gene, respectively. White flower with red/pink spots carrying two nonfunctional alleles of the *Riant* gene. (a) Pink flower derived from periclinal chimera, (b) pink flower with red somatic sectors derived from mericlinal chimera, (c) white flower with pink somatic sectors derived from mericlinal chimera, (d) white flower with red somatic sectors derived from sectorial chimera, (e) red flower carrying one functional and one nonfunctional allele of the *Riant* gene. (A colour version of this figure is available at *JXB* online.).

The type 1 variegated flower results from a sectorial chimera, whereas, both the type 2 and type 3 variegated flowers have arisen from a mericlinal chimera. These three types of variegated flowers occur at low frequency, with no branches that produce predominantly one of the three types of variegated flowers. This is consistent with a previous finding that sectorial and mericlinal chimeras are unstable (Pelsy, 2010; Filippis *et al.*, 2013). In contrast, periclinal chimeras are very stable giving rise to the pink-flower branch that bears predominantly pink flowers.

No pure white flowers are found on the trees of cv. HBH, and the petals carrying no functional *Riant* allele still have red- and pink-coloured spots. This is consistent with previous reports that knockout of the *GST* gene cannot completely inhibit anthocyanin accumulation in maize (Goodman *et al.*, 2004) and *Arabidopsis* (Li *et al.*, 2011; Sun *et al.*, 2012). The development of white with red/pink spotted flowers could be explained by the following reasons. First, mutated cells capable of accumulating anthocyanin appear at late stages of floral meristem development, and they are distributed within the L1 and/ or L2 layers. An invasion by cells from the inner L2 layers into the outer L1 layer, termed 'displacement', has been reported in grapevine (Hocquigny *et al.*, 2004; Walker *et al.*, 2006; Pelsy, 2010). Thus, these mutated cells will be mixed with the wild-type cells incapable of accumulating anthocyanin in petal tissue, resulting in white and red/pink spotted flowers. Second, functional redundancy in the *GST* gene family has been reported in grapevine (Conn *et al.*, 2008). Thus, it is unclear whether other anthocyanin carrier(s) could complement anthocyanin accumulation, resulting in variegated colouration.

This study reveals that the *Riant* gene encoding GST is essential for flower colouration in peach. Mutations involving small indels frequently occur in the last exon of the *Riant* gene, which causes the variegated flower in peach cv. HBH. However, the mechanism underlying the small indel formation requires further study.

ACKNOWLEDGEMENTS

This project was supported by funds received from the National High Technology Research and Development Program of China (Grant No. 2011AA100206) and the National Natural Science Foundation of China. We would also like to thank Dr Andrew Charles Allan for his critical review of the manuscript.

REFERENCES

1.   Ahsan N Lee DG Alam I et al. 2008. Comparative proteomic study of arsenic-induced differentially expressed proteins in rice roots reveals glutathione plays a central role during As stress. Proteomics 8, 3561–3576.

2.   Alfenito MR Souer E Goodman CD Buell R Mol J Koes R Walbot V .

1998. Functional complementation of anthocyanin sequestration in the vacuole by widely divergent glutathione S-transferases. Plant Cell 10, 1135–1149.

3. Bogs J Downey MO Harvey JS Ashton AR Tanner GJ Robinson SP . 2005. Proanthocyanidin synthesis and expression of genes encoding leucoanthocyanidin reductase and anthocyanidin reductase in developing grape berries and grapevine leaves. Plant Physiology 139, 652–663.

4. Boss PK Davies C Robinson SP . 1996. Anthocyanin composition and anthocyanin pathway gene expression in grapevine sports differing in berry skin colour. Australian Journal of Grape and Wine Research 2, 163–170.

5. Bradford MM . 1976. A rapid and sensitive method for the quantitation of microgram quantities of protein utilizing the principle of protein-dye binding. Analytical Biochemistry 72, 248–254.

6. Brand U Hobe M Simon R . 2001. Functional domains in plant shoot meristems. Bioessays 23, 134–141.

7. Carles CC Fletcher JC . 2003. Shoot apical meristem maintenance, the art of a dynamic balance. Trends in Plant Science 8, 394–401.

8. Chaparro JX Werner DJ Whetten RW O'Malley DM . 1995 a. Inheritance, genetic interaction, and biochemical characterization of anthocyanin phenotypes in peach. Journal of Heredity 86, 32–37.

9. Chaparro JX Werner DJ Whetten RW O'Malley DM . 1995 b. Characterization of an unstable anthocyanin phenotype and estimation of somatic mutation rates in peach. Journal of Heredity 86, 186–193.

10. Chen Y Mao Y Liu H Yu F Li S Yin T . 2014. Transcriptome analysis of differentially expressed genes relevant to variegation in peach flowers. PLoS One 9, e90842.

11. Cheng J Wei G Zhou H Gu C Vimolmangkang S Liao L Han Y . 2014. Unraveling the mechanism underlying the glycosylation and methylation of anthocyanins in peach. Plant Physiology 166, 1044–1058.

12. Clough SJ Bent AF . 1998. Floral dip, a simplified method for Agrobacterium mediated transformation of Arabidopsis thaliana. Plant Journal 16, 735–743.

13. Conn S Curtin C Bezier A Franco C Zhang W . 2008. Purification, molecular cloning, and characterization of glutathione S-transferases. (GSTs) from pigmented Vitis vinifera L. cell suspension cultures as putative anthocyanin transport proteins. Journal of Experimental Botany 59, 3621–3634.

14. Conn S Franco C Zhang W . 2010. Characterization of anthocyanic vacuolar inclusions in Vitis vinifera L. cell suspension cultures. Planta 231, 1343–1360.

15. Cutanda-Perez MC Ageorges A Gomez C Vialet S Terrier N Romieu C Torregrosa L . 2009. Ectopic expression of VlmybA1 in grapevine activates a narrow set of genes involved in anthocyanin synthesis and transport. Plant Molecular Biology 69, 633–648.

16. De S Babu MM . 2010. A time-invariant principle of genome evolution. Proceedings of the National Academy of Sciences 107, 13004–13009.

17. Debeaujon I Peeters AJ Léon-Kloosterziel KM Koornneef M . 2001. The TRANSPARENT TESTA12 gene of Arabidopsis encodes a multidrug secondary transporter-like protein required for flavonoid sequestration in vacuoles of the seed coat endothelium. Plant Cell 13, 853–871.

18. Filippis I Lopez-Cobollo R Abbott J Butcher S Bishop GJ . 2013. Using a periclinal chimera to unravel layer-specific gene expression in plants. Plant Journal 75, 1039–1049.

19. Francisco RM Regalado A Ageorges A et al. 2013. ABCC1, an ATP binding cassette protein from grape berry, transports anthocyanidin 3-O-glucosides. Plant Cell 25, 1840–1854.

20. Garcia-Diaz M Kunkel TA . 2006. Mechanism of a genetic glissando: structural biology of indel mutations. Trends in Biochemical Sciences 31, 206–214.

21. Gomez C Terrier N Torregrosa L et al. 2009. Grapevine MATE-type proteins act as vacuolar H+-dependent acylated anthocyanin transporters. Plant Physiology 150, 402–415.

22. Gomez C Conejero G Torregrosa L Cheynier V Terrier N Ageorges A . 2011. In vivo grapevine anthocyanin transport involves vesicle-mediated trafficking and the contribution of anthoMATE transporters and GST. Plant Journal 67, 960–970.

23. Goodman CD Casati P Walbot V . 2004. A multidrug resistance-associated protein involved in anthocyanin transport in Zea mays. Plant Cell 16, 1812–1826.

24. Goodman MF . 2002. Error-prone repair DNA polymerases in prokaryotes and eukaryotes. Annual review of biochemistry 71, 17–50.

25. Gou JY Felippes FF Liu CJ Weigel D Wang JW . 2011. Negative regulation of anthocyanin biosynthesis in Arabidopsis by a miR156-targeted SPL transcription factor. Plant Cell 23, 1512–1522.

26. Grotewold E . 2004. The challenges of moving chemicals within and out of cells: insights into the transport of plant natural products. Planta 219, 906–909.

27. Grotewold E . 2006. The genetics and biochemistry of floral pigments. Annual Review of Plant Biology , 57, 761–780.

28. Grotewold E Davis K . 2008. Trafficking and sequestration of anthocyanins. Natural Product Communications 3, 1251–1258.

29. Habu Y Hisatomi Y Iida S . 1998. Molecular characterization of the mutable flaked allele for flower variegation in the common morning glory. Plant Journal 16, 371–376.

30. Han Y Vimolmangkang S Soria-Guerra RE Rosales-Mendoza S Zheng D Lygin AV Korban SS . 2010. Ectopic expression of apple F3'H genes contributes to anthocyanin accumulation in the Arabidopsis tt7 mutant grown under nitrogen stress. Plant Physiology 153, 806–820.

31. Hellens RP Allan AC Friel EN Bolitho K Grafton K Templeton MD Karunairetnam S Gleave AP Laing WA . 2005. Transient expression vectors for functional genomics, quantification of promoter activity and RNA silencing in plants. Plant Methods 1, 13–14.

32. Hocquigny S Pelsy F Dumas V Kindt S Heloir MC Merdinoglu D . 2004. Diversification within grapevine cultivars goes through chimeric states. Genome 47, 579–589.

33. Hurkman WJ Tanaka CK . 1986. Solubilization of plant membrane proteins for analysis of two-dimensional gel electrophoresis. Plant Physiology 81, 802–806.

34. Inagaki Y Hisatomi Y Suzuki T Kasahara K Lida S . 1994. Isolation of a suppressor-mutator / enhancer-like Transposable element, Tpnl, from Japanese bearing variegated flowers morning glory. Plant Cell 6, 375–383.

35. Itoh Y Higeta D Suzuki A Yoshida H Ozeki Y . 2002. Excision of transposable elements from the chalcone isomerase and dihydroflavonol 4-reductase genes may contribute to the variegation of the yellow-flowered carnation (Dianthus caryophyllus). Plant and Cell Physiology 43, 578–585.

36. Kitamura S Shikazono N Tanaka A . 2004. TRANSPARENT TESTA 19 is involved in the accumulation of both anthocyanins and proanthocyanidins in Arabidopsis. Plant Journal 37, 104–114.

37. Koseki M Goto K Masuta C Kanazawa A . 2005. The star-type color pattern in petunia hybrida 'Red Star' flowers is induced by sequence-

specific degradation of chalcone synthase RNA. Plant and Cell Physiology 46, 1879–1883.

38. Larsen ES Alfenito MR Briggs WR Walbot V . 2003. A carnation anthocyanin mutant is complemented by the glutathione S-transferases encoded by maize Bz2 and petunia An9. Plant Cell Reports 21, 900–904.

39. Lazarow K Du M Weimer R Kunze R . 2012. A hyperactive transposase of the maize transposable element Activator(Ac). Genetics 191, 747–756.

40. Li X Gao P Cui D Wu L Parkin I Saberianfar R Menassa R Pan H Westcott N Gruber MY . 2011. The Arabidopsis tt19-4 mutant differentially accumulates proanthocyanidin and anthocyanin through a 3' amino acid substitution in glutathione S-transferase. Plant Cell and Environment 34, 374–388.

41. Mandal AKA Chakrabarty D Datta SK . 2000. In vitro isolation of solid novel flower colour mutants from induced chimeric ray florets of chrysanthemum. Euphytica 114, 9–12.

42. Marrs KA Alfenito MR Lloyd AM Walbot V . 1995. A glutathione S-transferase involved in vacuolar transfer encoded by the maize gene Bronze-2. Nature 375, 397–400.

43. Martínez-Gómez P Sánchez-Pérez R Dicenta F Howad W Arús P Gradziel TM . 2007. Almond. In: Kole C , ed. Genome Mapping and Molecular Breeding in Plants. Fruits and Nuts . Berlin Heidelberg: Springer-Verlag, 229–242.

44. Montgomery SB Goode DL Kvikstad E et al. 2013. The origin, evolution, and functional impact of short insertion-deletion variants identified in 179 human genomes. Genome Research 23, 749–761.

45. Mueller LA Goodman CD Silady RA Walbot V . 2000. AN9, a petunia glutathione S-transferase required for anthocyanin sequestration, is a flavonoid-binding protein. Plant Physiology 123, 1561–1570.

46. Nishizaki Y Matsuba Y Okamoto E Okamura M Ozeki Y Sasaki N . 2011. Structure of the acyl-glucose-dependent anthocyanin 5-O-glucosyltransferase gene in carnations and its disruption by transposable elements in some varieties. Molecular Genetics and Genomics 286, 383–394.

47. Noda K Glover BJ Linstead P Martin C . 1994. Flower colour intensity depends on specialized cell shape controlled by a Myb-related transcription factor. Nature 369, 661–664.

48. Owens DK Alerding AB Crosby KC Bandara AB Westwood JH Winkel BS . 2008. Functional analysis of a predicted flavonol synthase gene family in Arabidopsis. Plant Physiology 147, 1046–1061.

49. Pelsy F . 2010. Molecular and cellular mechanisms of diversity within grapevine varieties. Heredity 104, 331–340.

50. Pooma W Gersos C Grotewold E . 2002. Transposon insertions in the promoter of the Zea mays a1 gene differentially affect transcription by the Myb factors P and C1. Genetics 161, 793–801.

51. Poustka F Irani NG Feller A Lu Y Pourcel L Frame K Grotewold E . 2007. A trafficking pathway for anthocyanins overlaps with the endoplasmic reticulum-to-vacuole protein-sorting route in Arabidopsis and contributes to the formation of vacuolar inclusions. Plant Physiology 45, 1323–1335.

52. Quattrocchio F Wing J Woude K Souer E Vetten N Mol J Koes R . 1999. Molecular analysis of the anthocyanin2 gene of Petunia and its role in the evolution of flower color. Plant Cell 11, 1433–1444.

53. Roerink SF Koole W Stapel LC Romeijn RJ Tijsterman M . 2012. A broad requirement for TLS polymerases η and κ, and interacting sumoylation and nuclear pore proteins, in lesion bypass during C. elegans embryogenesis. PLoS Genetics 8, e1002800

54. Schiefelbein JW Furtek DB Dooner HK Nelson OE . 1988. Two mutations in a maize bronze-1 allele caused by transposable elements of the Ac-Ds family alter the quantity and quality of the gene product. Genetics 120, 767–777.

55. Schwinn K Venail J Shang Y Mackay S Alm V Butelli E. Oyama R Bailey P Davies K Martin C . 2006. A Small Family of MYB-Regulatory Genes Controls Floral Pigmentation Intensity and Patterning in the Genus Antirrhinum. Plant Cell 18, 831–851.

56. Spelt C Quattrocchio F Mol J Koes R . 2000. anthocyanin1 of petunia encodes a basic Helix-Loop-Helix protein that directly activates transcription of structural anthocyanin genes. Plant Cell 12, 1619–1631.

57. Stewart RN Dermen H . 1979. Ontogeny in monocotyledons as revealed by studies of the developmental anatomy of periclinal chloroplast chimeras. American Journal of Botany 66, 47–58.

58. Sun Y Li H Huang JR . 2012. Arabidopsis TT19 functions as a carrier to transport anthocyanin from the cytosol to tonoplasts. Molecular Plant 5, 387–400.

59. Tanaka Y Sasaki N Ohmiya A . 2008. Biosynthesis of plant pigments: anthocyanins, betalains and carotenoids. Plant Journal 54, 733–749.

60. Tanay A Siggia ED . 2008. Sequence context affects the rate of short insertions and deletions in flies and primates. Genome Biology 9, R37.

61. Telias A Lin-Wang K Stevenson DE Cooney JM Hellens RP Allan AC Hoover EE Bradeen JM . 2011. Apple skin patterning is associated with differential expression of MYB10. BMC Plant Biology 11, 93.

62. The International Peach Genome Initiative. 2013. The high-quality draft genome of peach (Prunus persica) identifies unique patterns of genetic diversity, domestication and genome evolution. Nature Genetics 45, 487–494.

63. Uematsu C Katayama H Makino I Inagaki A Arakawa O Martin C . 2014. Peace, a MYB-like transcription factor, regulates petal pigmentation in flowering peach 'Genpei' bearing variegated and fully pigmented flowers. Journal of Experimental Botany 65, 1081–1094.

64. van Houwelingen A Souer E Spelt K Kloos D Mol J Koes R . 1998. Analysis of flower pigmentation mutants generated by random transposon mutagenesis in Petunia hybrida. Plant Journal 13, 39–50.

65. Walker AR Lee E Robinson SP . 2006. Two new grape cultivars, bud sports of Cabernet Sauvignon bearing pale-coloured berries, are the result of deletion of two regulatory genes of the berry colour locus. Plant Molecular Biology 62, 623–635.

66. Waters LS Minesinger BK Wiltrout ME D'Souza S Woodruff RV Walker GC . 2009. Eukaryotic Translesion polymerases and their roles and regulation in DNA damage tolerance. Microbiology and Molecular Biology Reviews 73, 134–154.

67. Winkel-Shirley B . 2001. Flavonoid biosynthesis.A colorful model for genetics, biochemistry, cell biology, and biotechnology. Plant Physiology 126 (2), 485–493.

68. Yamamoto T Yamaguchi M Hayashi T . 2005. An Integrated Genetic Linkage Map of Peach by SSR, STS, AFLP and RAPD. Journal of The Japanese Society for Horticultural Science 74, 204–213.

69. Zhang H Wang L Deroles S Bennett R Davies K . 2006. New insight into the structures and formation of anthocyanic vacuolar inclusions in flower petals. BMC Plant Biology 17, 6–29.

70. Zhao J Dixon RA . 2009. MATE transporters facilitate vacuolar uptake of epicatechin 3'-O-glucoside for proanthocyanidin biosynthesis in Medicago truncatula and Arabidopsis. Plant Cell 21, 2323–2340.

71. Zhao J Dixon RA . 2010. The 'ins' and 'outs' of flavonoid transport. Trends in Plant Science 15, 72–80.

Chapter 11

A SPONTANEOUS DOMINANT-NEGATIVE MUTATION WITHIN A 35S::ATMYB90 TRANSGENE INHIBITS FLOWER PIGMENT PRODUCTION IN TOBACCO

Jeff Velten[1], Cahid Cakir[1], Christopher I. Cazzonelli[2]

[1]Plant Stress and Water Conservation Laboratory, United States Department of Agriculture - Agricultural Research Service, Lubbock, Texas, United States of America

[2]Australian Research Council - Centre of Excellence in Plant Energy Biology, Research School of Biology, Australian National University, Canberra, Australian Capital Territory, Australia

ABSTRACT

Background

In part due to the ease of visual detection of phenotypic changes, anthocyanin pigment production has long been the target of genetic and molecular research in plants. Specific members of the large family of plant myb transcription factors have been found to play critical roles in regulating expression of anthocyanin biosynthetic genes and these genes continue to serve as important tools in dissecting the molecular mechanisms of plant gene regulation.

Findings

A spontaneous mutation within the coding region of an Arabidopsis 35S::*AtMYB90* transgene converted the activator of plant-wide anthocyanin production to a dominant-negative allele (PG-1) that inhibits normal pigment production within tobacco petals. Sequence analysis identified a single base change that created a premature nonsense codon, truncating the encoded myb protein. The resulting mutant protein lacks 78 amino acids from the wild type C-terminus and was confirmed as the source of the white-flower phenotype. A putative tobacco homolog of*AtMYB90* (*NtAN2*) was isolated and found to be expressed in flower petals but not leaves of all tobacco plants tested. Using transgenic tobacco constitutively expressing the *NtAN2* gene confirmed the NtAN2 protein as the likely target of PG-1-based inhibition of tobacco pigment production.

Conclusions

Messenger RNA and anthocyanin analysis of PG-1Sh transgenic lines (and PG-1Sh x purple 35S::*NtAN2* seedlings) support a model in which the mutant myb transgene product acts as a competitive inhibitor of the native tobacco NtAN2 protein. This finding is important to researchers in the field of plant transcription factor analysis, representing a potential outcome for experiments analyzing *in vivo* protein function in test transgenic systems that over-express or mutate plant transcription factors.

INTRODUCTION

Anthocyanins represent a broad family of plant pigments that contribute to flower and fruit pigmentation [1], plant stress response [2], [3] and have been implicated as helpful nutrients that contribute to improved human health [4]. The production of anthocyanins and related pigments in plants has been the target of extensive genetic and molecular research and represents one of the better understood plant gene regulatory systems. Specific members of the Myb family of plant transcription factors have been found to play critical roles in controlling the expression of genes associated with anthocyanin production, often in conjunction with members of the basic helix-loop-helix (bHLH) and WD40 families of trans factors (e.g. [5], [6],[7], [8], [9], [10], [11], [12], [13]). A classic example of this form of gene regulation was originally identified through genetic mapping of maize mutations affecting seed-coat color. Many of these maize mutant alleles mapped to the C1 (MYB) [14], [15], R (bHLH) [16], [17], or PAC1 (WD40) [18] loci [19]. More recently, other examples of plant MYB genes in the R2R3 family [20], [21] have been found to play significant roles in controlling pigment production in flowers, fruit and vegetative tissues of several plant species [9], [22]. Transgenic ectopic over-expression of several of these MYB genes has been shown to dramatically impact anthocyanin accumulation, in many cases affecting pigmentation within plant species other than those from which the MYB transgenes originated [13], [23], [24], [25], [26], [27], [28],[29], [30], [31], [32], [33], [34]. Ectopic expression of either of two closely related Arabidopsis MYB genes, *AtMYB75* (*PAP1*) and *AtMYB90* (*PAP2*) in *Nicotiana tabacum* produced striking levels of anthocyanin pigmentation in most parts of transgenic plants, providing a clear visual indicator of transgene activity [35]. A similar dark purple 35S::*AtMYB90* transgenic tobacco line was created in this laboratory (Myb-27, Fig. 1 & 2) and used as test material in a visual screen for molecular mechanisms that can alter transgene expression levels and/or patterns during *in vitro* de-differentiated growth, and subsequent *de novo* shoot production, processes that are normally part of plant genetic transformation protocols. A single plant line (PG-1) regenerated from purple Myb-27 callus, was initially identified by a complete loss of the darkly pigmented phenotype of the parental line. Upon reaching maturity, the PG-1 line was found to display a white flower phenotype that differed from the dark purple flowers of MYB-27 and the lightly pigmented red flowers of wild-type tobacco [*N. tabacum*, cv SR1 [36]]. Genetic and molecular analysis of the PG-1 line indicate that both the loss of hyper-pigmentation and the white flower phenotype are the result of a spontaneous dominant-negative nonsense mutation within the coding region of the *AtMYB90* transgene. The observed dominant-negative white flower phenotype seen with the *PG-1* allele is similar to that reported in transgenic tobacco lines expressing the maize *C1-I* mutant allele [37]; and a wild type strawberry myb (*FaMYB1*[38]). The structure and properties of the *PG-1* dominant-negative mutation demonstrate a mechanism for manipulating Myb gene structure that can provide useful insight into the mechanisms by which MYB transcription factors function to regulate gene expression in plants.

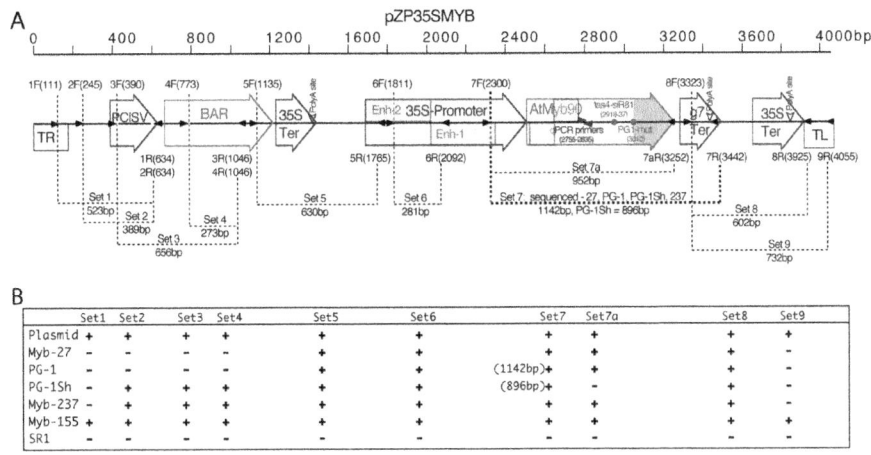

Figure 1: PCR scan across the T-DNA construct introduced into Myb-27.

A. Map of the T-DNA containing a 35S::*AtMYB90* transgene introduced into *N. tabacum*to create the Myb-27 purple plant line: 'TR', right T-DNA border; 'PCISV-Pro', PCISV promoter; 'BAR-Coding', basta resistance gene; '35S-Ter', CaMV 35S termination signal; '2xEnh35S-Pro', CaMV 35S promoter with duplicated enhancer region; '*AtMYB90*-Coding', Arabidopsis MYB90 gene; 'g7-Ter', termination signal from gene-7 of octopine T-DNA; 'TL', left T-DNA border. The small black arrows show PCR primers (primer identifiers listed above [forward] and below [reverse] each arrow) used to confirm the structure of the 35S::*AtMYB90* transgene in plant samples. Primer sets used are indicated by dashed lines (PCR product size, bp, in parenthesies). Set 7 indicates the area of the *Myb-27* and *PG-1 alleles* that was PCR amplified from transgenic plants and sequenced, with the red spot in *AtMYB90*-coding showing the location of the PG-1 nonsense (AAT->TAG, K172*) mutation (shaded area of the *AtMYB90* coding region indicates the amino acids missing from PG-1 and the DNA segment deleted in PG-1Sh). B. PCR results are alligned with the corresponding primer sets indicated in part A (numbered 1–9), with '+' indicating a positive PCR band of the predicted size, and '-' signifying no PCR product. The plasmid DNA used as a positive control, pZP35SMYB, is the binary construct used to generate the Myb-27 transgenic plant line. The remaining templates (total plant leaf DNA) are from the purple Myb-27 line, the white-flower PG-1 line, the white-flower PG-1Sh line and two additional independently derived purple transgenic tobacco lines (Myb-155 and Myb-237).

doi:10.1371/journal.pone.0009917.g001

Figure 2: Photos displaying the phenotypes of transgenic plant lines used in this study.

A. The Myb-27 transgenic plant line, wild type *N. tabaccum* cv SR1, Myb-27 callus with induced green and purple shoots and the NtAN2-1-59 line (35S::*NtAN2*). B. Flowers from the purple Myb-27 line, wild type *N. tabaccum* cv SR1, the dominant-negative white flower mutant PG-1 line, the shortened *Myb-27*, PG-1Sh (ransgenic line 32), the NtAN2 hairpin RNA (transgenic line 29) and the NtAN2-1-59 line. Flowers on the right were hand sectioned longitudinally to show internal components.

doi:10.1371/journal.pone.0009917.g002

RESULTS

Myb-27: Production and Properties of the 35S::*AtMYB90* Transgenic Lines; Callus Propagation; and *de novo* Shoot Induction

The *AtMYB90* coding region, under control of a CaMV 35S promoter [39] and the T-DNA gene-7 transcription termination/polyadenylation signal sequence ([40], Fig. 1A), was introduced into tobacco (*N. tabacum* cv SR1) and resulting transgenic shoots screened visually for ectopic anthocyanin production. The Myb-27 line was selected as a purple shoot from callus associated with the initial Agrobacterium-treated tobacco leaf explants. Subsequent phosphinothricin treatment of R1 Myb-27 seedlings indicated that the line was not herbicide resistant, consistent with PCR scans spanning the introduced T-DNA (Fig. 1B). Other transgenic lines also chosen for their purple phenotypes (e.g. Myb-237 and Myb-155) were found to harbor functional glufosinate resistance genes (Fig. 1B). The transgenic line, Myb-27, was selected for additional analysis based upon its dominant, heavily pigmented phenotype (Fig. 2A). Although the purple Myb-27 plants grow more slowly than their wild-type tobacco parent under low light conditions ($\sim$60 uMol quanta m^{-2} s^{-1}), they otherwise display no obvious developmental or morphological changes. Actively growing cultured callus derived from surface sterilized hemizygous Myb-27 leaf material was found to display extensive anthocyanin pigmentation and was capable of producing new shoots, most of which displayed anthocyanin pigment patterns and levels similar to the parent Myb-27 plant (Fig. 2A).

Myb-27 Plants Regenerated from Callus can Revert to a Wild-Type, Green, Phenotype

Of $\sim$100 plantlets regenerated and rooted from hemizygous purple Myb-27 callus, 4 completely lacked ectopic purple pigmentation (Fig. 2A). These 4 green regenerants were subsequently screened by PCR for the presence of the 35S::*AtMYB90* transgene (primer set 7a, Fig. 1A). Only one plant, designated line PG-1, gave a positive PCR signal, with the other three green plants apparently having lost the transgene during callus growth and/or plant regeneration. After reaching maturity the PG-1 line was found to display a white flower phenotype, producing flower petals that not only lacked the dark pigmentation of Myb-27 flowers, but also failed to produce the normal lightly pigmented red petals seen in wild-type tobacco (Fig. 2B).

The *PG-1* Locus Contains a Single-Base, Dominant-Negative, nonsense Mutation within the *AtMYTB90* Transgene

Plants grown from seed of the selfed R_0 PG-1 plant displayed an approximately 3:1 ratio of white to pink flowered plants (29 white, 11 pink), results consistent with the original PG-1 transgenic plant being hemizygous for a single, dominant-negative, white-flower locus. The dominant-negative character of the *PG-1* allele was confirmed by crossing the PG-1 R_0 plant to wild-type tobacco, producing an approximate 1:1 ratio of white (18) to red (21) flower phenotypes in the resulting seedlings.

PCR analysis using primers targeting additional sites within the T-DNA used to create the Myb-27, and subsequent PG-1, transgenic lines failed to indicate any gross rearrangements of the PG-1 T-DNA relative to that present in Myb-27 plants (Fig. 1B). DNA isolated from Myb-27, PG-1 and Myb-237 lines was used to produce PCR products covering the area flanked by primer set 7 (extending from the 35S promoter to the g7 termination signal, Fig. 1A). Sequence derived from these PCR products indicated that, relative to the wild-type Myb-27 *AtMYB90*allele, the *PG-1* allele contains a single base change within the myb coding region. This mutation, an A to T transversion, converts an AAG (lysine) codon to a TAG (ocher) nonsense triplet at the 172nd codon (Fig. 3), and is predicted to produce a truncated AtMYB90 protein that lacks the C-terminal 78 amino acids of the 249 amino acid AtMYB90 protein (Fig. 4). The A to T mutation also creates a new XbaI cleavage site (Fig. 4), allowing direct detection of the*PG-1* allele by XbaI digestion of PCR products from flanking primers, followed by electrophoretic separation of the resulting two DNA fragments. The new XbaI site was used to confirm the presence of the *PG-1* allele in all experiments involving PG-1 plant lines.

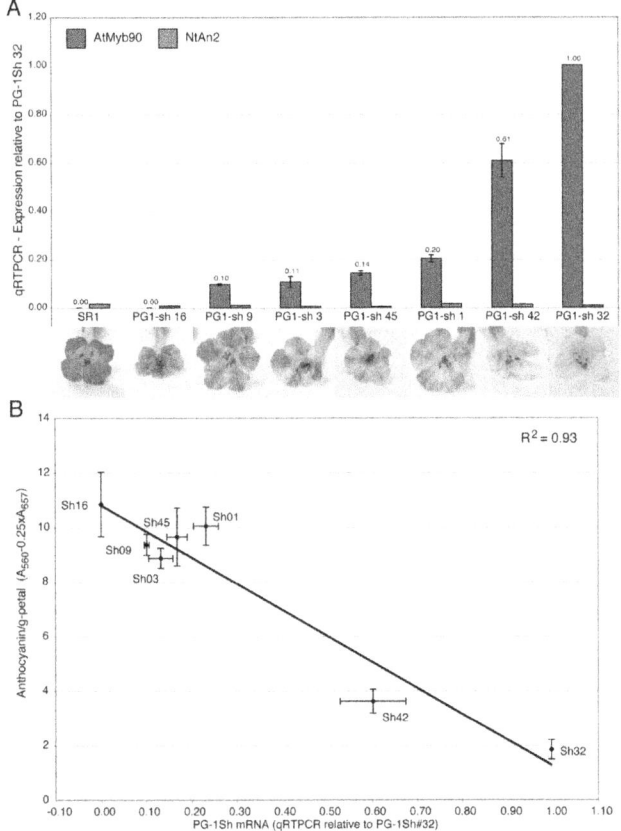

Figure 3: Analysis of anthocyanin levels and *AtMYB(90)* **expression in PG-1Sh transgenic lines.**

A. Flower total RNA was used for qRTPCR determination of mRNA levels from the PG-1Sh transgene (purple) and the endogenous tobacco homolog, *NtAN2* (blue). All values (shown above the PG-1Sh bars) are reported relative to the mRNA level for the PG-1Sh transgene in line #32 and are the mean of 3 to 4 biological reps. The PG-1Sh transgene appears to be inactive in line #16. Photos of representative flowers from each plant line are shown below the graph. B. Spectrophotometically determined anthocyanin levels in flowers (n=3 to 4) from the same transgenic lines were plotted against the relative PG-1Sh mRNA amounts shown in part A. PG-1Sh mRNA levels show an inverse correlation with anthocyanin content (R^2=0.94), while an identical plot of anthocyanin content against *NtAN2* mRNA levels using the same flower RNA samples showed no correlation with pigmentation (R^2=0.02).

doi:10.1371/journal.pone.0009917.g003

```
        CCAACCACGT CTTCAAAGCA AGTGGATTGA TGTGATATCT CCACTGACGT      50

        AAGGGATGAC GCACAATCCC ACTATCCTTC GCAAGACCCT TCCTCTATAT     100
                          +1r---------->
        AAGGAAGTTC ATTTCATTTG GAGAGGACAC GCTGAAATCA CCAGTCTCTC     150

        TCTACAAATC TATCTCTCTC GAGCTTTCGC AGATCTGTCG ATCGACCATG     200

        GAGGGTTCGT CCAAAGGGTT GAGGAAAGGT GCATGGACTG CTGAAGAAGA     250

        TAGTCTCTTG AGGCTATGTA TTGATAAGTA TGGAGAAGGC AAATGGCATC     300

        AAGTTCCTTT GAGAGCTGGG CTAAATCGAT GCAGAAAGAG TTGTAGACTA     350

        AGATGGTTGA ACTATTTGAA GCCAAGTATC AAGAGAGGAA GACTTAGCAA     400

        TGATGAAGTT GATCTTCTTC TTCGCCTTCA TAAGCTTCTA GGAAATAGGT     450

        GGTCCTTGAT TGCTGGTCGA TTGCCTGGTC GGACCGCTAA TGATGTCAAA     500

        AATTACTGGA ACACCCATCT GAGTAAAAAA CATGAGTCTT CGTGTTGTAA     550

        GTCTAAAATG AAAAAGAAAA ACATTATTTC CCCTCCTACA ACACCGGTCC     600

Tas4-siR81(-)          3'-aC GGAGCUGGAG CUAGGAAGu-5'
        AAAAAATCGG TGTTTTTAAG CCTCGACCTC GATCCTTCTC TGTTAACAAT     650
NtAN2 AgAAgATCac caTaTTcAga CCTCGgCCTC GAaCCTTCTC aaAgAcaAAT
      LysIlu-- ----PheArg ProArgProA rgThrPheSe r

        GGTTGCAGCC ATCTCAATGG TCTGCCAGAA GTTGATTTAA TTCCTTCATG     700

        CCTTGGACTC AAGAAAAATA ATGTTTGTGA AAATAGTATC ACATGTAACA     750
PG-1         TC TAGA(XbaI)

        AAGATGATGA GAAAGATGAT TTTGTGAATA ATCTAATGAA TGGAGATAAT     800

        ATGTGGTTGG AGAATTTACT GGGGGAAAAC CAAGAAGCTG ATGCGATTGT     850

        TCCTGAAGCG ACGACAGCTG AACATGGGGC CACTTTGGCG TTTGACGTTG     900

        AGCAACTTTG GAGTCTGTTT GATGGAGAGA CTGTTGAACT TGATTAGTGT     950

        TTCATGCATG GATCCTCTAG GTAGATGAGC TAAGCTAGCT ATATCATCAA    1000

        TTTATGTATT ACACATAATA TCGCACTCAG TCTTTCATCT ACGGCAATGT    1050

        ACCAGCTGAT ATAATCAGTT ATTGAAATAT TTCTGAATTT AAACTTGCAT    1100

        CAATAAATTT ATGTTTTTGC TTGGACTATA ATACCTGACT TGT           1143
```

Figure 4: DNA sequence of the *AtMYB90* **region within PG-1 and Myb-27 transgenic plants.**

The *AtMYB90* coding region is indicated by bold text (Red is Repeat 2, and Blue, Repeat 3) and the predicted transcription start site by a dashed arrow. PCR primers used to amplify the sequenced segment from total plant DNA are indicated by arrows. The PG-1 mutated codon is boxed (A to T mutation produces a new XbaI cut site). A TAS4-siR81(-) tasiRNA recognition site [43] is indicated (grey box). In the area of the recognition site the coresponding *NtAN2* DNA and predicted amino acid sequences are shown below (divergent bases, lower case). Areas of significant DNA homology between the *AtMYB90* and *NtAN2* sequences are underlined.

doi:10.1371/journal.pone.0009917.g004

The Predicted PG-1 Protein can Produce a White-Flower Phenotype in Tobacco

To test the hypothesis that the predicted shortened PG-1 protein is responsible for the observed white-flower phenotype, a new 35S::*AtMYB90* variant (PG-1 Short, or *PG-1Sh*) was generated and introduced into tobacco plants. The PG-1Sh construct lacks DNA encoding the 78 C-terminal amino acids downstream from the site of the *PG-1* mutant stop codon (Fig. 1A), and should produce the same shortened AtMYB90 protein as is predicted for the *PG-1* mutant allele. Transgenic tobacco lines expressing the PG-1Sh transgene displayed a range of flower color phenotypes, including plants with completely white flowers similar to those seen with the PG-1 line (Fig. 2B). Quantitative reverse-transcriptase PCR (qRTPCR) using mRNA from flowers of PG-1Sh lines chosen for their broad range in flower pigmentation indicated that expression of the PG-1Sh transgene was inversely proportional ($R^2=0.93$) to flower anthocyanin pigment levels (Fig. 3A&B). These results support a model in which the *PG-1* or *PG-1Sh* gene product interferes competitively with the normal functioning of an endogenous tobacco myb factor controlling anthocyanin production.

Cloning and Expression of a Putative Tobacco Homolog of *AtMYB90*

Alignment of the *AtMYB90* sequence against those contained in the tobacco transcription factor sequence database, TOBFAC, (<http://compsysbio.achs. virginia.edu/tobfac/>, [41]) identified a tobacco myb gene (gnl|tobfac|R2R3-MYB_141) with sequence similarity to the *AtMYB90* coding region. A PCR primer targeting the N-terminus of the predicted R2R3-MYB_141 coding region was designed and used to amplify and clone a cDNA for this putative tobacco *AtMYB90* homolog (PCR from start codon to a poly-A adaptor sequence, primers in Table 1)). The cloned tobacco Myb cDNA was sequenced and found to match that of a tobacco homolog (*NtAN2*) of the Petunia *AN2* myb gene recently added to the NCBI Genbank (FJ472647). In the spirit of standardized nomenclature we will refer to our tobacco myb homolog as *NtAN2*.

Table 1: PCR primers

Set ID.	Forward Primer (5'->3')	Reverse Primer (5'->3')	Product (bp)
1	GGTTTACCCGCCAATATATCC	GACGCGTCGACGTCTTCTCGATCGTGTCGATCAATAC	523
2	CGGGCCTCTTCGCTATTAC	GACGCGTCGACGTCTTCTCGATCGTGTCGATCAATAC	389
3	GATCTTGAGCCAATCAAAGAGGAGTGATGTAGAC	AGCCCGATGACAGCGAC	656
4	GTACCGAGCCGCAGGAAC	AGCCCGATGACAGCGAC	273
5	TGGCATGACGTGGGTTTC	CCCTCTGGTCTTCTGAGACTGTATC	630
6	GATTCCATTGCCCAGCTATC	CCCTCTGGTCTTCTGAGACTGTATC	281
7	CCAACCACGTCTTCAAAGCA	ATCAAGTTCAACAGTCTCTCCATCA	1142/896 [1]
7a	CCAACCACGTCTTCAAAGCA	ACAAGTCAGGTATTATAGTCCAAGC	952
8	ACATAATATCGCACTCAGTCTTTCATC	TGCGAACGTTTTTAATGTACTG	602
9	ACATAATATCGCACTCAGTCTTTCATC	CGAGTGGTGATTTTGTGCCGA	732
NtAN2 (PCR)	ATGAATATTTGTACTAATAAGTCGTCGTCAG	AAAGATTAAATCCTACGTCTGCCTCATAAG	549
NtAN2 (cDNA)	TACCAAGACCATGGATATTTGTACT	ACAGGATCCTATCAACTGAAAAGTG	683
AtMybQ1 (qPCR)	GACTGCTGAAGAAGATAGTCTCTTG	GCCCAGCTCTCAAAGGAACTTGATG	104
NtMybQ1 (qPCR)	AGGCCACATATAAAGAGAGGGAGACT	AATAAGTGACCATCTGTTGCCTAAC	107
icMGB (qPCR)	TCGCTAATGTGAGGACAGTGTA	ATCATCCATGTGCGTGGGACAGCAT	108
35S:: NtAN2	CACAATCCCACTATCCTTCG	AATAAGTGACCATCTGTTGCCTAAC	411
35S:: PG1Sh32	CACAATCCCACTATCCTTCG	TGTTTTTCTTTTTCATTTTAGACTT	511
NtAN2 In-Ex	AATGTAATTCTACTTATTGTAACAGGTACTTATC	CTTATGAAGCCTCAAAATGATGATCTAC	305

[1]Two product sizes are indicated for Set 7 with the smaller number being associated with the PG-1Sh deletion construct.
doi:10.1371/journal.pone.0009917.t001

doi:10.1371/journal.pone.0009917.t001

A protein BLAST search using the *NtAN2* sequence identified *AtMyb113, 75, 90* and *114*genes (BLAST scores: 205, 194, 183, and 180) as the Arabidopsis proteins most closely related to *NtAN2*. All of these Arabidopsis Myb genes have been implicated in regulation of Anthrocyanin production and the next closest Arabidopsis gene in the search, transparent testa 2 (*TT2, AtMYB123*) is associated with proanthocyanin production in the seed coat. Consistent with a role as an activator of anthocyanin production in tobacco, qRTPCR analysis of *NtAN2* mRNA (primers listed in Table 1) detected *NtAN2* expression in flowers but none in leaf tissue (leaf Ct>35, at least 1000 fold less than flower mRNA levels [Ct~23]). Further support for *NtAN2*'s role as a myb activator of anthocyanin production was provided by generation of transgenic *N. tabacum* (SR1) plants expressing a 35S::*NtAN2* transgene (the 35S::*NtAN2* construct substitutes the *NtAN2* coding region for that of *AtMYB90* in Fig. 1A). Several *NtAN2*-expressing R_0 lines (12 of 71) displayed extensive ectopic purple pigmentation similar to patterns observed in tobacco lines expressing the 35S::*AtMYB90* transgene (e.g.Fig. 2A and 2B). Finally, transgenic tobacco plants expressing a double-stranded hairpin construct targeting the entire *NtAN2* coding region for RNAi (ihpNtAN2, a 35S::antisense-intron-sense hairpin within the pKO vector, [42]) was able to produce white flowers similar to those of PG-1 plants (2 of 12 lines showed a white flower phenotype, with the remaining lines displaying varying levels of pigment reduction, Fig. 2B and 3A). These findings are consistent with those reported by Pattanaik et al, at the ASPB Plant Biology Symposium, 2009 <http://abstracts.aspb.org/pb2009/

public/P30/P30031.html>, and strongly suggest that *NtAN2*is a likely target for the interference with anthocyanin production seen in plants expressing the*PG-1* allele or PG-1Sh transgenes.

qRTPCR analysis of *NtAN2* gene expression in flowers from the set of representative PG-1Sh plants analyzed for PG-1Sh mRNA (Fig. 3A) did not indicate any correlation between flower*NtAN2* mRNA levels and anthocyanin pigmentation (R^2=0.01). These results strongly suggest that PG-1Sh-associated interference in pigment production does not result from transgene-induced alterations in *NtAN2* transcription or from post transcriptional gene silencing of the*NtAN2* gene, leaving competitive protein-protein interaction as the most likely mechanism for the observed white flower phenotype.

Alignment of the *NtAN2* cDNA with that of *AtMYB90* showed very little sequence similarity outside of that occurring within the 5′ repeats that are definitive of the R2R3 family of plant myb genes (Fig. 4). The only clear exception was a small region of sequence similarity just downstream from the R2R3 repeats (at ~625 bp) which, interestingly, overlaps the area of the*AtMYB90* transcript targeted by an Arabidopsis trans-acting small interfering RNA [tasiRNA, specifically TAS4-siR81(−)] [43]. The tobacco sequence is not a perfect complement to the TAS4-siR81 (2 mismatches and a G::T pairing) and there is as yet no direct evidence suggesting that the observed sequence similarity reflects evolutionary conservation of a functional mRNA::siRNA interaction. In fact, alignment of the predicted amino acid sequences (Probcons, [44]) from the *NtAN2* and *AtMYB90* genes at the TAS4-siRN81 target site indicated only highly conservative amino acid substitutions (Arginine for Lysine and Threonine for Serine, Fig. 5) within a conserved nine amino acid segment. It is thus conceivable that the observed sequence similarity at the TAS4-siRN81 site is the result of an evolutionarily conserved protein function.

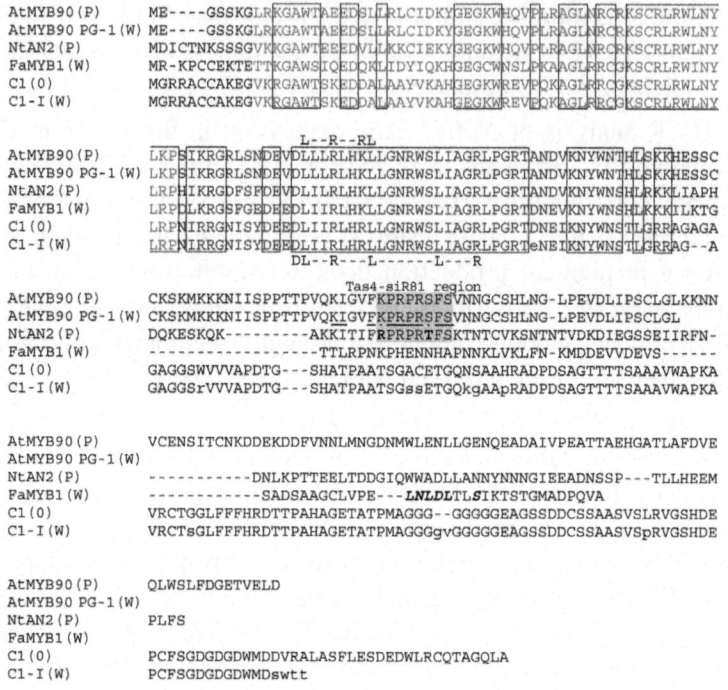

Figure 5: Protein sequence allignment (ProbCon, [44]) of R2R3 Myb proteins demonstrated to produce a reduction in anthocyanin phenotypes when expressed in transgenic tobacco.

Ectopic over-expression of the indicated Myb genes produces extensive purple pigmentation (P), white flower phenotype (W) or no phenotypic change (0). The R2 repeat is indicated as Red text and R3 repeat as Blue. Amino acid sequences that align in all proteins are boxed, differences between C1 andC1-I are shown as lower case. Sequences associated with Myb-bHLH interaction (L--R--RL [49], DL--R---L------L---R[50]) are indicated above and below the aligned sequences. The amino acids encoded by the mRNA region of *AtMYB90* mRNA targeted by TAS4-siR81(-) are indicated by a grey box (bold indicates conservative amino acid differences between the *NtAN2* and*AtMYB90* sequences in that area). *Bold-italic* amino acids indicate the conserved C2 domain proposed to be important to *FaMYB1* repressor function [56]. Due to a native nonsense mutation the protein sequence of the *AtMYB114* allele in the Columbia ecotype is predicted to end 32 amino acids upstream from the PG-1 nonsense mutation (at the F residue just prior to the TAS4-siR81 (-) grey boxed region).

doi:10.1371/journal.pone.0009917.g005

The PG-1Sh version of *AtMYB90* also Impacts Anthocyanin Production in Transgenic 35S::*NtAN2* Plants

To confirm functional *in vivo* interaction between the PG1 and *NtAN2* gene products, PG-1Sh #32 transgenic plants were crossed with a 35S::*NtAN2* transgenic line (NtAN2-1-59) that displays enhanced anthocyanin production (Fig. 2A and 2B). The phenotypes (anthocyanin pigmentation) and genotypes (determined by gene-specific PCR, Table 1) of resulting F1 seedlings were compared (Fig. 6). As expected, plants containing only the 35S::*NtAN2*transgene displayed enhanced anthocyanin production within their leaves (Fig. 6). Seedlings containing both the 35S::*NtAN2* and *PG-1Sh* transgenes showed dramatically reduced anthocyanin production in leaves, in most cases appearing phenotypically identical to leaves from wildtype SR1 seedlings or plants containing only the *PG-1Sh* construct (Fig. 6). These data confirm the ability of the *PG1* gene product to interfere with *NtAN2* function in tissues other than flower petals, and indicate that the observed interference is independent of the promoter associated with *NtAN2* expression (the native *NtAN2* promoter drives expression in tobacco flower petals, while the virally derived CaMV-35S promoter controls *NtAN2* expression in NtAN2-1-59 transgenic leaves).

Pigmentation phenotypes

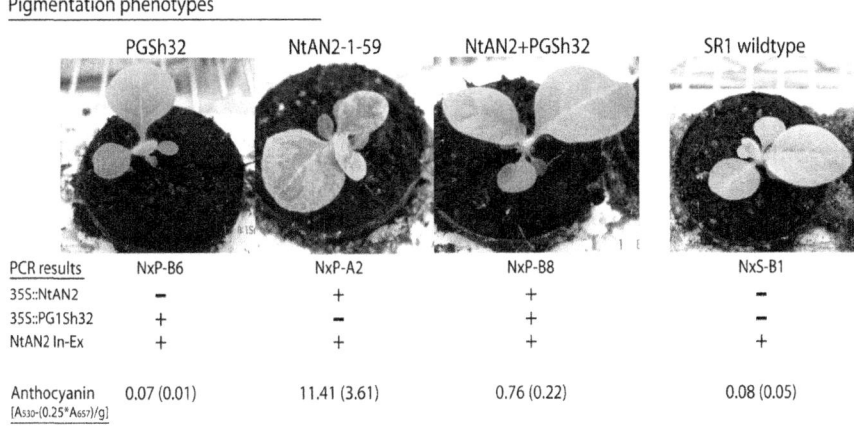

PCR results	NxP-B6	NxP-A2	NxP-B8	NxS-B1
35S::NtAN2	−	+	+	−
35S::PG1Sh32	+	−	+	−
NtAN2 In-Ex	+	+	+	+
Anthocyanin [A₅₃₀-(0.25*A₆₅₇)/g]	0.07 (0.01)	11.41 (3.61)	0.76 (0.22)	0.08 (0.05)

Figure 6: Representative anthocyanin pigmentation phenotypes for all possible transgene genotypes resulting from NtAN2-1-59 x PG-1Sh #32 crosses.

The genotypes for each seedling ('NxP': NtAN2-1-59 x PG1Sh32, 'NxS': NtAN2-1-59 x SR1) are indicated below the photos as determined by PCR using primers that specifically target each transgene construct (35S::*NtAN2* or 35S::*PG-1Sh*). Primers targeting an intron-exon junction within the native tobacco *NtAN2* gene (NtAN2 In-Ex,Table 1) were used as a positive PCR control. Relative anthocyanin levels, determined using leaf tissues from each genotype, are listed below the PCR results (standard error for each measurement [n=3 to 10] is shown in parentheses).

doi:10.1371/journal.pone.0009917.g006

DISCUSSION

A single-base nonsense mutation within the coding region of an active Arabidopsis *AtMYB90*transgene (the *PG-1* allele) was found to convert the R2R3-myb gene from a transcriptional activator of plant-wide anthocyanin biosynthesis to a dominant-negative allele that was able to interfere with normal tobacco pigment production within flower petals. Confirmation that the*PG-1* gene product is responsible for the observed white-flower phenotype was provided by expression in transgenic tobacco of a truncated *AtMYB90* gene (*PG-1Sh*) engineered to produce the same shortened myb protein as that predicted for the mutant *PG-1* allele. The *PG-1Sh* transgenic lines displayed a range of flower pigmentation phenotypes, including white flowers similar to those seen with *PG-1* plants. Furthermore, anthocyanin content in representative *PG-1Sh* flowers was found to be inversely proportional to *PG-1Sh* transgene expression levels (Fig. 3A & 3B), supporting a negative function for the *PG-1Sh* gene product.

Based upon the highly pigmented phenotype of the Myb-27 tobacco line, the AtMYB90 protein is able to interact with those native tobacco transcription factors and promoters required to activate transcription of anthocyanin biosynthetic genes. This ability of an anthocyanin-associated myb factor to function in a non-native plant system is not unique, as similar pigmented phenotypes have been seen with ectopic over-expressed Myb transgenes in several heterologous plant species (e.g.: Maize *C1* expressed in tobacco; [33], Apple *MdMYB1*expressed in Arabidopsis [25]; Daisy *GMYB10* expressed in tobacco [23]; Arabidopsis*AtMYB75* expressed in petunia [26], tobacco [35] or tomato [31]; Sweet potato *IbMYB1*expressed in Arabidopsis [29]; Grape *VvMYB5a* expressed in tobacco [45]; and *Medicago truncatula LAP1* in legumes and tobacco [46]). The predicted PG-1 and PG-1Sh protein is a shortened version of the *AtMYB90* gene product, retaining the highly conserved R2R3 domains but lacking 78 amino acids at the C-terminus (Fig. 5).

Based on our results, the truncated PG-1 protein has lost the ability to induce pigment production but retained sufficient function to allow it to interfere with the tobacco anthocyanin regulatory system active in flower petals. The observed interference in flower anthocyanin biosynthesis does not appear to be the result of altered transcription or message stability (e.g. RNAi) of the presumed functional tobacco myb homolog (*NtAN2*) since steady-state *NtAN2*mRNA levels show no correlative relationship with *PG-1Sh* mRNA content or anthocyanin levels in transgenic flowers displaying a wide range of pigmentation (Fig. 3A).

A literature search identified two other examples of myb-based genes that effectively eliminate flower pigment production when over-expressed in tobacco, the *C1-I* allele from maize [37] and a wild-type strawberry myb gene (*FaMYB1* [38]). It was proposed that *FaMYB1* may act directly as a transcriptional repressor [38], while the mutant transcriptional activator, C1-I, was assumed to act as a competitor to a native tobacco Myb protein, replacing the native protein within specific transcription initiation complexes [37], [38], [47]. The high ratios of *PG-1Sh* to*NtAN2* expression seen in the least pigmented *PG-1Sh* transgenic flowers (~40-fold PG-1Sh mRNA excess in the mostly white flower line #42 or ~120-fold excess in the white-flower line #32], Fig. 3A), support a model that proposes competition between the 'inactive' PG-1 and 'active' NtAN2 proteins for a common site within anthocyanin-associated transcription complexes. A similar competitive inhibition of transcription complexes may explain the loss of pigmentation associated with over-expression of *AtMYB60* in lettuce [48]. The ability of an active PG-1Sh gene (PG-1Sh #32) to dramatically reduce anthocyanin production when crossed into the purple 35S::*NtAN2* transgenic line, NtAN2-1-59 (Fig. 6), further supports a model of protein competition since the observed interference occurs in non-flower tissues and affects *NtAN2* activity controlled by a promoter unrelated to that which regulates expression of the native *NtAN2* gene in flower petals.

Alignment of predicted C1, C1-I, FaMYB1, AtMYB90/PG-1 and NtAN2 protein sequences indicates that sequence similarity is primarily limited to the highly conserved R2R3 DNA-binding domains common to this family of plant myb genes (Fig. 5). All of the aligned anthocyanin-associated myb proteins do, however, share sequence motifs (Fig. 5) linked to myb-bHLH binding (L--R--RL [49], DL--R---L------L---R [50]). The presence of the conserved bHLH binding motif is consistent with possible competition between the dominant-negative PG1 gene product and NtAN2 protein for association with one or more tobacco bHLH proteins. Just downstream from the R2R3 domains there is a noticeable short segment of protein similarity between the *AtMYB90* and *NtAN2* sequences, KI--F[K/R]PRP[R/T]FS. This sequence overlaps with an active tasiRNA target site identified in the *AtMYB90* mRNA (TAS4-siR81−, [43]) and it is not clear whether the common amino acids represent a conserved protein domain or reflect a possible homologous tobacco tasiRNA target within the *NtAN2*message. Our current results do not directly support any interaction between the *PG-1* and*NtAN2* genes at the level of mRNA regulation.

The simplest model for a competitive interaction between the PG-1 and NtAN2 myb proteins assumes that the 78 C-terminal amino acids missing in the *PG-1* product contain, or overlap with, a transcriptional regulatory domain required for gene activation. Although sequences downstream from the conserved R2R3 domains are generally assumed to contain protein sequences responsible for transcription activation and/or repression, very few specific motifs or functional domains have been confirmed in plant myb proteins (e.g. [11], [51], [52], [53]). Support for this model of plant myb protein function comes from work in which fusion of a 12 amino acid EAR repressor motif to the 3' end of the AtMYB75 protein transformed the transcriptional activator into a gene specific repressor [54]. A search for conserved protein motifs in the AtMYB90, C1, NtAN2 and FaMYB1 protein sequences (online MEME analysis,[55]) failed to identify any motifs outside those already identified by protein alignment, specifically the R2, R3 domains, and for *AtMYB90* and *NtAN2*, the TAS4 target region. Specifically, the short conserved 'C2' motif (LNL[D/E]L-[G/S] [38], [56]), which contains the core EAR motif (LXLXL, [57]), present in the proposed myb repressor, FaMYB1 was not identified in any of the other myb protein sequences examined.

The PG-1 allele is the result of a spontaneous single-base mutation within a *AtMYB90*transgene that acts as a dominant-negative 'repressor' of pigment production in tobacco flowers. The *AtMYB114* gene present in the Arabidopsis Columbia ecotype (*AtMYB114* is one of three Arabidopsis genes with very high sequence similarity to the *AtMYB90* gene) contains a premature stop codon located 31 amino acids upstream from the PG1 mutation, and over-production of the AtMYB114 (Col) truncated myb protein was recently shown to negatively impact anthocyanin production in Arabidopsis [13]. Similar dominant-negative mutations that produce truncated Myb proteins have been identified as naturally occurring alleles of the maize C1 gene [58], [59]. Both gene systems demonstrate a potential evolutionary mechanism that can convert myb transcriptional activators into repressors. In the case of PG-1, repression of tobacco anthocyanin production appears to be the result of competitive inhibition of one or more tobacco myb proteins. This mechanism is different from that proposed for plant myb proteins that contain a functional repressor domain such as the conserved C2 domain [56]implicated in the regulatory function of *AtMYB4* [60] and *FaMYB1* [38], and should be considered as a possibility when plant myb genes are over-expressed to test their function *in vivo* [48]. The authors are unaware of any documented examples of native plant gene regulatory systems that use competitive inhibition by an 'inactive' R2R3 myb protein to down-regulate gene expression. It is, however, important that the potential for such regulatory mechanism be kept in mind when dissecting plant gene control pathways that make use of myb genes.

MATERIALS AND METHODS

Gene Constructs and Stable Plant Transformation

Plasmids were prepared using standard cloning techniques [61] and appropriate DNA segments sequenced to confirm final constructs. When possible, different promoter, terminator, reporter and selectable marker cassettes were used within constructs to reduce the potential for recombination within plasmids. The 35S::*AtMYB90* constructs (T-DNA depicted inFig. 1A) used the pPZP200 vector [62] modified to contain a glufosinate-resistance plant selectable marker near the T-DNA right border. The plant resistance construct consists of the bar gene coding region (552 bp) encoding phosphinothricin acetyl transferase (Accession number: AX235900), regulated by the peanut chlorotic streak virus promoter (240 to +1 bp) [63]and CaMV 35S transcript termination signal.

Transformation of tobacco (*N. tabacum* cv SR1) was accomplished using the *Agrobacterium tumefaciens* line EHA105 [64]. Plasmid constructs were electroporated into EHA105 as previously described [65] and transformation of tobacco carried out by the conventional leaf disc method [66], [67]. Regenerated transgenic shoots were rooted on MS-agar medium [68]containing B5 vitamins [69] and 500 µg/ml Claforan (sodium cefotaxime, Hoechst).

Callus was produced *de novo* from Myb-27 leaf tissue by placing surface sterilized material on MS-agar media supplemented with plant hormones (MS Salt; B5 Vitamins; Sucrose 2% [w/v]; indol-3-acetic acid (0.5 mg/mL); benzlaminopurine (0.5 mg/mL). After 2–3 weeks shoot production was induced by transfer of actively growing purple callus to the same media lacking indol-3-acetic acid. Shoots that displayed altered anthocyanin pigmentation levels or patterns were excised above the callus and moved to the same media lacking hormones for root induction and eventually transferred to soil.

PCR and Quantitative RT-PCR

Routine PCR used MJ Research PTC-100 thermocyclers (95°C-8 Min, 30 cycles-[94C-45 Sec, 56°C-30 Sec, 72°C-60 Sec], 74°C-5 Min) and reagents from Applied Biosystems®. Primer sets and product sizes are listed in Table 1.

Quantitative reverse transcriptase PCR (qRT-PCR, primers listed in Table 1) was performed using a LightCycler® 480 System and SYBR green kits (LightCycler® DNA Master SYBR Green I) from Roche Applied Science according to protocols provided by the manufacturer (2-step; 60°–72°, read once per second, ramp at 4.4°C/s up & 2.2°C/s down). Total RNA was prepared using either Ambion mirVana™ RNA isolation kits and suggested protocols or using Tri-Reagent® reagent from Ambion®. To control for potential variability in the biochemical processes that precede qRTPCR reactions, total RNA samples (5 μg each) were spiked with a synthetic control internal control (IC) mRNA (250 pg/reaction) produced in vitro using T7 RNA polymerase (using Ambion® MEGAscript® and MEGAclear™ kits) acting on a PCR product template (IC2r, Genebank Accession # GQ215228). Spiked samples were treated with RNAse-free DNAase (TURBO® DNase, from Ambion®) and cleaned post reaction as per manufacturer's instructions. Reverse transcription was performed using RETROscript® from Ambion® (following the manufacturer's protocols). Relative RNA values were calculated using formulas for ΔΔCt, the Pfaffl method [70], and according to Norgard, et al [71], applied to qRT-PCR data from total RNA samples (triplicate technical assays and the indicated number of biological replicates).

Spectrophotometric Anthocyanin Assay

Anthocyanin levels were determined by extraction of soluble anthocyanins as described by Martin et al [72], and spectrophotometic measurement at 530 nm and 657 nm. The formula used for relative anthocyanin content is: A_{530}-$(0.25 \times A_{657})$/g tissue extracted.

ACKNOWLEDGMENTS

Our appreciation goes out to: Drs. John Burke, Junping Chen and Zhanguo Xin for critical reading of the manuscript; Ryan Mize, Natalie Bizzell and Gracie Mahan for technical assistance and for keeping our precious plants alive; Kay McCrary and DeeDee Laumbach for tobacco transformation; and Nancy Layland for excellent technical support. The source plasmid for the *AtMYB90* cDNA was graciously provided by Mendel Biotechnology. Mention of a commercial or proprietary product does not constitute an endorsement by the USDA. USDA offers its programs to all eligible persons regardless of race, color, age, sex, or national origin.

AUTHOR CONTRIBUTIONS

Conceived and designed the experiments: JV. Performed the experiments: CC CIC. Analyzed the data: JV CIC. Contributed reagents/materials/analysis tools: JV. Wrote the paper: JV.

REFERENCES

1.   Yoshida K, Mori M, Kondo T (2009) Blue flower color development by anthocyanins: from chemical structure to cell physiology. Nat Prod Rep 26: 884–915.

2. Winkel-Shirley B (2002) Biosynthesis of flavonoids and effects of stress. Curr Opin Plant Biol 5: 218–223.

3. Ververidis F, Trantas E, Douglas C, Vollmer G, Kretzschmar G, et al. (2007) Biotechnology of flavonoids and other phenylpropanoid-derived natural products. Part II: Reconstruction of multienzyme pathways in plants and microbes. Biotechnol J 2: 1235–1249.

4. Crozier A, Jaganath IB, Clifford MN (2009) Dietary phenolics: chemistry, bioavailability and effects on health. Nat Prod Rep 26: 1001–1043.

5. Sablowski RW, Moyano E, Culianez-Macia FA, Schuch W, Martin C, et al. (1994) A flower-specific Myb protein activates transcription of phenylpropanoid biosynthetic genes. Embo J 13: 128–137.

6. Jin H, Martin C (1999) Multifunctionality and diversity within the plant MYB-gene family. Plant Mol Biol 41: 577–585.

7. Springob K, Nakajima J, Yamazaki M, Saito K (2003) Recent advances in the biosynthesis and accumulation of anthocyanins. Nat Prod Rep 20: 288–303.

8. Ramsay NA, Glover BJ (2005) MYB-bHLH-WD40 protein complex and the evolution of cellular diversity. Trends Plant Sci 10: 63–70.

9. Allan AC, Hellens RP, Laing WA (2008) MYB transcription factors that colour our fruit. Trends Plant Sci 13: 99–102.

10. Pattanaik S, Xie CH, Yuan L (2008) The interaction domains of the plant Myc-like bHLH transcription factors can regulate the transactivation strength. Planta 227: 707–715.

11. Du H, Zhang L, Liu L, Tang XF, Yang WJ, et al. (2009) Biochemical and molecular characterization of plant MYB transcription factor family. Biochemistry (Mosc) 74: 1–11.

12. Cominelli E, Gusmaroli G, Allegra D, Galbiati M, Wade HK, et al. (2008) Expression analysis of anthocyanin regulatory genes in response to different light qualities in Arabidopsis thaliana. J Plant Physiol 165: 886–894.

13. Gonzalez A, Zhao M, Leavitt JM, Lloyd AM (2008) Regulation of the anthocyanin biosynthetic pathway by the TTG1/bHLH/Myb transcriptional complex in Arabidopsis seedlings. Plant J 53: 814–827.

14. Coe EH (1962) Spontaneous Mutation of the Aleurone Color Inhibitor in Maize. Genetics 47: 779–783.

15. Paz-Ares J, Ghosal D, Wienand U, Peterson PA, Saedler H (1987) The regulatory c1 locus of Zea mays encodes a protein with homology to myb

proto-oncogene products and with structural similarities to transcriptional activators. Embo J 6: 3553–3558.

16. Stadler LJ (1946) Spontaneous Mutation at the R Locus in Maize. I. the Aleurone-Color and Plant-Color Effects. Genetics 31: 377–394.

17. Ludwig SR, Habera LF, Dellaporta SL, Wessler SR (1989) Lc, a member of the maize R gene family responsible for tissue-specific anthocyanin production, encodes a protein similar to transcriptional activators and contains the myc-homology region. Proc Natl Acad Sci U S A 86: 7092–7096.

18. Carey CC, Strahle JT, Selinger DA, Chandler VL (2004) Mutations in the pale aleurone color1 regulatory gene of the Zea mays anthocyanin pathway have distinct phenotypes relative to the functionally similar TRANSPARENT TESTA GLABRA1 gene in Arabidopsis thaliana. Plant Cell 16: 450–464.

19. Mol J, Grotewald E, Koes R (1998) How genes paint flowers and seeds. Trends in Plant Science 3: 212–217.

20. Meissner RC, Jin H, Cominelli E, Denekamp M, Fuertes A, et al. (1999) Function search in a large transcription factor gene family in Arabidopsis: assessing the potential of reverse genetics to identify insertional mutations in R2R3 MYB genes. Plant Cell 11: 1827–1840.

21. Stracke R, Werber M, Weisshaar B (2001) The R2R3-MYB gene family in Arabidopsis thaliana. Curr Opin Plant Biol 4: 447–456.

22. Grotewold E (2006) The genetics and biochemistry of floral pigments. Annu Rev Plant Biol 57: 761–780.

23. Elomaa P, Uimari A, Mehto M, Albert VA, Laitinen RA, et al. (2003) Activation of anthocyanin biosynthesis in Gerbera hybrida (Asteraceae) suggests conserved protein-protein and protein-promoter interactions between the anciently diverged monocots and eudicots. Plant Physiol 133: 1831–1842.

24. Mathews H, Clendennen SK, Caldwell CG, Liu XL, Connors K, et al. (2003) Activation tagging in tomato identifies a transcriptional regulator of anthocyanin biosynthesis, modification, and transport. Plant Cell 15: 1689–1703.

25. Takos AM, Jaffe FW, Jacob SR, Bogs J, Robinson SP, et al. (2006) Light-induced expression of a MYB gene regulates anthocyanin biosynthesis in red apples. Plant Physiol 142: 1216–1232.

26. Matousek J, Vrba L, Skopek J, Orctova L, Pesina K, et al. (2006) Sequence analysis of a "true" chalcone synthase (chs_H1) oligofamily from hop

(Humulus lupulus L.) and PAP1 activation of chs_H1 in heterologous systems. J Agric Food Chem 54: 7606–7615.

27. Shen LY, Petolino JF (2006) Pigmented Maize Seed via Tissue-specific Expression of Anthocyanin Pathway Gene Transcription Factors. Molec breeding 18: 57–67.

28. Espley RV, Hellens RP, Putterill J, Stevenson DE, Kutty-Amma S, et al. (2007) Red colouration in apple fruit is due to the activity of the MYB transcription factor, MdMYB10. Plant J 49: 414–427.

29. Mano H, Ogasawara F, Sato K, Higo H, Minobe Y (2007) Isolation of a regulatory gene of anthocyanin biosynthesis in tuberous roots of purple-fleshed sweet potato. Plant Physiol 143: 1252–1268.

30. Deluc L, Bogs J, Walker AR, Ferrier T, Decendit A, et al. (2008) The transcription factor VvMYB5b contributes to the regulation of anthocyanin and proanthocyanidin biosynthesis in developing grape berries. Plant Physiol 147: 2041–2053.

31. Zuluaga DL, Gonzali S, Loreti E, Pucciariello C, Degl'Innocenti E, et al. (2008) Arabidopsis thaliana MYB75/PAP1 transcription factor induces anthocyanin production in transgenic tomato plants. Functional Plant Biology 35: 606–618.

32. Cutanda-Perez MC, Ageorges A, Gomez C, Vialet S, Terrier N, et al. (2008) Ectopic expression of VlmybA1 in grapevine activates a narrow set of genes involved in anthocyanin synthesis and transport. Plant Mol Biol 69: 633–648.

33. Lloyd AM, Walbot V, Davis RW (1992) Arabidopsis and Nicotiana anthocyanin production activated by maize regulators R and C1. Science 258: 1773–1775.

34. Orzaez D, Medina A, Torre S, Fernandez-Moreno JP, Rambla JL, et al. (2009) A visual reporter system for virus-induced gene silencing in tomato fruit based on anthocyanin accumulation. Plant Physiol 150: 1122–1134.

35. Borevitz JO, Xia Y, Blount J, Dixon RA, Lamb C (2000) Activation tagging identifies a conserved MYB regulator of phenylpropanoid biosynthesis. Plant Cell 12: 2383–2394.

36. Maliga P, Sz-Breznovits A, Marton L, Joo F (1975) Non-Mendelian streptomycin-resistant tobacco mutant with altered chlorplasts and mitochondria. Nature 255: 401–402.

37. Chen B, Wang X, Hu Y, Wang Y, Lin Z (2004) Ectopic expression of a c1-I allele from maize inhibits pigment formation in the flower of transgenic tobacco. Mol Biotechnol 26: 187–192.

38. Aharoni A, De Vos CH, Wein M, Sun Z, Greco R, et al. (2001) The strawberry FaMYB1 transcription factor suppresses anthocyanin and flavonol accumulation in transgenic tobacco. Plant J 28: 319–332.

39. Odell JT, Nagy F, Chua NH (1985) Identification of DNA sequences required for activity of the cauliflower mosaic virus 35S promoter. Nature 313: 810–812.

40. Velten J, Schell J (1985) Selection-expression plasmid vectors for use in genetic transformation of higher plants. Nucleic Acids Res 13: 6981–6998.

41. Rushton PJ, Bokowiec MT, Laudeman TW, Brannock JF, Chen X, et al. (2008) TOBFAC: the database of tobacco transcription factors. BMC Bioinformatics 9: 53.

42. Cazzonelli CI, Velten J (2004) Analysis of RNA-mediated gene silencing using a new vector (pKNOCKOUT) and an in planta Agrobacterium transient expression system. Plant molecular biology reporter 22: 347–359.

43. Rajagopalan R, Vaucheret H, Trejo J, Bartel DP (2006) A diverse and evolutionarily fluid set of microRNAs in Arabidopsis thaliana. Genes Dev 20: 3407–3425.

44. Do CB, Mahabhashyam MS, Brudno M, Batzoglou S (2005) ProbCons: Probabilistic consistency-based multiple sequence alignment. Genome Res 15: 330–340.

45. Deluc L, Barrieu F, Marchive C, Lauvergeat V, Decendit A, et al. (2006) Characterization of a grapevine R2R3-MYB transcription factor that regulates the phenylpropanoid pathway. Plant Physiol 140: 499–511.

46. Peel GJ, Pang Y, Modolo LV, Dixon RA (2009) The LAP1 MYB transcription factor orchestrates anthocyanidin biosynthesis and glycosylation in Medicago. Plant J 59: 136–149.

47. Franken P, Schrell S, Peterson PA, Saedler H, Wienand U (1994) Molecular analysis of protein domain function encoded by the myb-homologous maize genes C1, Zm 1 and Zm 38. Plant J 6: 21–30.

48. Park JS, Kim JB, Cho KJ, Cheon CI, Sung MK, et al. (2008) Arabidopsis R2R3-MYB transcription factor AtMYB60 functions as a transcriptional repressor of anthocyanin biosynthesis in lettuce (Lactuca sativa). Plant Cell Rep 27: 985–994.

49. Grotewold E, Sainz MB, Tagliani L, Hernandez JM, Bowen B, et al. (2000) Identification of the residues in the Myb domain of maize C1 that specify the interaction with the bHLH cofactor R. Proc Natl Acad Sci U S A 97: 13579–13584.

50. Zimmermann IM, Heim MA, Weisshaar B, Uhrig JF (2004) Comprehensive identification of Arabidopsis thaliana MYB transcription factors interacting with R/B-like BHLH proteins. Plant J 40: 22–34.

51. Sainz MB, Goff SA, Chandler VL (1997) Extensive mutagenesis of a transcriptional activation domain identifies single hydrophobic and acidic amino acids important for activation in vivo. Mol Cell Biol 17: 115–122.

52. Urao T, Noji M, Yamaguchi-Shinozaki K, Shinozaki K (1996) A transcriptional activation domain of ATMYB2, a drought-inducible Arabidopsis Myb-related protein. Plant J 10: 1145–1148.

53. Jiang C, Gu X, Peterson T (2004) Identification of conserved gene structures and carboxy-terminal motifs in the Myb gene family of Arabidopsis and Oryza sativa L. ssp. indica. Genome Biol 5: R46.

54. Hiratsu K, Matsui K, Koyama T, Ohme-Takagi M (2003) Dominant repression of target genes by chimeric repressors that include the EAR motif, a repression domain, in Arabidopsis. Plant J 34: 733–739.

55. Bailey TL, Elkan C (1994) Fitting a mixture model by expectation maximization to discover motifs in biopolymers. Proc Int Conf Intell Syst Mol Biol 2: 28–36.

56. Kranz HD, Denekamp M, Greco R, Jin H, Leyva A, et al. (1998) Towards functional characterisation of the members of the R2R3-MYB gene family from Arabidopsis thaliana. Plant J 16: 263–276.

57. Ikeda M, Ohme-Takagi M (2009) A novel group of transcriptional repressors in Arabidopsis. Plant Cell Physiol 50: 970–975.

58. Singer T, Gierl A, Peterson PA (1998) Three new dominant C1 suppressor alleles in Zea mays. Genet Res 71: 127–132.

59. Paz-Ares J, Ghosal D, Saedler H (1990) Molecular analysis of the C1-I allele from Zea mays: a dominant mutant of the regulatory C1 locus. Embo J 9: 315–321.

60. Jin H, Cominelli E, Bailey P, Parr A, Mehrtens F, et al. (2000) Transcriptional repression by AtMYB4 controls production of UV-protecting sunscreens in Arabidopsis. Embo J 19: 6150–6161.

61. Sambrook J, Russell D (2001) Molecular Cloning: a laboratory manual. New York: Cold Spring Harbor Laboratory.

62. Hajdukiewicz P, Svab Z, Maliga P (1994) The small, versatile pPZP family of Agrobacterium binary vectors for plant transformation. Plant Mol Biol 25: 989–994.

63. Maiti IB, Shepherd RJ (1998) Isolation and expression analysis of peanut chlorotic streak caulimovirus (PCISV) full-length transcript (FLt)

promoter in transgenic plants. Biochem Biophys Res Commun 244: 440–444.

64. Hood EE, Gelvin SB, Melchers LS, Hoekema A (1993) New Agrobacterium helper plasmids for gene transfer to plants. Transgenic Research 218: 208–218.

65. Walkerpeach C, Velten J (1994) Agrobacterium-mediated gene transfer to plant cells cointegrate and binary vector systems. In: Gelvin S, Schilperoort R, editors. Plant Molecular Biology Manual. Second ed. Dordrecht: Kluwer Academic. pp. B1:1–B1:19.

66. Horsch R, Fry J, Hoffman N, Neidermeyer J, Rogers S, et al. (1988) Leaf disc transformation. In: Gelvin S, Schilperoort R, editors. Plant Molecular Biology Manual. Belgium: Kluwer Academic Publishers. pp. 1–9. First ed.

67. Svab Z, Hajdukiewicz P, Maliga P (1995) Generation of transgenic tobacco plants by cocultivation of leaf disks with Agrobacterium pPZP binary vectors. In: Maliga P, editor. Methods in plant molecular biology: A laboratory course manual. 1 ed. Plainview, NY: Cold Spring Harbor Laboratory Press. pp. 55–77.

68. Murashige T, Skoog F (1962) A revised medium for rapid growth and bioassays with tobacco tissue cultures. Physiol Plant 15: 473–497.

69. Gamborg OL, Miller RA, Ojima K (1968) Nutrient requirements of suspension cultures of soybean root cells. Exp. Exp Cell Res 50: 151–158.

70. Pfaffl MW (2001) A new mathematical model for relative quantification in real-time RT-PCR. Nucleic Acids Res 29: 2003–2007.

71. Nordgard O, Kvaloy JT, Farmen RK, Heikkila R (2006) Error propagation in relative real-time reverse transcription polymerase chain reaction quantification models: the balance between accuracy and precision. Anal Biochem 356: 182–193.

72. Martin T, Oswald O, Graham IA (2002) Arabidopsis seedling growth, storage lipid mobilization, and photosynthetic gene expression are regulated by carbon:nitrogen availability. Plant Physiol 128: 472–481.

Chapter 12

THE ABUNDANCE OF PINK-PIGMENTED FACULTATIVE METHYLOTROPHS IN THE ROOT ZONE OF PLANT SPECIES IN INVADED COASTAL SAGE SCRUB HABITAT

Irina C. Irvine[1,2], Christy A. Brigham[2] , Katharine N. Suding[3] , Jennifer B. H. Martiny[1]

[1]Department of Ecology and Evolutionary Biology, University of California Irvine, Irvine, California, United States of America

[2]Santa Monica Mountains National Recreation Area, United States National Park Service, Thousand Oaks, California, United States of America

[3]Department of Environmental Science, Policy and Management, University of California, Berkeley, California, United States of America

ABSTRACT

Pink-pigmented facultative methylotrophic bacteria (PPFMs) are associated with the roots, leaves and seeds of most terrestrial plants and utilize volatile C_1 compounds such as methanol generated by growing plants during cell division. PPFMs have been well studied in agricultural systems due to their importance in crop seed germination, yield, pathogen resistance and drought stress tolerance. In contrast, little is known about the PPFM abundance and diversity in natural ecosystems, let alone their interactions with non-crop species. Here we surveyed PPFM abundance in the root zone soil of 5 native and 5 invasive plant species along ten invasion gradients in Southern California coastal sage scrub habitat. PPFMs were present in every soil sample and ranged in abundance from 10^2 to 10^5 CFU/g dry soil. This abundance varied significantly among plant species. PPFM abundance was 50% higher in the root zones of annual or biennial species (many invasives) than perennial species (all natives). Further, PPFM abundance appears to be influenced by the plant community beyond the root zone; pure stands of either native or invasive species had 50% more PPFMs than mixed species stands. In sum, PPFM abundance in the root zone of coastal sage scrub plants is influenced by both the immediate and surrounding plant communities. The results also suggest that PPFMs are a good target for future work on plant-microorganism feedbacks in natural ecosystems.

INTRODUCTION

Methylotrophic bacteria utilize single carbon (C_1) compounds for energy and assimilation and are an important component of the global carbon cycle [1], [2]. One group of methylotrophs, pink-pigmented facultative methylotrophic bacteria (PPFMs), is distinguished based on their formation of pink to red colonies on selective isolation media. Classified within the genus*Methylobacterium*, PPFMs are facultative methylotrophs, using both single and multicarbon compounds.

PPFMs are associated with the roots, leaves and seeds of most terrestrial plants, and many are thought to be phytosymbionts [3]. The bacteria use C_1 compounds, such as methanol, generated by growing plants during cell division. In return, they can positively affect plant growth and survival [3]. Much of the evidence for these positive effects derives from agricultural systems, where PPFMs have been shown to improve seed germination, crop yield, pathogen resistance and drought stress tolerance [4], [5], [6].

There are at least two mechanisms by which PPFMs can positively affect plants, particularly in dry climates. First, PPFMs excrete auxins and cytokinins, plant growth hormones that influence germination and root growth and play critical roles in a plant›s response to water stress [7], [8]. In dry conditions, plants that send their roots deep quickly after germination may gain a competitive advantage over more shallowly rooted species. Second, PPFMs exude osmoprotectants (sugars and alcohols) on the surface of host plants [3]. This matrix may help protect the plants from desiccation and high temperatures.

In a recent study, we investigated the effect of PPFMs on several coastal sage scrub (CSS) plant species [9]. CSS is a low shrubland community that once dominated the Mediterranean-type climate regions of coastal California. Most native CSS plants are perennial, drought-deciduous species that can persist during dry, hot (+33°C) summers. Invasive species, many of which are annuals, have significantly fragmented and diminished the quality of CSS habitat[10]. In a combination of laboratory and field experiments, we found that PPFM or methanol addition stimulated germination and/or growth of two native species (*Artemisia californica* and*Nasella pulchra*), but not that of three invasive species [9]. These results suggest that PPFMs may have a greater benefit to native than invasive CSS species.

Given these interactions, we aimed to better understand the distribution of PPFMs in the CSS habitat, particularly in the root zones of native and invasive plant species. Despite their potential importance for plant growth and survival, most of what is known about PPFM distribution and diversity comes from agricultural settings. PPFMs and more broadly, *Methylobacterium* species, often make up the majority of cultivable heterotrophic bacteria in the phyllosphere (on leaf surfaces) of many plant families [11], [12], [6], [13], [14]. Several studies report that PPFM densities on leaves differ among crop species [15], [12], [16], suggesting that plant species vary in the quality of habitat that they provide for PPFMs. In addition, PPFM abundance and composition also appears to vary in the rhizosphere (next to the roots) of various crop species [13], [16], [17]. In natural ecosystems, PPFMs have been detected on the phyllosphere of temperate and tropical plant species worldwide [18], [19]. However, few studies have shown that methylobacteria or PPFM abundance varies among plant species in these ecosystems [18], [20], [21]. Further, we are unaware of any studies that investigate the abundance or composition of PPFMs in the rhizosphere or soils of natural terrestrial ecosystems.

Here we surveyed PPFM abundance from the root zones of a variety of 10 common native and invasive species in CSS. We addressed three questions: (1) Does PPFM abundance in the root zone differ by CSS plant species? (2) If so, does this abundance vary by native or invasive plant species? (3) Is PPFM abundance affected by the surrounding plant community (other plant species in the area)?

We estimated PPFM abundance from plant root zone soil using the most probable number (MPN) technique. Because PPFMs are defined by their appearance on a selective media, quantification by culturing is appropriate for this group. We predicted that PPFM root zone abundance would vary among plant species in natural systems, as observed in crop species. Further, we expected that PPFM abundance under native plant species would be higher than those under invasive plant species, hypothesizing that native plants adapted to the CSS habitat would be associated with more PPFMs to mitigate drought and heat stress. Finally, given the local scale of potential plant-PPFM interactions, we expected that root zone abundance would be determined by the immediate plant species and not the surrounding community (other native or invasive plant species).

MATERIALS AND METHODS

Study Site Description and Sample Collection

The study was conducted in the Santa Monica Mountains National Recreation Area (SMMNRA, U.S. National Park Service) in southern California, USA. The Santa Monica Mountains have a Mediterranean-type climate with most precipitation in the cool winter months (December through February, rainfall typical ranges between 250–330 mm/yr) followed by long, hot and dry conditions for the remainder of the year. The SMMNRA is bounded by the heavily populated Los Angeles and Ventura counties (pop. ~10 million and ~750,000 respectively). Invasive species threaten critical habitat to about 100 sensitive, threatened or endangered plant and animal species in the park.

In early June 2007, we identified 10 plant invasion gradients (or "sites") in CSS within a 0.5 km² area of the Satwiwa/Rancho Sierra Vista region of the SMMNRA. We laid one transect (10–32 m in length) through each invasion gradient, such that one end of the transect fell in a stand of all native plant species and the other end in a stand of all invasive species. Every effort was made to select sites with qualitatively similar soil type (clay), soil moisture (dry), slope (zero to 5 degrees), and aspect (zero to north or northwest). The transects included native species that are common constituents of CSS assemblages within the SMMNRA:*Artemisia californica* (California sagebrush); *Eriogonum fasciculatum* (California buckwheat);*Nassella lepida* (foothill needle grass); *Nassella pulchra* (purple needle grass); and *Baccharis pilularis* (coyote brush). Invasive species in the transects included common, problematic invaders that are subject to ongoing control efforts at the SMMNRA: *Carduus pycnocephalus*(Italian thistle); *Conium maculatum* (poison hemlock); *Foeniculum vulgare* (fennel); *Hirschfeldia incana* (shortpod mustard); and *Phalaris aquatica* (Harding grass). Not all species occurred at each site, although there was overlap between sites.

We established five, ~1 m² plots along each of the 10 transects. One plot was placed at the "pure" native (0% invaded) end of the transect, and one plot was placed at the "pure" invasive (100% invaded) end. Three additional plots were placed in "mixed" stands along the transect, such that the plots contained approximately 20, 50, and 80% invasive plant cover as determined by eye.

We then sampled soil from the root zones (~100 g, 5 cm deep at the base of the root crown) of green (photosynthesizing) plants from the plots. We took samples from 3 individuals of each plant species in each plot. Thus, the number of samples from a plot depended on the number of species present in a plot. Further, not all species present at a site (Table 1) were present in all the mixed plots. In total, we collected 246 soil samples, each of ~100 g. As typical for early June, soils were dry at the time of collection. Samples were stored at 4°C at the University of California, Irvine for 60 days and processed in random order over a two week period.

Table 1: PPFM abundance associated with the root zone of the different plant species

Plant Species	Life History and Form	Native CSS or Invasive	Site	Mean CFU/g dry soil ± SEM
Artemisia californica	PS	Native	1,4,5,7,8	5595±1620
Baccharis pilularis	PS	Native	7,10	5809±2146
Eriogonum fasciculatum	PS	Native	2	7243±2916
Nassella lepida	PB	Native	2,9	12774±9753
Nassella pulchra	PB	Native	1,3,6	5564±1516
Carduus pycnocephalus	AH	Invasive	1,2,8	15282±4076
Conium maculatum	BH	Invasive	8	12743±7818
Foeniculum vulgare	PH	Invasive	3,5	3882±576
Hirschfeldia incana	AH	Invasive	6,7,8,10	12012±2941
Phalaris aquatica	PB	Invasive	4,9	2817±470

AH = annual herb, BH = biennial herb, PB = perennial bunchgrass, PH = perennial herb, PS = perennial shrub. Mean MPN ± SEM is the average across all sites where the plant species was found (3 samples/species at each zone of invasion both pure and mixed, N = 246).
doi:10.1371/journal.pone.0031026.t001

doi:10.1371/journal.pone.0031026.t001

Most Probable Number (MPN) Dilutions

In the lab, each soil sample was homogenized, sieved and divided into three sub-samples prior to serial dilution. To estimate PPFM abundance, a ten-fold dilution series (1:10 to 1:100,000) for each sub-sample was prepared using sterile, modified nitrate mineral salts (NMS) medium[22] with methanol (1.0% vol/vol) as the sole carbon source in the media. Cycloheximide (100 mg/L) was added to the media to prevent fungal growth. Homogenized soil (2 g) was suspended directly in NMS medium and then further diluted. The dilutions were plated in triplicate into 1.5 mL/well uncoated, sterile microtiter plates. The plates were covered and incubated at 30°C for 14 days. Wells with pink growth at 14 days were scored as positive for PPFMs.

PCR and Sequencing

To confirm that the pink bacteria in the MPN dilutions were members of the genus*Methylobacterium*, we performed polymerase chain reaction (PCR) amplification and sequencing on nine purified colonies. For each colony, we targeted the methanol dehydrogenase gene (*mxa*F) and the 16 S small sub-unit ribosomal DNA (16 S SSU). The*mxa*F gene was selected for amplification because it has been found in all *Methylobacterium*to date [23]. Nine positive wells from the MPN plates were chosen randomly and streaked on NMS agar (methanol 1.0% vol/vol, 100 mg cycloheximide/L media) and incubated as above. A colony from these plates was re-streaked for further purification. One colony from each of the second streak plates was transferred to NMS broth and incubated until log phase (30°C, shaken at 150 rpm for 7 days). Total genomic DNA was extracted (Promega Wizard® Genomic DNA Purification Kit) from each of the pure cultures. Primers used to target the *mxa*F gene (*mxa*F1003 forward and *mxa*F1541 reverse [24]) yielded a 538 bp amplicon. We also used universal eubacterial primers to target the highly conserved 16 S small sub-unit ribosomal DNA (pA forward & pH'reverse, [25]) yielding a 1500 bp amplicon. PCR conditions for the *mxa*F gene were as follows: (25-μl reaction vol.) initial denaturing 95°C (5 min.), anneal at 55.6°C (30 sec.), extension at 72°C (40 sec.), 30 cycles total with a final extension step of 72°C (5 min). [Final concentration: forward and reverse primers 250 nM, 1 unit Taq polymerase, MasterAmp™ 1× Premix F (Epicenter Biotechnologies - Wisconsin, USA)]. PCR conditions for the 16 S SSU rDNA gene were: initial denaturing 95°C (4 min), anneal at 55°C (40 sec), extension at 72°C (2 min) for 30 cycles with a final extension step of 72°C (5 min). [Final concentration: forward and reverse primers 125 nM, 1 unit Taq polymerase, MasterAmp™ 1× Premix F (as above)]. PCR products were visualized by gel electrophoresis and the target bands were excised and purified following the manufacturer's protocol (MinElute® Gel Extraction Kit - Qiagen Sciences, Maryland, USA) prior to sequencing with the forward primer (Agencourt - Massachusetts, USA). Sequences were compared to GenBank accessioned sequences using the blastn algorithm. The sequences have been submitted to GenBank under the accession numbers HQ219729-HQ219746.

Statistical Analysis

To capture variation in PPFM abundance across different sites, we sampled from a variety of locations. This sampling pattern also meant that the plant species did not occur at every site, such that we are not able to test for interactive effects of site and plant species on PPFM abundance. Thus, we first tested for overall site differences in PPFM abundance with a one-way ANOVA. For subsequent analyses, we combined the data from all sites. Additionally, we tested for differences in PPFM abundances among the three mixed stand levels (20, 50 & 80% invaded). Since there were no significant differences among these levels, we combined these samples into one "mixed" category for all further tests. We then performed a two-way ANOVA to test for differences in PPFM abundance by plant species and by the surrounding plant community (i.e., whether in a pure or mixed stand). Because there was no significant interaction between species and pure/mixed stands, we could then use a one-way ANOVA to test whether PPFM abundance varied among native vs. invasive (or alternatively, annual/biennial vs. perennial life history). For all analyses, we ln-transformed the estimates of PPFM abundance to improve normality for the statistical tests. We performed the tests using the JMP 7.02 software (SAS Institute, Inc.). We present the back-transformed means+one standard error of the mean in the figures.

RESULTS

PCR and Sequencing

As expected, the pink bacteria that we cultured all appeared to belong to the genus*Methylobacterium*. The *mxa*F sequences of the isolates were most similar to*Methylobacterium*. In particular, 2 of 9 isolates had a ≥99.2% pairwise similarity to*Methylobacterium extorquens* and 5 of 9 had a ≥98.8% similarity to the cultured*Methylobacterium* sp. F3.2 strain. The remaining sequences were most similar to *M. dichloromethanicum* and *M. rhodinium* (99.4%). The 16 S sequences yielded similar results; 4 of 9 sequences were most similar (≥99.2%) to *M. extorquens* and the remaining most similar (≥99.3%) to three cultured *Methylobacterium* species strains (Table 2).

Table 2: Comparison of most similar cultured isolate to GenBank Sequences (blastn) for 16 SSU ribosomal DNA and the methanol dehydrogenase gene, *mxa*F

	*mxa*F			16 S rDNA		
Isolate	Accession #	Description	% Pairwise Identity	Accession #	Description	% Pairwise Identity
1	FJ157958	*Methylobacterium* sp. F3.2	98.8	FJ157961	*Methylobacterium* sp. 1b.3	99.3
8	FJ157958	*Methylobacterium* sp. F3.2	98.8	AM910536	*Methylobacterium* sp. F38	100.0
9	GU353343	*Methylobacterium* sp. F3.2	98.8	CP001510	*M. extorquens* AM1	99.7
4	GU353343	*Methylobacterium* sp. F3.2	98.8	AM910536	*Methylobacterium* sp. F38	99.5
6	GU353343	*Methylobacterium* sp. F3.2	98.8	FJ157976	*Methylobacterium* sp. JT1	99.4
2	U70527	*M. rhodinum*	99.4	CP000908	*M. extorquens* PA1	99.2
3	EF562465	*M. dichloromethanicum* KACC 11438	99.4	AF531770	*M. extorquens*	99.5
7	AJ878068	*M. extorquens* DM4	99.2	AB175632	*M. extorquens*	99.6
10	AB455974	*M. extorquens* NRIC 0601	99.6	FJ157961	*Methylobacterium* sp. 1b.3	99.3

doi:10.1371/journal.pone.0031026.t002

doi:10.1371/journal.pone.0031026.t002

PPFM Abundance

PPFMs were present in every soil sample, from a minimum of 10^2 to a maximum of 10^5 CFU/g dry soil. There was no obvious fungal or non-target bacterial growth observed in the MPN cultures. PPFM abundance in the root zones varied significantly among sites (P=0.0017); however, due to the changing identity of plant species present at each site, we cannot determine whether these site differences were due to the particular plant species present or abiotic effects. To examine this question further, we tested whether PPFM abundance differed in the pure stands of the four species that occurred at more than two sites (*A. californica, N. pulchra, C. pycnocephalus, H. incana*). Abundance did not differ across sites for any of the species (one-way ANOVA; P=0.09–0.31), suggesting that plant species is more important to PPFM abundance than direct abiotic effects.

Overall, PPFM abundance differed significantly among plant species (P=0.0034; Table 1,Figure 1). The natives, *E. fasciculatum, B. pilularis, A. californica* and *N. pulchra* showed mean PPFM abundances ranging from 5,600–7,200 CFU/g dry soil. The highest mean PPFM abundance was found under the invasive thistle, *C. pycnocephalus* (1.5×10⁴ CFU/g dry soil), followed by the native *N. lepida* and invasives *C. maculatum* and *H. incana*.

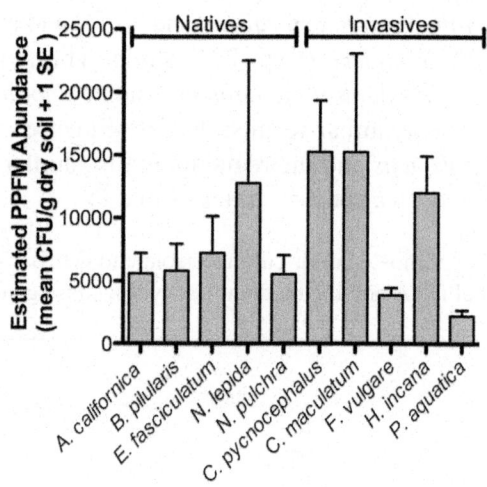

Figure 1: Average PPFM abundance for each species (pure and mixed stands pooled).

Two-way ANOVA, Factor: species ($F_{1,9}$=3.15, P=0.0034, N=82). Species with the same letters are not significantly different from each other (Tukey post hoc test α=0.05, Q=3.64). Error bars were constructed with one standard error of the mean for all soil samples taken under each plant species.

doi:10.1371/journal.pone.0031026.g001

The lowest mean PPFM abundance occurred under the two perennial invasives, *P. aquatica*and *F. vulgare*, 2.8–3.9×10³ CFU/g dry soil, respectively. PPFM abundance was 1.4 times higher in the root zones of invasive plant species than native plant species (9,808±1,733 SEM versus 6,878±1,715 SEM, respectively; P=0.025; Figure 2a). To examine whether this difference might be related to plant life history, we classified the species as annual/biennial or perennial and performed the same analysis. Indeed, annual/biennials had over 2 times more PPFMs than perennials (13,130±2,251 SEM versus 5,963±1,367 SEM, respectively; P<0.0001; Figure 2b).

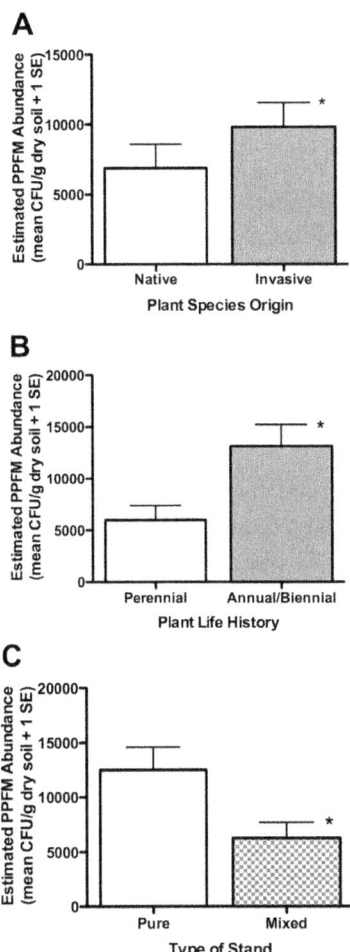

Figure 2: Average PPFM abundance in the root zone by species origin, life history, and pure and mixed stands.

(a) Native versus invasive species: one-way ANOVA, (mixed and pure stands pooled) $F_{1,80}$=5.21, P=0.0251, N=82; (b) Annual/biennial versus perennial life history: one-way ANOVA, (mixed and pure stands pooled) $F_{1,80}$=20.45, P<0.0001, N=82; (c) Pure versus mixed stands: two-way ANOVA, $F_{1,9}$=4.06, P=0.048, N=82. Error bars were constructed with one standard error of the mean.

doi:10.1371/journal.pone.0031026.g002

Finally, the number of PPFMs in the root zone of a particular species appeared to be affected by the broader, surrounding plant community. We found 50% fewer PPFMs in the mixed species stands than in pure stands (6,262±809 SEM versus 12,487±3342 SEM, respectively; P=0.0476; Figure 2c). There was no interactive effect of the focal plant species and the type (mixed/pure) of community stand on PPFM abundance (P=0.553).

DISCUSSION

PPFMs were ubiquitous among our soil samples from plant root zones in coastal sage scrub habitat. As far as we are aware, this study is first to demonstrate that PPFM abundance in the root zone varies among plant species in a natural ecosystem. This variation could be due in part to differences in root growth habitats among the plant species or in the amount and types of carbon compounds exuded from them. The result parallels prior work that shows that phyllosphere PPFM abundance varies among plant species in natural ecosystems [21], [18].

Contrary to our original hypothesis, native plant species generally had significantly fewer PPFMs than invasives. However, the pattern observed may be driven primarily by plant life history. All of the native species in this study were slow growing, drought-deciduous perennials. There were no native annual species in the transects (native annuals have a patchy distribution in CSS at SMMNRA), whereas three of the five invasive species were annuals. Supporting this interpretation, the two perennial invasives had the fewest PPFMs of all of the plants. Overall, annual/biennial species had twice as many PPFMs in their root zones than the perennial species. Perhaps drought deciduous perennials may not be as favorable to PPFMs because of their slower growth compared to annuals [15], [26], [16]. PPFMs consume the C_1 by-products of cell division, thus faster plant growth may result in faster PPFM growth and higher abundances. In future studies, the relative importance of native/invasive status versus life history on PPFM abundance should be tested in a system with better representation of the species' types.

The fact that we found on average 50% fewer PPFMs in the mixed stands (native and invasive) versus pure stands (only natives or only invasives) suggests that the surrounding plant community influences PPFM abundance in another plant›s root zone. Many studies have found that plants exude complex organics and extracellular enzymes from their roots to compete for space and resources [27], [28]. Some of these compounds have antimicrobial properties that could be affecting PPFMs [29]. Further, there is evidence that the amount and/or potency of phytotoxins released by invasive plant species changes when in competition with native species [30], [31]. Thus, the reduction in

PPFM abundance in mixed stands relative to pure stands could be due to the antimicrobial properties of phytotoxins.

Much work remains to be done to identify the particular environmental and biotic factors influencing PPFM abundance. For instance, the site level effects on PPFM abundance that we found are presumably due to plant composition, but this is affected by abiotic conditions. Though we attempted to minimize the abiotic conditions between sites (e.g., slope, aspect, soil type and soil moisture), minor differences in these variables (or other uncontrolled variables) may also be driving some of the variation in PPFM abundance. Further, the sampling design here did not allow us to test for the effect of plant species by site interactions on PPFM abundance. In addition, PPFM abundance is known to vary over the growing season[15]; therefore, it would be useful to compare these results (from late spring) to other times of year, such as during the plants› peak growing season (in the winter/spring) or when the perennial species are dormant (in the late summer/fall). Finally, given the effect of the surrounding plant composition on root zone PPFMs, it may also be important to consider the effect of the surrounding plant density.

The differential distribution of PPFMs among CSS plant species provides further motivation for controlled experimental studies on PPFMs in non-crop species. In particular, further work should examine how the genetic composition of PPFMs varies in natural ecosystems. In addition, experiments in CSS (e.g., Irvine et al. 2011) and other habitats are needed to test whether the relative strength of PPFM-plant interactions varies among native and invasive plant species. If so, PPFMs could play a role in structuring plant communities generally [32]. Plants and soil microorganisms are also known to alter the success of invasive plants into native communities; however, these studies have focused on nitrogen-fixers, mycorrhizal fungi, and soil pathogens [33], [34], [35], [36]. PPFMs might offer a promising direction for future investigations of plant-microorganism feedbacks and native plant community restoration.

ACKNOWLEDGMENTS

The authors wish to thank Steven Allison, Lucía Vivanco and our anonymous reviewers for their thoughtful comments on previous versions of manuscripts. We also thank Ashley Whelpley for her invaluable help preparing MPN dilutions. Finally, we thank the Santa Monica Mountains National Recreation Area for its support of PPFM research in the park.

AUTHOR CONTRIBUTIONS

Conceived and designed the experiments: ICI KNS. Performed the experiments: ICI. Analyzed the data: ICI JBHM. Contributed reagents/materials/analysis tools: ICI CB KNS JBHM. Wrote the paper: ICI JBHM.

REFERENCES

1.   Reeburgh WS, Whalen SC, Alperin MJ (1993) The role of methylotrophy in the global methane budget. In: Murrell JC, Kelly DP, editors. Microbial Growth on C1 Compounds. Andover, UK: Intercept. pp. 1–14.

2.   Lidstrom ME (2006) pp. 618–634. Aerobic Methylotrophic Prokaryotes.

3.   Trotsenko YA, Ivanova EG, Doronina NV (2001) Aerobic methylotrophic bacteria as phytosymbionts. Microbiology 70: 623–632.

4.   Kalyaeva MA, Zacharchenko NS, Doronina NV, Rukavtsova EB, Ivanova EG, et al. (2001) Plant growth and morphogenesis in vitro is promoted by associative methylotrophic bacteria. Russian Journal of Plant Physiology 48: 514–517.

5.   Madhaiyan M, Poonguzhali S, Senthilkumar M, Seshadri S, Chung HY, et al. (2004) Growth promotion and induction of systemic resistance in rice cultivar Co-47 (*Oryza sativa* L.) by *Methylobacterium* spp. Botanical Bulletin of Academia Sinica 45: 315–324.

6.   Madhaiyan M, Reddy BVS, Anandham R, Senthilkumar M, Poonguzhali S, et al. (2006) Plant growth-promoting *Methylobacterium* induces defense responses in groundnut (*Arachis hypogaea* L.) compared with rot pathogens. Current Microbiology 53: 270–276.

7.   Doronina NV, Ivanova EG, Trotsenko YA (2002) New evidence for the ability of methylobacteria and methanotrophs to synthesize auxins. Microbiology 71: 116–118.

8.   Madhaiyan M, Poonguzhali S, Lee HS, Hari K, Sundaram SP, et al. (2005) Pink-pigmented facultative methylotrophic bacteria accelerate germination, growth and yield of sugarcane clone Co86032 (*Saccharum officinarum* L.). Biology and Fertility of Soils 41: 350–358.

9.   Irvine IC, Witter MS, Brigham CA, Martiny JBH (2011) Relationships between methylobacteria and glyphosate with native and invasive plant species: implications for restoration. Restoration Ecology. (DOI 10.1111/j.1526-100X.2011.00850x).

10. Bowler PA (2000) Ecological restoration of coastal sage scrub and its potential role in habitat conservation plans. Environmental Management 26: S85–S96.

11. Corpe WA, Rheem S (1989) Ecology of the methylotrophic bacteria on living leaf surfaces. FEMS Microbiology Letters 62: 243–249.

12. Chanprame S, Todd JJ, Widholm JM (1996) Prevention of pink-pigmented methylotrophic bacteria (*Methylobacterium mesophilicum*) contamination of plant tissue cultures. Plant Cell Reports 16: 222–225.

13. Balachandar D, Raja P, Nirmala K, Rithyl TR, Sundaram SP (2008) Impact of transgenic Bt-cotton on the diversity of pink-pigmented facultative methylotrophs. World Journal of Microbiology & Biotechnology 24: 2087–2095.

14. Balachandar D, Raja P, Sundaram SP (2008) Genetic and metabolic diversity of Pink-Pigmented Facultative Methylotrophs in phyllosphere of tropical plants. Brazilian Journal of Microbiology 39: 68–73.

15. Omer ZS, Tombolini R, Gerhardson B (2004) Plant colonization by pink-pigmented facultative methylotrophic bacteria (PPFMs). FEMS Microbiology Ecology 47: 319–326.

16. Madhaiyan M, Poonguzhali S, Sa T (2007) Influence of plant species and environmental conditions on epiphytic and endophytic pink-pigmented facultative methylotrophic bacterial populations associated with field-grown rice cultivars. Journal of Microbiology and Biotechnology 17: 1645–1654.

17. Schauer S, Kutschera U (2008) Methylotrophic bacteria on the surfaces of field-grown sunflower plants: a biogeographic perspective. Theory in Biosciences 127: 23–29.

18. Romanovskaya VA, Stolyar SM, Malashenko YR (1996) Distribution of bacteria of the genus *Methylobacterium* in different ecosystems of Ukraine. Mikrobiologicheskii Zhurnal (Kiev) 58: 3–10.

19. Raja P, Balachandar D, Sundaram SP (2008) Genetic diversity and phylogeny of pink-pigmented facultative methylotrophic bacteria isolated from the phyllosphere of tropical crop plants. Biology and Fertility of Soils 45: 45–53.

20. Delmotte N, Knief C, Chaffron S, Innerebner G, Roschitzki B, et al. (2009) Community proteogenomics reveals insights into the physiology of phyllosphere bacteria. Proceedings of the National Academy of Sciences of the United States of America 106: 16428–16433.

21. Knief C, Frances L, Cantet F, Vorholt JA (2008) Cultivation-independent characterization of *Methylobacterium* populations in the plant phyllosphere by automated ribosomal intergenic spacer analysis. Applied and Environmental Microbiology 74: 2218–2228.

22. Burlage RS, Altlas R, Stahl D, Geesey G, Sayler G (1998) Techniques in Microbial Ecology. New York: Oxford University Press. 480 p.

23. Trotsenko YA, Murrell JC (2008) Metabolic aspects of aerobic obligate methanotrophy. Advances in Applied Microbiology 63: 183–229.

24. McDonald IR, Murrell JC (1997) The methanol dehydrogenase structural gene *mxaF* and its use as a functional gene probe for methanotrophs and methylotrophs. Applied and Environmental Microbiology 63: 3218–3224.

25. Edwards U, Rogall T, Blocker H, Emde M, Bottger EC (1989) Isolation and direct complete nucleotide determination of entire genes - characterization of a gene coding for 16s-ribosomal RNA. Nucleic Acids Research 17: 7843–7853.

26. Pirttila AM, Pospiech H, Laukkanen H, Myllyla R, Hohtola A (2005) Seasonal variations in location and population structure of endophytes in buds of Scots pine. Tree Physiology 25: 289–297.

27. Allison SD, Nielsen C, Hughes RF (2006) Elevated enzyme activities in soils under the invasive nitrogen-fixing tree *Falcataria moluccana*. Soil Biology & Biochemistry 38: 1537–1544.

28. Weidenhamer JD, Callaway RM (2010) Direct and indirect effects of invasive plants on soil chemistry and ecosystem function. Journal of Chemical Ecology 36: 59–69.

29. Bais HP, Prithiviraj B, Jha AK, Ausubel FM, Vivanco JM (2005) Mediation of pathogen resistance by exudation of antimicrobials from roots. Nature 434: 217–221.

30. Bains G, Kumar AS, Rudrappa T, Alff E, Hanson TE, et al. (2009) Native plant and microbial contributions to a negative plant-plant interaction. Plant Physiology 151: 2145–2151.

31. Thorpe AS, Thelen GC, Diaconu A, Callaway RM (2009) Root exudate is allelopathic in invaded community but not in native community: field evidence for the novel weapons hypothesis. Journal of Ecology 97: 641–645.

32. Reynolds HL, Packer A, Bever JD, Clay K (2003) Grassroots ecology: plant-microbe-soil interactions as drivers of plant community structure and dynamics. Ecology 84: 2281–2291.

33. Vitousek PM, Walker LR (1989) Biological invasion by *Myrica faya* in Hawai'i: plant demography, nitrogen fixation, ecosystem effects. Ecological Monographs 59: 247–265.

34. Hawkes CV, Wren IF, Herman DJ, Firestone MK (2005) Plant invasion alters nitrogen cycling by modifying the soil nitrifying community. Ecology Letters 8: 976–985.

35. Vogelsang KM, Bever JD (2009) Mycorrhizal densities decline in association with nonnative plants and contribute to plant invasion. Ecology 90: 399–407.

36. Mangla S, Inderjit , Callaway RM (2008) Exotic invasive plant accumulates native soil pathogens which inhibit native plants. Ecology 96: 58–67.

CITATION

CHAPTER 1

Lilian Cristina Baldon Aizza and Marcelo Carnier Dornelas, "A Genomic Approach to Study Anthocyanin Synthesis and Flower Pigmentation in Passionflowers," Journal of Nucleic Acids, vol. 2011, Article ID 371517, 17 pages, 2011. doi:10.4061/2011/371517.

CHAPTER 2

Hock-Eng Khoo K. Nagendra Prasad Kin-Weng Kong Yueming Jiang and Amin Ismail, Carotenoids and Their Isomers: Color Pigments in Fruits and Vegetables, doi:10.3390/molecules16021710.

CHAPTER 3

Hussein, M. and Alva, A. (2014) Effects of Zinc and Ascorbic Acid Application on the Growth and Photosynthetic Pigments of Millet Plants Grown under Different Salinity. Agricultural Sciences, 5, 1253-1260. doi: 10.4236/as.2014.513133.

CHAPTER 4

Schwieterman ML, Colquhoun TA, Jaworski EA, Bartoshuk LM, Gilbert JL, et al. (2014) Strawberry Flavor: Diverse Chemical Compositions, a Seasonal Influence, and Effects on Sensory Perception. PLoS ONE 9(2): e88446. doi:10.1371/journal.pone.0088446.

CHAPTER 5

Paulina Kuczynska, Malgorzata Jemiola-Rzeminska and Kazimierz Strzalka,Photosynthetic Pigments in Diatoms, doi:10.3390/md13095847.

CHAPTER 6

Huang J, Wei C, Zhang Y, Blackburn GA, Wang X, Wei C, et al. (2015) Meta-Analysis of the Detection of Plant Pigment Concentrations Using Hyperspectral Remotely Sensed Data. PLoS ONE 10(9): e0137029. doi:10.1371/journal.pone.0137029

CHAPTER 7

Beverley J. Glover and Heather M. Whitney, Structural colour and iridescence in plants: the poorly studied relations of pigment colour, doi: 10.1093/aob/mcq007.

CHAPTER 8

G. Pratta, G. Rodríguez, R. Zorzoli, L. Picardi and E. Valle, "Biodiversity in a Tomato Germplasm for Free Amino Acid and Pigment Content of Ripening Fruits," American Journal of Plant Sciences, Vol. 2 No. 2, 2011, pp. 255-261. doi: 10.4236/ajps.2011.22027.

CHAPTER 9

Congming Lu, Qingtao Lu, Jianhua Zhang and Tingyun Kuang, Characterization of photosynthetic pigment composition, photosystem II photochemistry and thermal energy dissipation during leaf senescence of wheat plants grown in the field, doi: 10.1093/jexbot/52.362.1805.

CHAPTER 10

Jun Cheng, Liao Liao, Hui Zhou, Chao Gu, Lu Wang and Yuepeng Han, A small indel mutation in an anthocyanin transporter causes variegated colouration of peach flowers, doi: 10.1093/jxb/erv419.

CHAPTER 11

Velten J, Cakir C, Cazzonelli CI (2010) A Spontaneous Dominant-Negative Mutation within a 35S::AtMYB90 Transgene Inhibits Flower Pigment Production in Tobacco. PLoS ONE 5(3): e9917. doi:10.1371/journal.pone.0009917.

CHAPTER 12

Irvine IC, Brigham CA, Suding KN, Martiny JBH (2012) The Abundance of Pink-Pigmented Facultative Methylotrophs in the Root Zone of Plant Species in Invaded Coastal Sage Scrub Habitat. PLoS ONE 7(2): e31026. doi:10.1371/journal.pone.0031026.

INDEX